Fluid Dynamics

T0177827

Fluid Dynamics

Part 3: Boundary Layers

Anatoly I. Ruban

Department of Mathematics
Imperial College London

OXFORD
UNIVERSITY PRESS

Great Clarendon Street, Oxford, OX2 6DP,
United Kingdom

Oxford University Press is a department of the University of Oxford.
It furthers the University's objective of excellence in research, scholarship,
and education by publishing worldwide. Oxford is a registered trade mark of
Oxford University Press in the UK and in certain other countries

© Anatoly I. Ruban 2018

The moral rights of the author have been asserted

First Edition published in 2018

Reprinted 2018

Impression: 2

All rights reserved. No part of this publication may be reproduced, stored in
a retrieval system, or transmitted, in any form or by any means, without the
prior permission in writing of Oxford University Press, or as expressly permitted
by law, by licence or under terms agreed with the appropriate reprographics
rights organization. Enquiries concerning reproduction outside the scope of the
above should be sent to the Rights Department, Oxford University Press, at the
address above

You must not circulate this work in any other form
and you must impose this same condition on any acquirer

Published in the United States of America by Oxford University Press
198 Madison Avenue, New York, NY 10016, United States of America

British Library Cataloguing in Publication Data

Data available

Library of Congress Control Number: 2013953774

ISBN 978–0–19–968175–4

DOI 10.1093/oso/9780199681754.001

Printed and bound by
CPI Group (UK) Ltd, Croydon, CR0 4YY

Preface

This is Part 3 of a book series on fluid dynamics that comprises the following four parts:

Part 1. Classical Fluid Dynamics

Part 2. Asymptotic Problems of Fluid Dynamics

Part 3. Boundary Layers

Part 4. Hydrodynamic Stability Theory

The series is designed to give a comprehensive and coherent description of fluid dynamics, starting with chapters on classical theory suitable for an introductory undergraduate lecture course, and then progressing through more advanced material up to the level of modern research in the field. Our main attention is on high-Reynolds-number flows, both incompressible and compressible. Correspondingly, the target reader groups are undergraduate and MSc students reading mathematics, aeronautical engineering, or physics, as well as PhD students and established researchers working in the field.

In Part 1, we started with discussion of fundamental concepts of fluid dynamics, based on the *continuum hypothesis*. We then analysed the forces acting inside a fluid, and deduced the Navier–Stokes equations for incompressible and compressible fluids in Cartesian and curvilinear coordinates. These were employed to study the properties of a number of flows that are represented by the so-called *exact solutions* of the Navier–Stokes equations. This was followed by detailed discussion of the theory of inviscid flows for incompressible and compressible fluids. When dealing with incompressible inviscid flows, a particular attention was paid to two-dimensional potential flows. These can be described in terms of the *complex potential*, allowing for the full power of the theory of functions of complex variable to be employed. We demonstrated how the method of conformal mapping can be used to study various flows of interest, such as flows past *Joukovskii aerofoils* and separated flows. For the later the *Kirchhoff model* was adopted. The final chapter of Part 1 was devoted to compressible flows of a perfect gas, including supersonic flows. Particular attention was given to the theory of characteristics, which was used, for example, to analyse the *Prandtl–Meyer flow* over a body surface with a bend or a corner. The properties of shock waves were also discussed in detail for steady and unsteady flows.

In Part 2 we introduced the reader to *asymptotic methods*. Also termed the *perturbation methods*, they are now an inherent part of fluid dynamics. We started with a discussion of the mathematical aspects of the asymptotic theory. This was followed by an exposition of the results of application of the asymptotic theory to various fluid-dynamic problems. The first of these was the *thin aerofoil theory* for incompressible and subsonic flows, steady and unsteady. In particular, it was shown that this theory allowed to reduce the task of calculating the lift force to a simple integral. We then turned to supersonic flows. We first analysed the linear approximation to the governing

Euler equations, which led to a remarkably simple relationship between the slope of the aerofoil surface and the pressure, known as the *Ackeret formula*. We then considered the second-order *Buzemann approximation*, and performed analysis of a rather slow process of attenuation of the perturbations in the far field. Part 2 also contains a detailed discussion of the properties of inviscid *transonic* and *hypersonic flows*. We concluded Part 2 with analysis of viscous low-Reynolds-number flows. Two classical problems of the low-Reynolds-number flow theory were considered: the flow past a sphere and the flow past a circular cylinder. In both cases, the flow analysis led to a difficulty, known as *Stokes paradox*. We showed how this paradox could be resolved using the formalism of matched asymptotic expansions.

The present Part 3 is concerned with the high-Reynolds-number fluid flows. In Chapter 1 we consider a class of flows that can be described in the framework of *classical boundary-layer theory*. These include the Blasius flow past a flat plate and the Falkner–Skan solutions for the flow past a wedge. We also discuss Chapman's shear-layer flow and Schlichting's solution for the laminar jet. Among other examples are Tollmien's solution for the viscous wake behind a rigid body and the periodic boundary layer on the surface of a rapidly rotating cylinder. This is followed by a discussion of the properties of compressible boundary layers, including hypersonic boundary layers. The latter are influenced by the interaction with inviscid part of the flow, and involve extremely strong heating of the gas near the body surface.

We then turn our attention to separated flows. Chapter 2 considers two examples of the boundary-layer separation. The first is the so-called *self-induced separation* of the boundary layer in supersonic flow. The theory of self-induced separation is based on the concept of *viscous-inviscid interaction*, which was put forward as a result of detailed experimental examination of the separation process. Experiments revealed a number of unexpected effects. Most disputed was the effect of upstream influence through the boundary layer. Since it could not be explained in the framework of the classical boundary-layer theory, it was suggested that near the separation, the boundary layer comes into 'strong interaction' with inviscid part of the flow. In Chapter 2 we present mathematical theory of the viscous-inviscid interaction, known as the *triple-deck theory*, following the original papers of Neiland (1969a) and Stewartson and Williams (1969) where this theory was formulated based on the asymptotic analysis of the Navier–Stokes equations at large values of the Reynolds number.

The second problem discussed in Chapter 2 is the classical problem of the boundary-layer separation from a smooth body surface in high-Reynolds-number flow of an incompressible fluid. The asymptotic theory for this class of separated flows was put forward by Sychev (1972). He demonstrated that, in order to resolve the long-standing contradiction associated with Goldstein's (1948) singularity at the separation point, one has to account for a strong interaction between the boundary layer and inviscid flow near the separation. Interestingly enough, this interaction develops naturally in response to the singularity in the classical Kirchhoff model for inviscid separated flows.

Chapter 3 is devoted to the flow of an incompressible fluid near the trailing edge of a flat plate. According to the classical boundary-layer theory, the flow past a flat plate is described by the Blasius solution. However, it has been found to fail near the trailing edge where the boundary layer again comes into interaction with the inviscid part of the flow. Stewartson (1969) and Messiter (1970) were the first to demonstrate that in the vicinity of the trailing edge, the flow is described by the triple-deck theory. The reader will see that the viscous-inviscid interaction changes the flow near the trailing edge, and leads to a noticeable increase of the viscous drag force acting on the plate.

Chapter 4 concerns subsonic and supersonic flows near corner points of the body surface, where increasing the corner angle allows us to observe the transition from an attached flow to a flow with local separation in the boundary layer.

Finally, in Chapter 5 we present the *Marginal Separation theory*. It is a special version of the triple-deck theory developed by Ruban (1981, 1982) and Stewartson *et al.* (1982) for the purpose of describing the boundary-layer separation at the leading edge of an aerofoil. The separation assumes the form of the so-called 'short separation bubble' that is observed to 'burst' when the angle of attack reaches a critical value. As a result, the lift force decreases dramatically—this phenomenon is known as the *leading-edge stall*.

The material presented in this book is based on lecture courses given by the author at the Moscow Institute of Physics and Technology, the University of Manchester, and Imperial College London.

Contents

Introduction

The notion of the *boundary layer* was first introduced by Prandtl (1904) in his seminal paper read before the 3rd International Mathematics Congress in Heidelberg. In this work, Prandtl was concerned with some paradoxical results of the theory of high-Reynolds-number flows. If the fluid viscosity is very small, then it might seem obvious that the Navier–Stokes equations governing the fluid motion could be replaced by their inviscid counterparts, the Euler equations. The latter are known to describe various properties of fluid motion perfectly well. In particular, the Euler equations may be used to calculate the lift force produced by an aerofoil. However, they fail completely in calculating the drag. According to the inviscid flow theory, the drag of an aerofoil in a steady flow is always zero. This result is known as the *d'Alembert paradox*. It is applicable not only to aerofoils, but also to other body shapes. For example, the solution of the Euler equations for the flow of an incompressible fluid past a circular cylinder is written in terms of the complex potential $w(z)$ as[1]

$$w(z) = V_\infty \left(z + \frac{a^2}{z} \right).$$

Here V_∞ is the free-stream velocity, a is the cylinder radius, and $z = x + iy$, with x and y being Cartesian coordinates. Figure I.1(a) shows the streamline pattern plotted using this solution. It is contrasted in Figure I.1(b) with an experimental visualization of the flow past a circular cylinder. Clearly, there is a significant difference between the theory and the real flow. While the Euler equations admit the solution where the fluid particles go smoothly around the cylinder and flow remains fully attached to the cylinder surface, the real flow develops a separation that leads to a formation of recirculation eddies behind the cylinder.

To explain these observations, Prandtl (1904) suggested that, at large values of the Reynolds number, the flow field around a rigid body has to be divided into two distinct regions: the main part of the flow and a thin boundary layer that forms along the body surface. The main part of the flow may be treated as inviscid. However, for all Reynolds numbers, no matter how large, the viscous forces maintain their importance in the boundary layer. According to Prandtl, it is due to specific behaviour of the flow in the boundary layer that separation takes place.

An immediate outgrowth of Prandtl's work was the development of the *classical boundary-layer theory* that inspired much of research of high-Reynolds-number flows in the first half of the twentieth century. These studies were mainly concerned with situation where the boundary layer remained attached to the body surface. An exposition of the results of this theory is given in Chapter 1 of this book. Then in Chapter 2 we turn to the analysis of separated flows. Prandtl described the separation process

---

[1]See equation (3.4.28) on page 166 in Part 1 of this book series.

Fluid Dynamics: Part 3: Boundary Layers. © Anatoly I. Ruban, 2018. Published 2018 by Oxford University Press. 10.1093/oso/9780199681754.001.0001

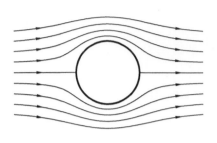

(a) Theoretical predictions.

(b) Experimental observations; the flow is from right to left.

Fig. I.1: Comparison of theoretical predictions for the flow past a circular cylinder with experimental observations by Prandtl and Tietjens (1934).

as follows. Since the flow in the boundary layer has to satisfy the no-slip condition on the body surface, the fluid velocity decreases from the value dictated by the inviscid theory at the outer edge of the boundary layer to zero on the body surface. The slow-moving fluid near the body surface is very sensitive to pressure variations along the boundary layer. If the pressure decreases downstream, then the boundary layer remains attached to the body surface. This situation is observed on the front part of the body; see Figure I.1. However, further downstream the pressure starts to rise, and the boundary layer tends to separate from the body surface. The reason for separation is that the kinetic energy of fluid particles inside the boundary layer appears to be less than it is at the outer edge of the boundary layer. This means that, while the pressure rise in the outer flow may be quite significant, the fluid particles inside the boundary layer may not be able to get over it. Even a small increase of pressure may cause the fluid particles near the wall to stop and then turn back to form a reverse flow region characteristic of separated flows.

It might seem surprising that the clear understanding of the physical processes leading to the separation, could not be converted into a rational mathematical theory for more than half a century. The fact is that the classical boundary-layer theory, as presented by Prandtl, is based on the so-called *hierarchical approach*, when the outer inviscid flow is calculated first, ignoring the existence of the boundary layer, and then the flow in the boundary layer is analysed. By the late 1940s, it became clear that such a strategy leads to a mathematical contradiction associated with singular behaviour of the boundary layer. For the steady flow past a solid wall, this singularity is referred to as the *Goldstein singularity*. This singularity was first described by Landau and Lifshitz (1944), who demonstrated that the shear stress on the body surface upstream of separation drops as the square root $\sqrt{s}$ of the distance s from the separation, and the velocity component normal to the surface tends to infinity being inversely proportional to $\sqrt{s}$. This result was later confirmed based on more rigorous mathematical analysis by Goldstein (1948). Goldstein also proved that the singularity at separation precludes a possibility of continuing the solution beyond the separation point into the region of reverse flow.

Simultaneously, another important development took place. During the 1940s and 1950s, a significant body of experimental evidence was produced which showed that the boundary-layer theory in its classical form was insufficient for describing the separation phenomenon. Most disputable was the effect of upstream influence through the boundary layer in a supersonic flow prior to separation. It could not be explained in the framework of the classical boundary-layer theory, and led to a concept of *viscous-inviscid interaction*; see, for example, Chapman *et al.* (1958). The asymptotic theory of viscous-inviscid interaction, known now as the *triple-deck theory*, was formulated simultaneously by Neiland (1969a) and Stewartson and Williams (1969) in application to the phenomenon of *self-induced separation* of the boundary layer in a supersonic flow, and by Stewartson (1969) and Messiter (1970) in application to incompressible fluid flow near the trailing edge of a flat plate. Two years later, using the concept of viscous-inviscid interaction, Sychev (1972) was able to resolve the contradiction associated with Goldstein's singularity at the separation point, and created the asymptotic theory of laminar boundary-layer separation from a smooth body surface, e.g., a circular cylinder, in an incompressible fluid flow.

Soon after that, it became clear that the viscous-inviscid interaction plays a key role in many fluid-dynamic phenomena. Chapters 2–5 provide an exposition of application of the theory to various separated flows. We will also use the theory in Part 4 of this book series. The reader will see that the viscous-inviscid interaction is instrumental in the development of instabilities of the boundary layer and transition to turbulence.

1

Classical Boundary-Layer Theory

The boundary-layer theory deals with fluid flows with small viscosity. To demonstrate how the boundary-layer equations are deduced, we begin with analysis of the flow of an incompressible fluid past a flat plate.

1.1   Flow Past a Flat Plate

Let us consider two-dimensional steady viscous flow past a flat plate aligned with the oncoming flow. We shall suppose that the fluid is incompressible, and, therefore, the density ρ and kinematic viscosity coefficient ν remain constant all over the flow field. We denote the plate length by L and introduce Cartesian coordinates $(\hat{x}, \hat{y})$ with $\hat{x}$ measured along the flat plate surface from its leading edge O, and $\hat{y}$ in the perpendicular direction as shown in Figure 1.1. With the velocity components denoted by $(\hat{u}, \hat{v})$ and the pressure by $\hat{p}$, the Navier–Stokes equations are written as[1]

$$\left.\begin{aligned} \hat{u}\frac{\partial \hat{u}}{\partial \hat{x}} + \hat{v}\frac{\partial \hat{u}}{\partial \hat{y}} &= -\frac{1}{\rho}\frac{\partial \hat{p}}{\partial \hat{x}} + \nu\left(\frac{\partial^2 \hat{u}}{\partial \hat{x}^2} + \frac{\partial^2 \hat{u}}{\partial \hat{y}^2}\right), \\ \hat{u}\frac{\partial \hat{v}}{\partial \hat{x}} + \hat{v}\frac{\partial \hat{v}}{\partial \hat{y}} &= -\frac{1}{\rho}\frac{\partial \hat{p}}{\partial \hat{y}} + \nu\left(\frac{\partial^2 \hat{v}}{\partial \hat{x}^2} + \frac{\partial^2 \hat{v}}{\partial \hat{y}^2}\right), \\ \frac{\partial \hat{u}}{\partial \hat{x}} + \frac{\partial \hat{v}}{\partial \hat{y}} &= 0. \end{aligned}\right\} \qquad (1.1.1)$$

The boundary conditions for these equations are the no-slip conditions that should be applied on both sides of the plate,

$$\hat{u} = \hat{v} = 0 \quad \text{at} \quad \hat{y} = 0\pm, \ \hat{x} \in [0, L], \qquad (1.1.2)$$

and the free-stream conditions

$$\left.\begin{aligned} \hat{u} &\to V_\infty, \\ \hat{v} &\to 0, \\ \hat{p} &\to p_\infty \end{aligned}\right\} \quad \text{as} \quad \hat{x}^2 + \hat{y}^2 \to \infty. \qquad (1.1.3)$$

Here V_∞ and p_∞ are the fluid velocity and pressure in the free-stream, respectively.

---

[1]See equations (1.7.6) on page 62 in Part 1. We assume here that body force $\mathbf{f}$ is negligible.

Fluid Dynamics: Part 3: Boundary Layers. © Anatoly I. Ruban, 2018. Published 2018 by Oxford University Press. 10.1093/oso/9780199681754.001.0001

Fig. 1.1: Problem layout.

In order to express equations (1.1.1) and boundary conditions (1.1.2), (1.1.3) in non-dimensional form, we perform the following transformation of the variables:

$$\left.\begin{array}{cc} \hat{x} = L\,x, & \hat{y} = L\,y, \\[2mm] \hat{u} = V_\infty u, & \hat{v} = V_\infty v, \qquad \hat{p} = p_\infty + \rho\, V_\infty^2 p. \end{array}\right\} \tag{1.1.4}$$

This turns the Navier–Stokes equations (1.1.1) into

$$u\frac{\partial u}{\partial x} + v\frac{\partial u}{\partial y} = -\frac{\partial p}{\partial x} + \frac{1}{Re}\left(\frac{\partial^2 u}{\partial x^2} + \frac{\partial^2 u}{\partial y^2}\right), \tag{1.1.5a}$$

$$u\frac{\partial v}{\partial x} + v\frac{\partial v}{\partial y} = -\frac{\partial p}{\partial y} + \frac{1}{Re}\left(\frac{\partial^2 v}{\partial x^2} + \frac{\partial^2 v}{\partial y^2}\right), \tag{1.1.5b}$$

$$\frac{\partial u}{\partial x} + \frac{\partial v}{\partial y} = 0. \tag{1.1.5c}$$

The no-slip conditions (1.1.2) become

$$u = v = 0 \quad \text{at} \quad y = 0\pm, \ \ x \in [\,0,1\,], \tag{1.1.6}$$

and the free-stream conditions (1.1.3) assume the form

$$\left.\begin{array}{l} u \to 1, \\ v \to 0, \\ p \to 0 \end{array}\right\} \quad \text{as} \quad x^2 + y^2 \to \infty. \tag{1.1.7}$$

The Reynolds number Re is defined as

$$Re = \frac{V_\infty L}{\nu},$$

and our task will be to find the asymptotic solution to the problem (1.1.5)–(1.1.7) as $Re \to \infty$.

Clearly, we are dealing here with a singular perturbation problem.[2] In fact, the basic ideas that lie in the foundation of the method of matched asymptotic expansions

---

[2]See Sections 1.3 and 1.4 in Part 2 of this book series.

were first put forward by Prandtl (1904) in his seminal paper read before the 3rd International Mathematics Congress in Heidelberg, where he introduced the notion of the *boundary layer*. Prandtl argued that at large values of the Reynolds number *Re*, almost the entire flow field can be treated as inviscid. Still, at any *Re*, no matter how large, there always exists a thin layer adjacent to the body surface, termed the boundary layer, where the role of internal viscosity remains of primary importance.

1.1.1 Asymptotic analysis of the flow

We shall start with the *outer region*, where both coordinates x and y are of the order of the plate length. In the non-dimensional variables, this is written as

$$x = O(1), \quad y = O(1), \quad Re \to \infty.$$

The outer region is shown in Figure 1.2 as region 1, and in what follows, we shall refer to it as the *external flow region*.

The solution of the Navier–Stokes equations (1.1.5) in this region will be sought in the form of the asymptotic expansions[3]

$$\left. \begin{aligned} u(x, y; Re) = u_0(x, y) + \cdots, \qquad v(x, y; Re) = v_0(x, y) + \cdots, \\ p(x, y; Re) = p_0(x, y) + \cdots. \end{aligned} \right\} \tag{1.1.8}$$

Substituting (1.1.8) into (1.1.5) and setting $Re \to \infty$, we find that functions u_0, v_0, and p_0 satisfy the Euler equations:

$$\left. \begin{aligned} u_0 \frac{\partial u_0}{\partial x} + v_0 \frac{\partial u_0}{\partial y} &= -\frac{\partial p_0}{\partial x}, \\ u_0 \frac{\partial v_0}{\partial x} + v_0 \frac{\partial v_0}{\partial y} &= -\frac{\partial p_0}{\partial y}, \\ \frac{\partial u_0}{\partial x} + \frac{\partial v_0}{\partial y} &= 0. \end{aligned} \right\} \tag{1.1.9}$$

Unlike the Navier–Stokes equations (1.1.5), the Euler equations (1.1.9) do not involve the second order derivatives of the velocity components, and therefore they cannot

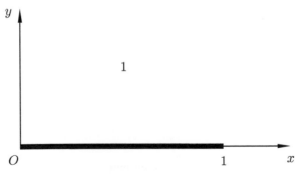

Fig. 1.2: Inviscid flow (region 1).

[3]Here we restrict our attention to the leading-order terms.

be solved with the entire set of the boundary conditions (1.1.6), (1.1.7). According to the inviscid flow theory, the Euler equations are compatible with the free-stream conditions

$$\left.\begin{array}{l} u_0 \to 1, \\ v_0 \to 0, \\ p_0 \to 0 \end{array}\right\} \quad \text{as} \quad x^2 + y^2 \to \infty, \tag{1.1.10}$$

and the impermeability condition on the rigid body surface. For the flat plate it is written as

$$v_0 = 0 \quad \text{at} \quad y = 0\pm, \ x \in [0, 1]. \tag{1.1.11}$$

An infinitely thin flat plate does not produce any perturbations in an inviscid flow and, indeed, by direct substitution, it is easily verified that

$$u_0 = 1, \quad v_0 = 0, \quad p_0 = 0 \tag{1.1.12}$$

is a solution to the boundary-value problem (1.1.9)–(1.1.11).

As was expected, the outer solution (1.1.12) does not satisfy the condition for u in (1.1.6). Hence, an inner region occupying a small vicinity of the plate surface should be brought into consideration. This new region is shown as region 2 in Figure 1.3. Since the boundary layer is thin, we have to scale the coordinate normal to the wall:

$$y = \delta(Re)Y. \tag{1.1.13}$$

Here $\delta(Re)$ is supposed small, that is

$$\delta(Re) \to 0 \quad \text{as} \quad Re \to \infty.$$

As the boundary layer extends along the entire surface of the flat plate, no scaling of the longitudinal coordinate x is required. Consequently, the asymptotic analysis of the Navier–Stokes equations (1.1.5) in region 2 has to be based on the limit

$$x = O(1), \quad Y = \delta^{-1}y = O(1), \quad Re \to \infty. \tag{1.1.14}$$

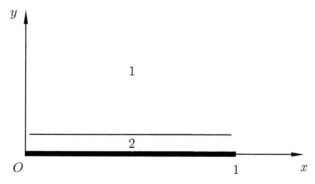

Fig. 1.3: The boundary layer (region 2).

Correspondingly, the leading-order terms of the asymptotic expansions of u, v, and p in this region will be sought in the form

$$u(x, y; Re) = U_0(x, Y) + \cdots, \qquad v(x, y; Re) = \sigma(Re)V_0(x, Y) + \cdots, \left.\right\}$$

$$p(x, y; Re) = \chi(Re)P_0(x, Y) + \cdots. \tag{1.1.15}$$

Here it has been taken into account that the longitudinal velocity component u decreases from $u = 1$ at the outer edge of the boundary layer to $u = 0$ on the wall, which suggests that u is an order one quantity. The coefficients $\sigma(Re)$ and $\chi(Re)$ in the asymptotic expansions of v and p are not known in advance. We expect to find them in the course of analysis of the Navier–Stokes equations.

Let us start with continuity equation (1.1.5c). The substitution of (1.1.15) into this equation yields

$$\frac{\partial U_0}{\partial x} + \frac{\sigma(Re)}{\delta(Re)}\frac{\partial V_0}{\partial Y} = 0.$$

According to the principle of least degeneration, we have to choose[4]

$$\sigma(Re) = \delta(Re), \tag{1.1.16}$$

and the continuity equation takes the form

$$\frac{\partial U_0}{\partial x} + \frac{\partial V_0}{\partial Y} = 0. \tag{1.1.17}$$

Let us now consider the longitudinal momentum equation (1.1.5a). The substitution of (1.1.15) into (1.1.5a) yields

$$U_0\frac{\partial U_0}{\partial x} + V_0\frac{\partial U_0}{\partial Y} = -\chi(Re)\frac{\partial P_0}{\partial x} + \frac{1}{Re}\frac{\partial^2 U_0}{\partial x^2} + \frac{1}{\delta^2 Re}\frac{\partial^2 U_0}{\partial Y^2}. \tag{1.1.18}$$

We see that the second term on the right-hand side of (1.1.18) is small compared to any of the terms on the left-hand side, and should be neglected as $Re \to \infty$. As far as the third term is concerned, it can be retained in the equation by setting

$$\delta(Re) = Re^{-1/2}. \tag{1.1.19}$$

---

[4]Indeed, if we assumed $\delta(Re) \gg \sigma(Re)$, which could be interpreted as an attempt to consider a region with a thickness much larger than the thickness of the real boundary layer, then the continuity equation would degenerate to

$$\frac{\partial U_0}{\partial x} = 0.$$

Since upstream of the trailing edge $u = 1$, it would lead to a conclusion that $U_0 \equiv 1$, as if we still were dealing with the outer inviscid region.

If, on the other hand, $\sigma(Re)$ was assumed much larger than $\delta(Re)$, which could be interpreted as an overestimation of the order of magnitude of v in (1.1.15), then the continuity equation would degenerate to

$$\frac{\partial V_0}{\partial Y} = 0.$$

With $V_0 = 0$ on the plate surface, the solution of the above equation is

$$V_0 = 0,$$

which means that the asymptotic expansion for v in (1.1.15) does not really have a term with σ larger than δ.

This choice can be justified with the help of the following arguments. If one assumes that $\delta^2 Re \gg 1$, then the flow would appear inviscid, and the no-slip condition $U_0 = 0$ on the plate surface could not be satisfied. If, on the other hand, $\delta^2 Re \ll 1$, then equation (1.1.18) degenerates to

$$\frac{\partial^2 U_0}{\partial Y^2} = 0. \qquad (1.1.20)$$

It has to be solved with the no-slip condition on the plate surface,

$$U_0\Big|_{Y=0} = 0, \qquad (1.1.21)$$

and the condition

$$U_0\Big|_{Y=\infty} = 1, \qquad (1.1.22)$$

which follows from matching of the asymptotic expansion for u in (1.1.15) with the solution (1.1.12) in region 1. Clearly, the boundary-value problem (1.1.20)–(1.1.22) does not have a solution.

This confirms, once again, that the principle of least degeneration provides the only possible choice of $\delta(Re)$ as given by (1.1.19). Using (1.1.19), we can write the longitudinal momentum equation (1.1.18) as

$$U_0\frac{\partial U_0}{\partial x} + V_0\frac{\partial U_0}{\partial Y} = -\chi(Re)\frac{\partial P_0}{\partial x} + \frac{\partial^2 U_0}{\partial Y^2}. \qquad (1.1.23)$$

It remains to determine $\chi(Re)$, for which purpose we shall use the lateral momentum equation (1.1.5b). The substitution of (1.1.15), (1.1.13), (1.1.16), and (1.1.19) into (1.1.5b) yields

$$U_0\frac{\partial V_0}{\partial x} + V_0\frac{\partial V_0}{\partial Y} = -\frac{\chi(Re)}{Re^{-1}}\frac{\partial P_0}{\partial Y} + \frac{1}{Re}\frac{\partial^2 V_0}{\partial x^2} + \frac{\partial^2 V_0}{\partial Y^2}. \qquad (1.1.24)$$

When dealing with equations (1.1.23) and (1.1.24), we shall consider the following two options. If we assume that $\chi(Re)$ is small, that is, $\chi \to 0$ as $Re \to \infty$, then the equation (1.1.23) reduces to

$$U_0\frac{\partial U_0}{\partial x} + V_0\frac{\partial U_0}{\partial Y} = \frac{\partial^2 U_0}{\partial Y^2}. \qquad (1.1.25)$$

If, on the other hand, $\chi(Re) = O(1)$ or $\chi(Re) \gg 1$, then it follows from (1.1.24) that

$$\frac{\partial P_0}{\partial Y} = 0. \qquad (1.1.26)$$

A boundary condition for this equation may be deduced by comparing the asymptotic expansions for the pressure in regions 1 and 2; these are given by (1.1.8) and (1.1.15), respectively. Keeping in mind that the solution in region 1 is given by (1.1.12), one can see that for any $\chi(Re) = O(1)$ or $\chi(Re) \gg 1$, the matching condition is written as

$$P_0\Big|_{Y=\infty} = 0.$$

Solving (1.1.26) with this condition, we can conclude that $P_0 = 0$ everywhere inside the boundary layer, which means that the pressure gradient should be disregarded in the longitudinal momentum equation (1.1.23), reducing it again to (1.1.25).

Equation (1.1.25) involves two unknown functions, U_0 and V_0. Being combined with the continuity equation (1.1.17), it forms a close set of equations, which serve to determine the velocity field (U_0, V_0) in the boundary layer. Now, we need to formulate the boundary conditions for (1.1.25), (1.1.17). A choice of the boundary conditions, which are compatible with a particular partial differential equation, depends on the type of the equation. The latter is determined by the higher order derivatives in the equation considered. If we start with the momentum equation (1.1.25), then we can see that the higher order derivative of U_0 with respect to Y is represented by the viscous term, $\partial^2 U_0 / \partial Y^2$, on the right-hand side of (1.1.25). The equation also involves the first order derivative of U_0 with respect to x in the convective term $U_0 \, \partial U_0 / \partial x$ on the left-hand side of the equation. This suggests that the momentum equation (1.1.25) is *parabolic*.

A classical example of a parabolic equation is the heat-transfer equation. When applied, say, to a heat-conducting rod, it may be written as

$$\frac{\partial T}{\partial t} = a \frac{\partial^2 T}{\partial x^2}. \tag{1.1.27}$$

Here, a is a positive constant that represents the thermal conductivity of the material the rod is made of, T is the temperature to be found, t is the time, and x is the coordinate measured along the rod. The latter is assumed to occupy the interval $x \in [0, d]$ of the x-axis, as shown in Figure 1.4. In order to determine the temperature T at a point $= x_0$ inside the rod, at time $t = t_0$, it is necessary to specify the temperature distribution along the rod at initial instant $t = 0$, and also to formulate the thermal conditions at the rod ends for $t \in [0, t_0]$. These boundaries are shown by braces in Figure 1.4. The solution at point (t_0, x_0), obviously, does not depend on the boundary conditions at the rod ends at later times $t > t_0$.

Similarly, if the coefficient U_0 in the first term on the left-hand side of (1.1.25) is positive, then to make the solution of the momentum equation (1.1.25) unique, we

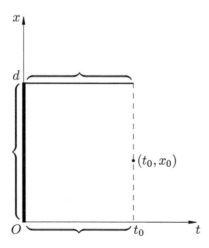

Fig. 1.4: The boundary conditions for the heat-transfer equation (1.1.27).

need, first of all, to specify initial distribution of U_0 at the leading edge of the flat plate. Since everywhere upstream of the plate the flow is unperturbed, we have

$$U_0 = 1 \quad \text{at} \quad x = 0, \ Y \in (0, \infty).$$

Secondly, corresponding to the boundary conditions at the rod ends, we need to formulate the conditions for U_0 on the plate surface ($Y = 0$) and at the outer edge of the boundary layer ($Y = \infty$). The first of these,

$$U_0 = 0 \quad \text{at} \quad Y = 0, \ x \in [0, 1],$$

is deduced by substituting the asymptotic expansion (1.1.15) for u into the no-slip condition (1.1.6). The boundary condition at the outer edge of the boundary layer

$$U_0 = 1 \quad \text{at} \quad Y = \infty$$

follows from matching with the solution (1.1.8), (1.1.12) in the outer flow region.

If the longitudinal velocity component U_0 were known, then the lateral velocity component V_0 could be found by integrating the continuity equation (1.1.17) with respect to Y. To perform this task, an initial condition for V_0 is needed. It is provided by the impermeability condition on the flat plate surface,

$$V_0 = 0 \quad \text{at} \quad Y = 0, \ x \in [0, 1].$$

Summarizing the above analysis, we see that the velocity components are represented in the boundary layer (region 2) in the form of asymptotic expansions

$$u = U_0(x, Y) + \cdots, \qquad v = Re^{-1/2} V_0(x, Y) + \cdots, \qquad Y = Re^{1/2} y, \qquad (1.1.28)$$

where functions U_0 and V_0 have to be found by solving the following boundary-value problem.

Problem 1.1 *Find solution to the equations*

$$U_0 \frac{\partial U_0}{\partial x} + V_0 \frac{\partial U_0}{\partial Y} = \frac{\partial^2 U_0}{\partial Y^2}, \qquad (1.1.29a)$$

$$\frac{\partial U_0}{\partial x} + \frac{\partial V_0}{\partial Y} = 0, \qquad (1.1.29b)$$

subject to the initial condition at the leading edge of the plate

$$U_0 = 1 \quad at \quad x = 0, \ Y \in (0, \infty), \qquad (1.1.29c)$$

the no-slip conditions on the plate surface

$$U_0 = V_0 = 0 \quad at \quad Y = 0, \ x \in [0, 1], \qquad (1.1.29d)$$

and the matching condition with the solution in the external inviscid flow

$$U_0 = 1 \quad at \quad Y = \infty. \qquad (1.1.29e)$$

1.1.2   Blasius solution

Asymptotic analysis presented in the previous section brings about a significant simplification to the mathematical description of the high-Reynolds-number flow past a flat plate. We see that the governing equations become less complicated, and the lateral momentum equation is no longer needed when determining the velocity field, (u, v). However, even more important is the fact that the *elliptic* Navier–Stokes equations (1.1.5) are reduced to the boundary-layer equations (1.1.29a), (1.1.29b), which are *parabolic*, and, therefore, possess the following property. Provided that the coefficient U_0 before $\partial U_0/\partial x$ in the first term on the left-hand side of (1.1.29a) is positive everywhere except on the wall, the solution of equations (1.1.29a), (1.1.29b) at any point (x_0, Y_0) in the boundary layer depends only on the boundary conditions upstream of this point, $x \leq x_0$. No variation of the boundary conditions downstream of the line $x = x_0$ can influence the solution at (x_0, Y_0). Hence, one can say that everywhere upstream of the trailing edge of the plate, the boundary layer 'does not know' that the plate is finite, and behaves in the same way as the boundary layer on a semi-infinite flat plate. Since the semi-infinite flat plate does not have a characteristic length scale, the solution to Problem 1.1 may be expected to have a *self-similar form*. This means that if distributions of velocity components U_0 and V_0 across the boundary layer were known at some location, say $x = 1$, then at any other location, U_0 and V_0 could be found using simple 'zooming'. Mathematically, this zooming is performed using affine transformations.

Let us suppose that

$$U_0 = F(x, Y), \qquad V_0 = G(x, Y) \tag{1.1.30}$$

is a solution to Problem 1.1, and, in particular, the distributions of U_0 and V_0 across the boundary layer at $x = 1$ are given by functions

$$F(1, Y) = f(Y), \qquad G(1, Y) = g(Y). \tag{1.1.31}$$

Let us then try to find an invariant affine transformation that leaves equations (1.1.29a), (1.1.29b), and boundary conditions (1.1.29c)–(1.1.29e) unchanged. We write

$$U_0 = A\tilde{U}_0, \qquad V_0 = B\tilde{V}_0, \qquad x = C\tilde{x}, \qquad Y = D\tilde{Y}, \tag{1.1.32}$$

where A, B, C, and D are positive constants.

The substitution of (1.1.32) into (1.1.29a) and (1.1.29b) gives

$$\frac{A^2}{C}\tilde{U}_0\frac{\partial\tilde{U}_0}{\partial\tilde{x}} + \frac{AB}{D}\tilde{V}_0\frac{\partial\tilde{U}_0}{\partial\tilde{Y}} = \frac{A}{D^2}\frac{\partial^2\tilde{U}_0}{\partial\tilde{Y}^2},$$

$$\frac{A}{C}\frac{\partial\tilde{U}_0}{\partial\tilde{x}} + \frac{B}{D}\frac{\partial\tilde{V}_0}{\partial\tilde{Y}} = 0,$$

while the boundary conditions (1.1.29c)–(1.1.29e) turn into

$$\begin{aligned} A\tilde{U}_0 &= 1 & \text{at} \quad \tilde{x} = 0, \\ \tilde{U}_0 = \tilde{V}_0 &= 0 & \text{at} \quad \tilde{Y} = 0, \\ A\tilde{U}_0 &= 1 & \text{at} \quad \tilde{Y} = \infty. \end{aligned}$$

To ensure that the equations and the boundary conditions remain unchanged, one has to set

$$\frac{A^2}{C} = \frac{AB}{D} = \frac{A}{D^2}, \qquad \frac{A}{C} = \frac{B}{D}, \qquad A = 1.$$

Solving these equations we find

$$A = 1, \qquad B = \frac{1}{\sqrt{C}}, \qquad D = \sqrt{C}, \tag{1.1.33}$$

with C remaining arbitrary.

Since the transformed boundary-value problem coincides with the original one, it also admits the solution (1.1.30), which now should be written as

$$\tilde{U}_0 = F(\tilde{x}, \tilde{Y}), \qquad \tilde{V}_0 = G(\tilde{x}, \tilde{Y}). \tag{1.1.34}$$

Using (1.1.32) and (1.1.33), we can express (1.1.34) in the form

$$U_0 = F\left(\frac{x}{C}, \frac{Y}{\sqrt{C}}\right), \qquad V_0 = \frac{1}{\sqrt{C}} G\left(\frac{x}{C}, \frac{Y}{\sqrt{C}}\right). \tag{1.1.35}$$

Here, parameter C may assume an arbitrary value, and therefore, may be considered as an additional independent variable. In particular, it may be chosen to coincide with x, which turns (1.1.35) into

$$U_0(x, Y) = F(1, \eta), \quad V_0(x, Y) = \frac{1}{\sqrt{x}} G(1, \eta), \quad \eta = \frac{Y}{\sqrt{x}}. \tag{1.1.36}$$

We see that for any $x \in (0, 1)$, the distributions of U_0 and V_0 across the boundary layer may be reduced to those at the trailing edge of the plate.

Using (1.1.31) in (1.1.36), we arrive at a conclusion that the solution to Problem 1.1 may be sought in the form

$$U_0 = f(\eta), \qquad V_0 = x^{-1/2} g(\eta), \tag{1.1.37}$$

where the independent variable

$$\eta = \frac{Y}{\sqrt{x}} \tag{1.1.38}$$

is referred to as the *similarity variable*.

In order to find functions $f(\eta)$ and $g(\eta)$, we have to substitute (1.1.37), (1.1.38) into the boundary-layer equations (1.1.29a), (1.1.29b), and the boundary conditions (1.1.29c)–(1.1.29e). It is easily seen that

$$\frac{\partial \eta}{\partial Y} = x^{-1/2}, \qquad \frac{\partial \eta}{\partial x} = -\frac{1}{2}\frac{Y}{x^{3/2}} = -\frac{1}{2}\frac{\eta}{x},$$

whence

$$\frac{\partial U_0}{\partial x} = -\frac{1}{2} x^{-1} \eta f', \quad \frac{\partial U_0}{\partial Y} = x^{-1/2} f', \quad \frac{\partial^2 U_0}{\partial Y^2} = x^{-1} f'', \quad \frac{\partial V_0}{\partial Y} = x^{-1} g'. \tag{1.1.39}$$

The substitution of (1.1.39) into the momentum equation (1.1.29a) results in

$$-\frac{1}{2}\eta f f' + g f' = f'',$$
(1.1.40)

with the continuity equation (1.1.29b) turning into

$$-\frac{1}{2}\eta f' + g' = 0.$$
(1.1.41)

It further follows from the no-slip conditions (1.1.29d) that

$$f(0) = g(0) = 0.$$
(1.1.42)

Since $\eta \to \infty$ as $x \to 0$, we find from the initial condition (1.1.29c) that

$$f \to 1 \quad \text{as} \quad \eta \to \infty.$$
(1.1.43)

Interestingly enough, the matching condition (1.1.29e) also leads to (1.1.43).

The boundary-value problem (1.1.40)–(1.1.43) serves to find $f(\eta)$ and $g(\eta)$. We can simplify this problem as follows. Let us write the continuity equation (1.1.41) as

$$g' = \frac{1}{2}\eta f' = \frac{1}{2}(\eta f)' - \frac{1}{2}f.$$
(1.1.44)

If we now introduce a new function $\varphi(\eta)$, such that

$$\varphi'(\eta) = f(\eta), \quad \varphi(0) = 0,$$
(1.1.45)

then the equation (1.1.44) may be integrated to yield

$$g = \frac{1}{2}\eta\varphi' - \frac{1}{2}\varphi.$$
(1.1.46)

Here the constant of integration has been set to zero in view of the boundary conditions (1.1.42), (1.1.45) imposed upon g and φ at $\eta = 0$.

It remains to substitute (1.1.46) into the momentum equation (1.1.40), and we will have the following ordinary differential equation for $\varphi(\eta)$:

$$\varphi''' + \frac{1}{2}\varphi\varphi'' = 0.$$
(1.1.47a)

It is known as the *Blasius equation*. The boundary conditions for (1.1.47a) are deduced by combining (1.1.45) with (1.1.42) and (1.1.43). We have

$$\varphi(0) = \varphi'(0) = 0, \quad \varphi'(\infty) = 1.$$
(1.1.47b)

While there is no analytical solution to the boundary-value problem (1.1.47), the function $\varphi(\eta)$ may be easily calculated using a numerical technique. Once $\varphi(\eta)$ is found,

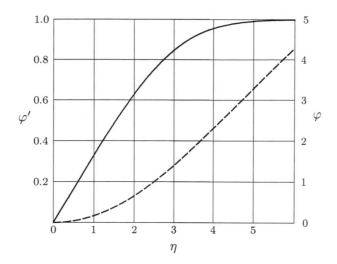

Fig. 1.5: Solution of the Blasius problem (1.1.47). The solid line shows the longitudinal velocity profile $\varphi'(\eta)$; the dashed line represents function $\varphi(\eta)$.

the velocity components in the boundary layer may be determined by substituting (1.1.45), (1.1.46) into (1.1.37),

$$U_0 = \varphi'(\eta), \qquad V_0 = \frac{1}{2\sqrt{x}}(\eta\varphi' - \varphi). \tag{1.1.48}$$

The results of numerical solution to the Blasius problem are shown in Figure 1.5.

For future reference, we need to study the asymptotic behaviour of $\varphi(\eta)$ near the wall ($\eta \to 0$) and at the outer edge of the boundary layer ($\eta \to \infty$). If η is small then, in view of (1.1.47b), the first non-zero term of the Taylor expansion of $\varphi(\eta)$ is written as

$$\varphi(\eta) = \frac{1}{2}\lambda\eta^2 + \cdots \quad \text{as} \quad \eta \to 0, \tag{1.1.49}$$

where $\lambda = \varphi''(0)$, and from the numerical solution of (1.1.47) it is known that $\lambda = 0.3321$.

Now let us turn to the asymptotic behaviour of $\varphi(\eta)$ at large values of η. Due to (1.1.47b), the leading-order term of the asymptotic expansion of $\varphi(\eta)$ is written as

$$\varphi(\eta) = \eta + \cdots \quad \text{as} \quad \eta \to \infty.$$

For the next term let us try a power function

$$\varphi(\eta) = \eta + A\eta^\alpha + \cdots . \tag{1.1.50}$$

Here, A is a constant, and to ensure that the second term in (1.1.50) is much smaller than the first term (as required by the definition of an asymptotic expansion), we have to assume that $\alpha < 1$.

The substitution of (1.1.50) into (1.1.47a) results in

$$A\,\alpha(\alpha-1)(\alpha-2)\eta^{\alpha-3} = -\frac{1}{2}A\,\alpha(\alpha-1)\eta^{\alpha-1}.$$

At large values of η, the right-hand side of this equation is much greater than the left-hand side. Therefore, we have to set

$$A\,\alpha(\alpha-1) = 0.$$

With $A \neq 0$, there are two possibilities to satisfy this equation:

$$\alpha = 0 \quad \text{and} \quad \alpha = 1.$$

Taking into account that $\alpha < 1$, we shall choose $\alpha = 0$, which turns (1.1.50) into

$$\varphi(\eta) = \eta + A + \cdots \quad \text{as} \quad \eta \to \infty. \tag{1.1.51}$$

It follows from the numerical solution of the Blasius problem (1.1.47) that A is negative, $A = -1.7208$. For that reason we shall write

$$\varphi(\eta) = \eta - A_- + \cdots \quad \text{as} \quad \eta \to \infty, \tag{1.1.52}$$

where a new constant, $A_- = -A = 1.7208$.

In conclusion, it should be noted that the self-similar form (1.1.37), (1.1.38) of the solution for the flat-plate flow was predicted by Prandtl (1904). Then, a detailed analysis of the behaviour of the fluid-dynamic function, as described above, was conducted by Blasius (1908).

Exercises 1

1. An alternative way to solve Problem 1.1 is as follows. Instead of dealing with velocity components, U_0 and V_0, one can introduce the stream function, $\Psi(x, Y)$. Its existence follows from the continuity equation (1.1.29b), and we can write

$$\frac{\partial \Psi}{\partial x} = -V_0, \qquad \frac{\partial \Psi}{\partial Y} = U_0.$$

   Your task is to try to find the solution of the boundary-value problem (1.1.29) in the form

$$\Psi(x, Y) = \sqrt{x}\,\varphi(\eta), \qquad \eta = \frac{Y}{\sqrt{x}}.$$

   Show that the function $\varphi(\eta)$ satisfies the Blasius equation (1.1.47a) and boundary conditions (1.1.47b).

2. The higher-order terms in the asymptotic expansions (1.1.49) and (1.1.52) may be found through iterative use of the Blasius equation (1.1.47a):

   (a) when dealing with the flow near the plate surface, write the Blasius equation in the form

$$\varphi''' = -\frac{1}{2}\varphi\varphi''$$

and, using (1.1.49) for the right-hand side, show that the two-term asymptotic expansion of $\varphi(\eta)$ near the plate surface is written as

$$\varphi(\eta) = \frac{1}{2}\lambda\eta^2 - \frac{\lambda^2}{2\cdot 5!}\eta^5 + \cdots \quad \text{as} \quad \eta \to 0.$$

(b) now rearrange the Blasius equation (1.1.47a) in the form

$$\frac{\varphi'''}{\varphi''} = -\frac{1}{2}\varphi$$

and, using (1.1.52) for the right-hand side, demonstrate that

$$\varphi = \eta - A_- + \frac{C}{(\eta - A_-)^2}\, e^{-(\eta - A_-)^2/4} + \cdots \quad \text{as} \quad \eta \to \infty, \qquad (1.1.53)$$

with C being a constant.

   Hint: You may find the integration by parts technique described in Section 1.1.2 in Part 2 of this book series useful.

3. The drag force produced by the boundary layers on the two sides of the flat plate (see Figure 1.1) is calculated as

$$D = 2\int\limits_0^L \hat{\tau}_w \, d\hat{x},$$

where $\hat{\tau}_w$ is the shear stress on the plate surface, also known as the *skin friction*. To find $\hat{\tau}_w$, consider equation (1.5.25) on page 46 in Part 1 of this book series. Choosing i and j in this equation to be 2 and 1, respectively, deduce that[5]

$$\hat{\tau}_w = p_{\hat{y}\hat{x}}\Big|_{\hat{y}=0} = \mu\left(\frac{\partial \hat{v}}{\partial \hat{x}} + \frac{\partial \hat{u}}{\partial \hat{y}}\right)\Bigg|_{\hat{y}=0} = \mu\frac{\partial \hat{u}}{\partial \hat{y}}\Big|_{\hat{y}=0}. \qquad (1.1.54)$$

Then show, using the Blasius solution, that the drag coefficient

$$C_D = \frac{D}{\frac{1}{2}\rho V_\infty^2 L} = \frac{8\lambda}{\sqrt{Re}}.$$

4. The Blasius solution for the boundary layer on a flat plate surface developes a singularity at the leading edge of the plate. Taking into account that the boundary-layer approximation remains valid as long as the relative thickness of the boundary layer is small, that is

$$\frac{\hat{y}}{\hat{x}} \ll 1,$$

find the characteristic size of a small region near the leading edge, where this condition is violated.

---

[5]Remember that when deriving equation (1.5.25) in Part 1 we used dimensional variables.

Represent the solution in this new region in the form

$$\left.\begin{aligned}
u(x, y; Re) &= U^\circ(X^\circ, Y^\circ) + \cdots, \\
v(x, y; Re) &= V^\circ(X^\circ, Y^\circ) + \cdots, \\
p(x, y; Re) &= P^\circ(X^\circ, Y^\circ) + \cdots,
\end{aligned}\right\} \tag{1.1.55}$$

where the order-one independent variables X°, Y° are defined by

$$x = Re^{-1}X^\circ, \qquad y = Re^{-1}Y^\circ.$$

Through substitution of the asymptotic expansions (1.1.55) into the Navier–Stokes equations (1.1.5), show that in the leading edge region the flow is described by the full set of the Navier–Stokes equations

$$U^\circ \frac{\partial U^\circ}{\partial X^\circ} + V^\circ \frac{\partial U^\circ}{\partial Y^\circ} = -\frac{\partial P^\circ}{\partial X^\circ} + \frac{\partial^2 U^\circ}{\partial X^{\circ 2}} + \frac{\partial^2 U^\circ}{\partial Y^{\circ 2}},$$

$$U^\circ \frac{\partial V^\circ}{\partial X^\circ} + V^\circ \frac{\partial V^\circ}{\partial Y^\circ} = -\frac{\partial P^\circ}{\partial Y^\circ} + \frac{\partial^2 V^\circ}{\partial X^{\circ 2}} + \frac{\partial^2 V^\circ}{\partial Y^{\circ 2}},$$

$$\frac{\partial U^\circ}{\partial X^\circ} + \frac{\partial V^\circ}{\partial Y^\circ} = 0,$$

with unit value of the 'local Reynolds number'.

1.2   Prandtl's Hierarchical Strategy

We shall now apply the ideas developed in Section 1.1 to the flow past a body with curvilinear surface S; see Figure 1.6. We shall suppose, as before, that the flow considered is two-dimensional and steady. We shall also suppose that the fluid is incompressible and the body force, $\mathbf{f}$, is negligible. In this case, the Navier–Stokes equations assume the form given by equations (1.1.1). To make these equations dimensionless, we shall use again the scalings (1.1.4). A choice of the characteristic length scale L in (1.1.4) depends on the shape of the body considered. For instance, when dealing with the flow past an aerofoil, it is convenient to choose L to coincide with the aerofoil chord. For the flow past a circular cylinder, the cylinder's radius may be chosen as the characteristic length L.

The substitution of (1.1.4) into (1.1.1) results in

$$\left.\begin{aligned}
u\frac{\partial u}{\partial x} + v\frac{\partial u}{\partial y} &= -\frac{\partial p}{\partial x} + \frac{1}{Re}\left(\frac{\partial^2 u}{\partial x^2} + \frac{\partial^2 u}{\partial y^2}\right), \\
u\frac{\partial v}{\partial x} + v\frac{\partial v}{\partial y} &= -\frac{\partial p}{\partial y} + \frac{1}{Re}\left(\frac{\partial^2 v}{\partial x^2} + \frac{\partial^2 v}{\partial y^2}\right), \\
\frac{\partial u}{\partial x} + \frac{\partial v}{\partial y} &= 0,
\end{aligned}\right\} \tag{1.2.1}$$

with the Reynolds number defined by

$$Re = \frac{V_\infty L}{\nu}.$$

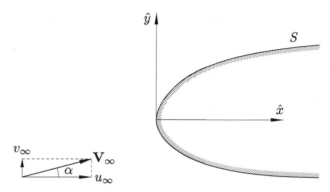

Fig. 1.6: Problem layout.

The free-stream boundary conditions for equations (1.2.1) are written in the non-dimensional variables as

$$
\left.\begin{array}{l}
u \to \cos \alpha, \\
v \to \sin \alpha, \\
p \to 0
\end{array}\right\} \quad \text{as} \quad x^2 + y^2 \to \infty. \tag{1.2.2}
$$

Here, α is the angle made by the free-stream velocity vector with the x-axis (see Figure 1.6). The no-slip conditions on the body surface

$$
u = v = 0 \quad \text{on} \quad S \tag{1.2.3}
$$

close the formulation of the boundary-value problem (1.2.1)–(1.2.3).

We shall assume that

$$
Re \to \infty,
$$

and we start the flow analysis with the *outer region*.

1.2.1 Outer inviscid flow region

The asymptotic analysis of the flow in the outer region is based on the limit

$$
x = O(1), \quad y = O(1), \quad Re \to \infty.
$$

Restricting for the moment our attention to the leading-order approximation, we write the asymptotic expansions for the three unknown functions u, v, and p in the form

$$
\left.\begin{array}{l}
u(x, y; Re) = u_0(x, y) + \cdots, \quad v(x, y; Re) = v_0(x, y) + \cdots, \\
p(x, y; Re) = p_0(x, y) + \cdots.
\end{array}\right\} \tag{1.2.4}
$$

The substitution of (1.2.4) into (1.2.1) leads to the Euler equations describing the inviscid flow motion:

$$\left.\begin{aligned}
u_0\frac{\partial u_0}{\partial x} + v_0\frac{\partial u_0}{\partial y} &= -\frac{\partial p_0}{\partial x}, \\
u_0\frac{\partial v_0}{\partial x} + v_0\frac{\partial v_0}{\partial y} &= -\frac{\partial p_0}{\partial y}, \\
\frac{\partial u_0}{\partial x} + \frac{\partial v_0}{\partial y} &= 0.
\end{aligned}\right\}
\qquad (1.2.5)$$

We know from the inviscid flow theory (see Section 3 in Part 1) that the Euler equations (1.2.5) are compatible with the free-stream conditions

$$\left.\begin{aligned}
u_0 &\to \cos\alpha, \\
v_0 &\to \sin\alpha, \\
p_0 &\to 0
\end{aligned}\right\}
\quad \text{as} \quad x^2 + y^2 \to \infty,
\qquad (1.2.6)$$

and the impermeability condition

$$\left(n_x u_0 + n_y v_0\right)\Big|_S = 0
\qquad (1.2.7)$$

on the body surface S. The latter states that body surface is impenetrable for the fluid, and, therefore, the velocity vector (u_0, v_0) should be tangent to the body surface, that is, perpendicular to the normal vector $\mathbf{n} = (n_x, n_y)$.

When solved, the boundary-value problem (1.2.5)–(1.2.7) gives the distributions of the velocity (u_0, v_0) and pressure p_0 in the flow field; these are dependent on the body shape.[6] Of particular interest are the tangential velocity and the pressure on body surface. We shall denote these as $U_e(s)$ and $P_e(s)$, respectively, with s being the arc length measured along the body surface from conveniently chosen point O, say, the front stagnation point (see Figure 1.7). The inviscid flow calculations invariably show that $U_e \neq 0$, except at the front stagnation point O. This means that the no-slip condition (1.2.3) is not satisfied by the solution in the outer region, and one has to introduce an *inner region* that occupies a thin layer adjacent to the body surface S. In what follows, we shall call it the *boundary layer*.

1.2.2   Boundary layer

When analysing the fluid motion in the boundary layer, it is convenient to use the body-fitted coordinates.[7] Remember that in these coordinates the position of a point M is defined by coordinates (s, n). The first of these is the distance measured along the body contour from an initial point O to point the M', which is obtained by dropping a perpendicular from M to the body surface. The second coordinate, n, is the distance between M and body surface; see Figure 1.7. When dealing with vector quantities we use the unit vector pair $(\mathbf{e}_1, \mathbf{e}_2)$ which is oriented such that $\mathbf{e}_1$ is parallel to the tangent

---

[6]Various examples of such solutions are given in Chapter 3 in Part 1 of this book series.

[7]See Section 1.8 in Part 1.

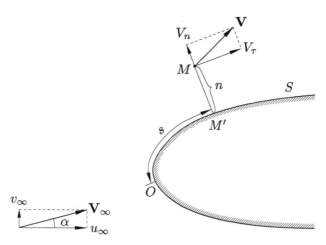

Fig. 1.7: Body-fitted coordinates.

to the body contour at point M', while $\mathbf{e}_2$ is directed along the n-axis. Correspondingly, the velocity vector is decomposed as

$$\mathbf{V} = V_\tau \mathbf{e}_1 + V_n \mathbf{e}_2,$$

with V_τ and V_n called the tangential and normal velocity components.

The non-dimensional Navier–Stokes equations (1.2.1) are written in these coordinates as[8]

$$\frac{V_\tau}{H_1}\frac{\partial V_\tau}{\partial s} + V_n\frac{\partial V_\tau}{\partial n} + \frac{\kappa V_\tau V_n}{H_1} = -\frac{1}{H_1}\frac{\partial p}{\partial s} + \frac{1}{Re}\left[\frac{1}{H_1}\frac{\partial}{\partial s}\left(\frac{1}{H_1}\frac{\partial V_\tau}{\partial s}\right)+\right.$$
$$\left.+ \frac{\partial^2 V_\tau}{\partial n^2} + \kappa\frac{\partial}{\partial n}\left(\frac{V_\tau}{H_1}\right) + \frac{\kappa}{H_1^2}\frac{\partial V_n}{\partial s} + \frac{1}{H_1}\frac{\partial}{\partial s}\left(\frac{\kappa V_n}{H_1}\right)\right], \quad (1.2.8a)$$

$$\frac{V_\tau}{H_1}\frac{\partial V_n}{\partial s} + V_n\frac{\partial V_n}{\partial n} - \frac{\kappa V_\tau^2}{H_1} = -\frac{\partial p}{\partial n} + \frac{1}{Re}\left[\frac{1}{H_1}\frac{\partial}{\partial s}\left(\frac{1}{H_1}\frac{\partial V_n}{\partial s}\right)+\right.$$
$$\left.+ \frac{\partial^2 V_n}{\partial n^2} + \kappa\frac{\partial}{\partial n}\left(\frac{V_n}{H_1}\right) - \frac{\kappa}{H_1^2}\frac{\partial V_\tau}{\partial s} - \frac{1}{H_1}\frac{\partial}{\partial s}\left(\frac{\kappa V_\tau}{H_1}\right)\right], \quad (1.2.8b)$$

$$\frac{1}{H_1}\frac{\partial V_\tau}{\partial s} + \frac{\partial V_n}{\partial n} + \frac{\kappa V_n}{H_1} = 0. \quad (1.2.8c)$$

Here, κ is the local curvature of the body contour, and H_1 is the Lamé coefficient given by

$$H_1 = 1 + \kappa(s)n. \quad (1.2.9)$$

Our task is to construct the solution of the equations (1.2.8) in the boundary layer. Since the boundary layer is supposed to be thin, we have to rescale the normal coordinate as

[8]See equations (1.8.49) on page 87 in Part 1.

$$n = \delta(Re)N, \tag{1.2.10}$$

where $\delta(Re)$ is such that

$$\delta(Re) \to 0 \quad \text{as} \quad Re \to \infty.$$

The asymptotic analysis of the Navier–Stokes equations (1.2.8) in the boundary layer is based on the limit

$$s = O(1), \quad N = \delta^{-1}n = O(1), \quad Re \to \infty.$$

The velocity components and the pressure will be sought in the form of the asymptotic expansions

$$\left. \begin{aligned} V_\tau(s, n; Re) = U_0(s, N) + \cdots, \qquad V_n = \sigma(Re)V_0(s, N) + \cdots, \\ p(s, n; Re) = P_0(s, N) + \cdots. \end{aligned} \right\} \tag{1.2.11}$$

Notice that, unlike in the flat plate problem, where the correspondent expansions are given by (1.1.15), we now know that the pressure at the outer edge of the boundary layer, $P_e(s)$, is an order one quantity. This is why $\chi(Re)$ has been chosen to be $\chi(Re) = O(1)$.

The substitution of (1.2.10) into (1.2.9) shows that in the boundary layer

$$H_1 = 1 + \delta(Re)\,\kappa(s)N,$$

which means that, when dealing with the leading-order equations, we can simply set $H_1 = 1$. Let us now consider the continuity equation (1.2.8c). By substituting (1.2.11) together with (1.2.10) into (1.2.8c), we find that

$$\frac{\partial U_0}{\partial s} + \frac{\sigma(Re)}{\delta(Re)}\frac{\partial V_0}{\partial N} + \sigma(Re)\kappa V_0 = 0. \tag{1.2.12}$$

Since $\delta(Re) \ll 1$, the third term in the above equation is small as compared to the second term, and should be disregarded. In order to avoid further degeneration in (1.2.12), we have to set

$$\sigma(Re) = \delta(Re), \tag{1.2.13}$$

which renders the continuity equation in the form

$$\frac{\partial U_0}{\partial s} + \frac{\partial V_0}{\partial N} = 0. \tag{1.2.14}$$

Now, we turn to the tangential momentum equation (1.2.8a). Using (1.2.11), (1.2.10) and (1.2.13), the three terms on the left-hand side of (1.2.8a) are calculated as

$$\frac{V_\tau}{H_1}\frac{\partial V_\tau}{\partial s} = U_0\frac{\partial U_0}{\partial s} + \cdots, \quad V_n\frac{\partial V_\tau}{\partial n} = V_0\frac{\partial U_0}{\partial N} + \cdots, \quad \frac{\kappa V_\tau V_n}{H_1} = \sigma\kappa U_0 V_0 + \cdots.$$

The first two terms remain finite, while the third term tends to zero as $Re \to \infty$, and, therefore, has to be disregarded in the leading-order approximation.

The first term on the right-hand side of (1.2.8a) represents the pressure gradient. Differentiating of the asymptotic expansion for the pressure in (1.2.11), we have

$$\frac{1}{H_1}\frac{\partial p}{\partial s} = \frac{\partial P_0}{\partial s} + \cdots .$$

Using (1.2.11) and (1.2.10), we find that the viscous terms are given by

$$\frac{1}{Re}\frac{1}{H_1}\frac{\partial}{\partial s}\left(\frac{1}{H_1}\frac{\partial V_\tau}{\partial s}\right) = \frac{1}{Re}\frac{\partial^2 U_0}{\partial s^2} + \cdots ,$$

$$\frac{1}{Re}\frac{\partial^2 V_\tau}{\partial n^2} = \frac{1}{Re\delta^2}\frac{\partial^2 U_0}{\partial N^2} + \cdots ,$$

$$\frac{1}{Re}\kappa\frac{\partial}{\partial n}\left(\frac{V_\tau}{H_1}\right) = \frac{1}{Re\delta}\kappa\frac{\partial U_0}{\partial N} + \cdots ,$$

$$\frac{1}{Re}\frac{\kappa}{H_1^2}\frac{\partial V_n}{\partial s} = \frac{\sigma}{Re}\kappa\frac{\partial V_0}{\partial s} + \cdots ,$$

$$\frac{1}{Re}\frac{1}{H_1}\frac{\partial}{\partial s}\left(\frac{\kappa V_n}{H_1}\right) = \frac{\sigma}{Re}\frac{\partial}{\partial s}(\kappa V_0) + \cdots .$$

The second of these is clearly much larger than the rest, and, therefore, this is the only viscous term that should be retained in the tangential momentum equation (1.2.8a). We have

$$U_0\frac{\partial U_0}{\partial s} + V_0\frac{\partial U_0}{\partial N} = -\frac{\partial P_0}{\partial s} + \frac{1}{Re\delta^2}\frac{\partial^2 U_0}{\partial N^2}.$$

The equations governing the flow in the boundary layer should admit solutions that satisfy the matching condition with the solution in the outer region and the no-slip condition on the body surface. This is only possible if the second order derivative $\partial^2 U_0/\partial N^2$ is retained in the above equation, which is achieved by setting

$$\delta = Re^{-1/2}. \tag{1.2.15}$$

As a result, the tangential momentum equation assumes the form

$$U_0\frac{\partial U_0}{\partial s} + V_0\frac{\partial U_0}{\partial N} = -\frac{\partial P_0}{\partial s} + \frac{\partial^2 U_0}{\partial N^2}. \tag{1.2.16}$$

It remains to analyse the normal momentum equation (1.2.8b). The dominant term on the left-hand side of this equation is

$$\frac{\kappa V_\tau^2}{H_1} = \kappa U_0^2 + \cdots . \tag{1.2.17}$$

The normal pressure gradient on the right-hand side of (1.2.8b) is calculated as

$$\frac{\partial p}{\partial n} = \frac{1}{\delta(Re)}\frac{\partial P_0}{\partial N} + \cdots = Re^{1/2}\frac{\partial P_0}{\partial N} + \cdots . \tag{1.2.18}$$

Finally, the dominant viscous term is

$$\frac{1}{Re}\frac{\partial^2 V_n}{\partial n^2} = Re^{-1/2}\frac{\partial^2 V_0}{\partial N^2} + \cdots .\tag{1.2.19}$$

Substituting (1.2.17), (1.2.18), and (1.2.19) into (1.2.8b), and multiplying all the terms by $Re^{-1/2}$, we have

$$-Re^{-1/2}\kappa U_0^2 = -\frac{\partial P_0}{\partial N} + Re^{-1}\frac{\partial^2 V_0}{\partial N^2}.$$

Clearly, setting $Re \to \infty$ in this equation results in

$$\frac{\partial P_0}{\partial N} = 0,\tag{1.2.20}$$

which means that the pressure does not change across the boundary layer.

Summarizing the results of the above analysis, we substitute (1.2.15) into (1.2.13) and then into (1.2.10) and (1.2.11), which leads to a conclusion that in the boundary layer

$$\left.\begin{aligned}V_\tau(s,n;Re) &= U_0(s,N) + \cdots , & V_n &= Re^{-1/2}V_0(s,N) + \cdots ,\\ p(s,n;Re) &= P_0(s,N) + \cdots , & n &= Re^{-1/2}N.\end{aligned}\right\}\tag{1.2.21}$$

To find functions U_0, V_0, and P_0, equations (1.2.16), (1.2.14) and (1.2.20) must be solved. Let us now formulate the boundary conditions for these equations. We shall start with the matching of the solution in the boundary layer with the solution in the inviscid outer region.

1.2.3   Matching procedure

The flow in the inviscid region is described by the Euler equations (1.2.5) subject to the free-stream condition (1.2.6) and impermeability condition on the body surface (1.2.7). As has been already mentioned, the solution of the boundary-value problem (1.2.5)–(1.2.7) gives the velocity and pressure at any point in the flow field, including the tangential velocity $U_e(s)$ and the pressure $P_e(s)$ on the body surface S. Let us now refine the information about the inviscid flow behaviour near this surface. When performing this task, it is convenient to express the Euler equations (1.2.5) in the body-fitted coordinates (s, n). This is done by setting $Re = \infty$ in the Navier–Stokes equations (1.2.8). Denoting the leading-order terms in the asymptotic expansion of V_τ and V_n in the outer region by $\breve{u}_0$ and $\breve{v}_0$, namely

$$\left.\begin{aligned}V_\tau(s,n;Re) &= \breve{u}_0(s,n) + \cdots , & V_n(s,n;Re) &= \breve{v}_0(s,n) + \cdots ,\\ p(s,n;Re) &= p_0(s,n) + \cdots ,\end{aligned}\right\}\tag{1.2.22}$$

we have

$$\frac{\breve{u}_0}{H_1}\frac{\partial \breve{u}_0}{\partial s} + \breve{v}_0\frac{\partial \breve{u}_0}{\partial n} + \frac{\kappa \breve{u}_0 \breve{v}_0}{H_1} = -\frac{1}{H_1}\frac{\partial p_0}{\partial s},\tag{1.2.23a}$$

$$\frac{\breve{u}_0}{H_1}\frac{\partial \breve{v}_0}{\partial s} + \breve{v}_0\frac{\partial \breve{v}_0}{\partial n} - \frac{\kappa \breve{u}_0^2}{H_1} = -\frac{\partial p_0}{\partial n},\tag{1.2.23b}$$

$$\frac{1}{H_1}\frac{\partial \breve{u}_0}{\partial s} + \frac{\partial \breve{v}_0}{\partial n} + \frac{\kappa \breve{v}_0}{H_1} = 0,\tag{1.2.23c}$$

with $H_1 = 1 + \kappa(s)n$.

In these coordinates, the impermeability condition on the body surface is written as

$$\check{v}_0 = 0 \quad \text{at} \quad n = 0, \ s \ge 0. \tag{1.2.24}$$

Therefore, at any point on the body surface, the tangential momentum equation (1.2.23a) assumes the form

$$\check{u}_0 \frac{\partial \check{u}_0}{\partial s} = -\frac{\partial p_0}{\partial s}. \tag{1.2.25}$$

We know that

$$\check{u}_0 \Big|_{n=0} = U_e(s), \qquad p_0 \Big|_{n=0} = P_e(s), \tag{1.2.26}$$

which, being substituted into (1.2.25), yields[9]

$$\frac{dP_e}{ds} = -U_e \frac{dU_e}{ds}. \tag{1.2.27}$$

Now we turn to the normal momentum equation (1.2.23b). Using (1.2.24) in (1.2.23b), and taking into account that $H_1 = 1$ at $n = 0$, we find that

$$\frac{\partial p_0}{\partial n} \Big|_{n=0} = \kappa \check{u}_0^2 = \kappa U_e^2.$$

Consequently, the first two terms of the Taylor expansion of p_0 near the body surface are written as

$$p_0 = P_e(s) + \kappa U_e^2 n + \cdots \quad \text{as} \quad n \to 0. \tag{1.2.28}$$

Finally, it follows from the continuity equation (1.2.23c) that

$$\frac{\partial \check{v}_0}{\partial n} \Big|_{n=0} = -\frac{dU_e}{ds},$$

whence

$$\check{v}_0 = -\frac{dU_e}{ds} n + \cdots \quad \text{as} \quad n \to 0. \tag{1.2.29}$$

We are now ready to perform the matching with the solution in the boundary layer. We start with the tangential velocity component V_τ. In the outer inviscid region the asymptotic expansion of V_τ is given by the first of equations (1.2.22),

$$V_\tau = \check{u}_0(s, n) + \cdots .$$

According to (1.2.21), in the boundary layer

$$V_\tau = U_0(s, N) + \cdots , \qquad N = Re^{1/2} n.$$

---

[9]Of course, the equation (1.2.27) may be also deduced from the Bernoulli equation:

$$\tfrac{1}{2} U_e(s) + P_e(s) = const.$$

Hence, the Prandtl matching rule states

$$\lim_{n \to 0} \breve{u}_0(s, n) = \lim_{N \to \infty} U_0(s, N). \tag{1.2.30}$$

Using the first equation in (1.2.26) on the left-hand side of (1.2.30), we can conclude that the sought boundary condition for U_0 is

$$U_0 = U_e(s) \quad \text{at} \quad N = \infty. \tag{1.2.31}$$

Similarly, the matching condition for the pressure is written as

$$P_0 = P_e(s) \quad \text{at} \quad N = \infty. \tag{1.2.32}$$

Integration of the equation (1.2.20) with the boundary condition (1.2.32) leads to a conclusion that

$$P_0 = P_e(s)$$

everywhere in the boundary layer. Taking this into account, and using (1.2.27), we can write the tangential momentum equation (1.2.16) as

$$U_0 \frac{\partial U_0}{\partial s} + V_0 \frac{\partial U_0}{\partial N} = U_e \frac{dU_e}{ds} + \frac{\partial^2 U_0}{\partial N^2}. \tag{1.2.33}$$

In order to study the flow in the boundary layer, the momentum equation (1.2.33) must be solved together with the continuity equation (1.2.14). In addition to the matching condition (1.2.31), the boundary-layer equations (1.2.33), (1.2.14) require the no-slip conditions on the body surface,

$$U_0 = V_0 = 0 \quad \text{at} \quad N = 0,$$

and an initial condition. The form of the latter depends on a particular problem considered. For the flow past a body with a rounded nose, we have

$$U_0 = 0 \quad \text{at} \quad s = 0.$$

Here, it is taken into account that the fluid velocity is zero at the front stagnation point, $s = 0$.

Thus, in the boundary layer, the following boundary-value problem must be solved:

Problem 1.2 *Find the solution to the equations*

$$U_0 \frac{\partial U_0}{\partial s} + V_0 \frac{\partial U_0}{\partial N} = U_e \frac{dU_e}{ds} + \frac{\partial^2 U_0}{\partial N^2}, \tag{1.2.34a}$$

$$\frac{\partial U_0}{\partial s} + \frac{\partial V_0}{\partial N} = 0, \tag{1.2.34b}$$

subject to the initial condition

$$U_0 = 0 \quad at \quad s = 0, \quad N \in [0, \infty), \tag{1.2.34c}$$

the no-slip conditions

$$U_0 = V_0 = 0 \quad at \quad N = 0, \quad s > 0 \tag{1.2.34d}$$

and the condition of matching with the solution in the external inviscid flow region

$$U_0 = U_e(s) \quad at \quad N = \infty, \quad s > 0. \tag{1.2.34e}$$

1.2.4 Displacement effect of the boundary layer

Given the distribution of the tangential velocity at the outer edge of the boundary layer $U_e(s)$, Problem 1.2 serves to determine the tangential and normal velocity components, U_0 and V_0, in the boundary layer. It should be noticed that, when formulating Problem 1.2, we did not need to perform the matching of the normal velocity component V_n. Still, it is interesting to see what this matching will lead to.

To determine the behaviour of V_0 near the outer edge of the boundary layer ($N \to \infty$), we write the continuity equation (1.2.34b) as

$$\frac{\partial V_0}{\partial N} = -\frac{\partial U_0}{\partial s},$$

and integrate it with respect to N, keeping in mind that $V_0 = 0$ at $N = 0$. We have

$$V_0 = -\int_0^N \frac{\partial U_0}{\partial s} \, dN. \tag{1.2.35}$$

The boundary condition (1.2.34e) shows that the integrand in (1.2.35) tends to dU_e/ds as N becomes large. In general, $dU_e/ds \neq 0$, which makes the integral divergent as $N \to \infty$. Keeping this in mind, we rearrange (1.2.35) as

$$V_0 = -\frac{dU_e}{ds}N + \int_0^N \left(\frac{dU_e}{ds} - \frac{\partial U_0}{\partial s} \right) dN = -\frac{dU_e}{ds}N + \frac{\partial}{\partial s}\left[U_e(s) \int_0^N \left(1 - \frac{U_0}{U_e} \right) dN \right].$$

We now see that the two-term asymptotic expansion of V_0 at the outer edge of the boundary layer has the form

$$V_0 = -\frac{dU_e}{ds}N + \frac{d}{ds}\left[U_e(s)\Delta^*(s) \right] + \cdots \quad as \quad N \to \infty, \tag{1.2.36}$$

where

$$\Delta^*(s) = \int_0^\infty \left(1 - \frac{U_0}{U_e} \right) dN. \tag{1.2.37}$$

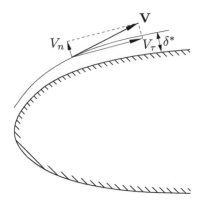

Fig. 1.8: Displacement thickness of the boundary layer, $\delta^*(s)$.

The function $\Delta^*(s)$, being multiplied by $Re^{-1/2}$, is known as the (dimensionless) *displacement thickness* of the boundary layer, $\delta^*(s)$. The reason for the name is explained as follows. Multiplying both sides of equation (1.2.36) by $Re^{-1/2}$, and using (1.2.21), we have

$$V_n = -\frac{dU_e}{ds}n + \frac{d}{ds}\left(U_e\delta^*\right),$$

or, equivalently,

$$V_n = (\delta^* - n)\frac{dU_e}{ds} + U_e\frac{d\delta^*}{ds}. \tag{1.2.38}$$

This shows that the line drawn at a distance $n = \delta^*(s)$ from the body contour (see Figure 1.8) is a streamline. Indeed, setting $n = \delta^*$ in (1.2.38) results in

$$\left.\frac{V_n}{U_e}\right|_{n=\delta^*} = \frac{d\delta^*}{ds}, \tag{1.2.39}$$

which proves that the velocity vector is tangent to the line $n = \delta^*(s)$. Equation (1.2.39) gives an 'improved boundary condition' for the Euler equations (1.2.5). It shows that the solution in the inviscid flow region can be made more accurate if, instead of using the impermeability condition (1.2.7) on the real body surface, one applies the impermeability condition on the surface of the so-called *effective body* which is displaced from the real body surface through a distance equal to the displacement thickness of the boundary layer, $\delta^*(s)$. Since $\delta^* \sim Re^{-1/2}$, one can expect the boundary layer to cause $O(Re^{-1/2})$ perturbations in the outer inviscid flow region. Consequently, the asymptotic expansions (1.2.22) have to be modified as

$$V_\tau(s, n; Re) = \breve{u}_0(s, n) + Re^{-1/2}\breve{u}_1(s, n) + \cdots, \tag{1.2.40a}$$

$$V_n(s, n; Re) = \breve{v}_0(s, n) + Re^{-1/2}\breve{v}_1(s, n) + \cdots, \tag{1.2.40b}$$

$$p(s, n; Re) = p_0(s, n) + Re^{-1/2}p_1(s, n) + \cdots. \tag{1.2.40c}$$

The substitution of (1.2.40) into the Navier–Stokes equations (1.2.8) results in the linearized Euler equations for $\breve{u}_1$, $\breve{v}_1$, and $\breve{p}_1$:

$$\frac{\breve{u}_0}{H_1}\frac{\partial \breve{u}_1}{\partial s} + \frac{1}{H_1}\frac{\partial \breve{u}_0}{\partial s}\breve{u}_1 + \breve{v}_0\frac{\partial \breve{u}_1}{\partial n} + \frac{\partial \breve{u}_0}{\partial n}\breve{v}_1 + \frac{\kappa}{H_1}(\breve{u}_0\breve{v}_1 + \breve{v}_0\breve{u}_1) = -\frac{1}{H_1}\frac{\partial p_1}{\partial s}, \quad (1.2.41a)$$

$$\frac{\breve{u}_0}{H_1}\frac{\partial \breve{v}_1}{\partial s} + \frac{1}{H_1}\frac{\partial \breve{v}_0}{\partial s}\breve{u}_1 + \breve{v}_0\frac{\partial \breve{v}_1}{\partial n} + \frac{\partial \breve{v}_0}{\partial n}\breve{v}_1 - \frac{2\kappa}{H_1}\breve{u}_0\breve{u}_1 = -\frac{\partial p_1}{\partial n}, \quad (1.2.41b)$$

$$\frac{1}{H_1}\frac{\partial \breve{u}_1}{\partial s} + \frac{\partial \breve{v}_1}{\partial n} + \frac{\kappa}{H_1}\breve{v}_1 = 0. \quad (1.2.41c)$$

These equations require the boundary conditions in the free stream and on the body surface. The first of these is formulated by substituting (1.2.40) into (1.2.2) and working with the $O(Re^{-1/2})$ terms. We find that

$$\breve{u}_1 = \breve{v}_1 = p_1 = 0 \quad \text{at} \quad s^2 + n^2 = \infty. \quad (1.2.42)$$

The condition on the body surface is represented in the leading-order approximation by the impermeability condition (1.2.7). To deduce its equivalent for the $O(Re^{-1/2})$ terms in (1.2.40), we need to perform the matching of the normal velocity component V_n in the inviscid flow region and in the boundary layer. We start with the inviscid flow where the normal velocity V_n is given by the asymptotic expansion (1.2.40b). Assuming n small, and using (1.2.29) in (1.2.40b), we have

$$V_n = -\frac{dU_e}{ds}n + Re^{-1/2}\breve{v}_1(s,0) + \cdots. \quad (1.2.43)$$

On the other hand, using (1.2.36) in the asymptotic expansion for V_n in (1.2.21), we find that at the outer edge of the boundary layer

$$V_n = -\frac{dU_e}{ds}n + Re^{-1/2}\frac{d}{ds}\big[U_e(s)\Delta^*(s)\big] + \cdots. \quad (1.2.44)$$

It remains to compare (1.2.43) with (1.2.44), and we can see that the sought boundary condition has the form

$$\breve{v}_1(s,0) = \frac{d}{ds}\big[U_e(s)\Delta^*(s)\big]. \quad (1.2.45)$$

The described procedure is known as the *Prandtl hierarchical strategy*. Remember that we started with flow analysis in the outer inviscid flow region. In the leading-order approximation, the fluid motion in this region is governed by the Euler equations (1.2.5). They should be solved with the free-stream conditions (1.2.6) and the impermeability condition (1.2.7) on the body surface. The latter ignores the existence of the boundary layer. Once the solution of the boundary-value problem (1.2.5)–(1.2.7) is constructed, and the pressure P_e and the tangential velocity U_e on the body surface are found, we can turn to the flow analysis in the boundary layer. The flow behaviour of the boundary layer is determined by solving Problem 1.2. As a part of the solution, the boundary layer displacement thickness Δ^* can be found. Once this is done, we can return to the inviscid flow region and improve the solution there by solving the linearized Euler equations (1.2.41), subject to the boundary conditions (1.2.42) and

(1.2.45). The next step is to improve the solution in the boundary layer. Adding the $O(Re^{-1/2})$ perturbations to (1.2.21), we have

$$\left.\begin{array}{l} V_\tau = U_0(s, N) + Re^{-1/2}U_1(s, N) + \cdots, \\ V_n = Re^{-1/2}V_0(s, N) + Re^{-1}V_1(s, N) + \cdots, \\ p = P_e(s) + Re^{-1/2}P_1(s, N) + \cdots \end{array}\right\} \quad \text{as} \quad Re \to \infty, \qquad (1.2.46)$$

where $N = Re^{1/2}n$. The equations for the correction functions, U_1, V_1, and P_1 can be derived, as usual, by substituting (1.2.46) into the Navier–Stokes equations (1.2.8). The improvement of the solutions in both regions may be continued in an obvious manner by including higher-order terms in (1.2.40) and (1.2.46).

It should be noted that the Prandtl hierarchical strategy presumes the flow to be free of singularities. However, we will see in Chapters 2–5 that many flows develop singular behaviour at various points in the flow field, which requires an appropriate adjustment of the solution strategy.

Exercises 2

1. When deriving the governing equations (1.1.29a), (1.1.29b) for the Blasius boundary layer, we did not need to know what $\chi(Re)$ in the asymptotic expansion (1.1.15) for the pressure p was. Now you are in a position to argue that the displacement effect of the boundary layer causes $O(Re^{-1/2})$ pressure perturbations. Hence, express the pressure in the boundary layer in the form of the asymptotic expansion

$$p(x, y; Re) = Re^{-1/2}P_1(x, Y) + \cdots \quad \text{as} \quad Re \to \infty. \qquad (1.2.47)$$

   Substitute (1.2.47) together with (1.1.28) into the lateral momentum equation (1.1.5b), and show that P_1 does not change across the boundary layer.

2. The flow behaviour in the boundary layer may be altered in a desired way (this is termed the *flow control*) by organizing suction of the fluid through the body surface. In particular, when dealing with the flow past a flat plate, one can assume that the plate is hollow, and that its interior is connected to a vacuum chamber, as shown in Figure 1.9. If the plate wall is perforated, then the fluid is forced to flow into the interior of the plate. The fluid flux through the perforation depends on the difference of the exterior and interior pressures, as well as on the density of the holes made in the wall.

Fig. 1.9: Blasius flow with suction.

   Assume that suction is organized such that on the plate surface

$$U_0 = 0, \quad V_0 = \frac{q}{\sqrt{x}} \quad \text{at} \quad Y = 0.$$

How should the Blasius problem (1.1.47a), (1.1.47b) be modified to accommodate for this new effect?

3. Consider now the case of uniform suction through flat plate surface,

$$U_0 = 0, \quad V_0 = -V_w \quad \text{at} \quad Y = 0,$$

with V_w being an order one positive constant.

Assume, subject to subsequent confirmation, that

$$\frac{\partial U_0}{\partial x} \to 0 \quad \text{as} \quad x \to \infty$$

in the boundary-layer equations (1.1.29a) and (1.1.29b), and find the *limiting velocity profile* across the boundary layer at large values of x.

4. Consider again the boundary layer on the surface of a flat plate.

(a) Through making use of (1.1.48), (1.1.52), and (1.2.37), show that the displacement thickness of the Blasius boundary layer is given by

$$\delta^* = Re^{-1/2} A_- \sqrt{x}. \tag{1.2.48}$$

(b) For the boundary layer on the flat plate surface the displacement thickness, δ^*, may be also defined as a function which satisfies the following equation:

$$\frac{d\delta^*}{dx} = \lim_{Y \to \infty} \frac{v}{u}. \tag{1.2.49}$$

Calculate the right-hand side in (1.2.49) using (1.1.28), (1.1.48), and (1.1.52), and show that the integration of (1.2.49) leads to (1.2.48).

5. Keeping in mind that not only $O(1)$ but also $O(Re^{-1/2})$ terms in (1.2.40) are governed by the inviscid equations (1.2.41), introduce the velocity potential φ such that $\mathbf{V} = \nabla\varphi$, and write the asymptotic expansion of φ in the outer inviscid region in the form

$$\varphi(x, y; Re) = \varphi_0(x, y) + Re^{-1/2}\varphi_1(x, y) + \cdots.$$

Formulate the equations and boundary conditions for φ_0 and φ_1.

Suggestion: You may find the discussion of the potential flow theory in Section 3.2 in Part 1 of this book series useful.

6. Substitute (1.2.46) into the lateral momentum equation (1.2.8b), and deduce that

$$\frac{\partial P_1}{\partial N} = \kappa U_0^2.$$

Integrate the above equation across the boundary layer, and show that on the body surface

$$p = P_e(s) + Re^{-1/2}\left[p_1(s, 0) + \kappa \int\limits_0^\infty (U_e^2 - U_0^2)dN\right] + O(Re^{-1}).$$

Here $p_1(s, 0)$ is the coefficient of the second term in the asymptotic expansion (1.2.40c) for the pressure in the inviscid flow region taken at $n = 0$.

7. Consider an incompressible viscous fluid flow between two coaxial circular cylinders of radii R_1 and R_2 rotating with angular velocities Ω_1 and Ω_2, respectively; see Figure 1.10. The cylinders are made permeable to allow the fluid to flow across the gap between them. The amount of fluid supplied (per unit time and unit length in the axial direction) into the gap through the inner cylinder is Q, with an equal amount removed through the outer cylinder.

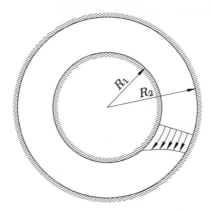

Fig. 1.10: Flow between two coaxial cylinders.

The Navier–Stokes equations admit an exact solution for this flow.[10] In particular, the circumferential velocity proves to be

$$\widehat{V}_\phi = \frac{C_1}{\hat{r}} + C_2\,\hat{r}^{1+q/\nu}, \tag{1.2.50}$$

where $\hat{r}$ is the radial coordinate measured from the cylinders' common centre, $q = Q/2\pi$, and

$$C_1 = \frac{V_1 - V_2\big(R_1/R_2\big)^{1+q/\nu}}{1 - \big(R_1/R_2\big)^{2+q/\nu}}R_1, \qquad C_2 = \frac{V_2 - V_1\big(R_1/R_2\big)}{1 - \big(R_1/R_2\big)^{2+q/\nu}}\frac{1}{R_2^{1+q/\nu}}. \tag{1.2.51}$$

with $V_1 = \Omega_1 R_1$ and $V_2 = \Omega_2 R_2$ being the circumferential velocities of the inner and outer cylinders. Your task is to study the flow behaviour in the limit when the fluid viscosity ν tends to zero. When performing this task you may use equations (1.2.50), (1.2.51) without proof.

Conduct your analysis in the following steps:

(a) First, consider the *outer region* that occupies almost the entire space between the cylinders. Assume that the radial fluid flux is in the direction from the inner cylinder to the outer cylinder; this corresponds to $Q = 2\pi q > 0$. Setting $q/\nu \to \infty$ in (1.2.50), (1.2.51), show that for any $\hat{r} < R_2$ the flow is represented by the 'potential vortex' solution

$$\widehat{V}_\phi = V_1\frac{R_1}{\hat{r}} + \cdots. \tag{1.2.52}$$

---

[10]For details, see Problem 7 in Exercises 6 in Part 1.

(b) Argue that equation (1.2.52) is not applicable near the outer cylinder where the *boundary layer* should form. Introduce the inner variable N such that

$$\hat{r} = R_2\left(1 - \frac{\nu}{q}N\right).$$

Express the circumferential velocity (1.2.50), (1.2.51) in terms of N and show, using the limit

$$\frac{q}{\nu} \to \infty, \qquad N = O(1),$$

that in the boundary layer

$$\widehat{V}_\phi = V_1\frac{R_1}{R_2} + \left(V_2 - V_1\frac{R_1}{R_2}\right)e^{-N} + \cdots.$$

Hint: It is known that for any finite constant κ,

$$\lim_{\varepsilon \to 0} \left(1 + \kappa\varepsilon\right)^{1/\varepsilon} = e^\kappa.$$

1.3 Solutions of Falkner and Skan

Falkner and Skan (1930) studied a class of solutions of the boundary-layer equations (1.2.34) with the inviscid velocity $U_e(s)$ in the form of the power function

$$U_e(s) = Cs^m, \tag{1.3.1}$$

with C and m being constants.

To this category belongs, for example, the flow past a body with a wedge-shaped nose. If we assume that the body is oriented with respect to the oncoming flow such that the front stagnation point finds itself at the wedge apex O (see Figure 1.11), then the inviscid flow solution can be constructed as suggested in Problems 1 and 2

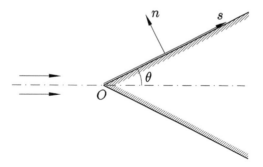

Fig. 1.11: Flow past a wedge.

in Exercises 10 in Part 1 of this book series. It appears that the tangential velocity on the wedge surface is described by equation (1.3.1) with the parameter m given by

$$m = \frac{\theta}{\pi - \theta}. \tag{1.3.2}$$

Here, θ is the half angle of the wedge apex. Factor C depends on the free-stream velocity as well as on the entire body shape, not just its nose. In what follows, we shall assume that $C > 0$.

Due to the symmetry of the flow, we can restrict our attention to the boundary layer on the upper surface of the wedge. The fluid motion in the boundary layer is governed by equations (1.2.34), with s measured along the wedge surface from its apex and $n = Re^{-1/2}N$ in the perpendicular direction. We start with the continuity equation (1.2.34b). It shows that there exists the stream function $\Psi(s, N)$ such that

$$\frac{\partial \Psi}{\partial s} = -V_0, \qquad \frac{\partial \Psi}{\partial N} = U_0. \tag{1.3.3}$$

Since no characteristic length scale is involved in the formulation of the boundary-layer problem (1.2.34), the solution for $\Psi(s, N)$ may be sought in the self-similar form

$$\Psi = s^\alpha f(\zeta), \qquad \zeta = \frac{N}{s^\beta}. \tag{1.3.4}$$

With (1.3.3), the continuity equation (1.2.34b) is satisfied automatically. Therefore, in order to deduce an equation for $f(\zeta)$ we need to substitute (1.3.4) into the momentum equation (1.2.34a). We start with differentiation of the similarity variable ζ:

$$\frac{\partial \zeta}{\partial s} = -\beta N s^{-\beta-1} = -\beta \frac{\zeta}{s}, \qquad \frac{\partial \zeta}{\partial N} = \frac{1}{s^\beta}.$$

Now we calculate the velocity components:

$$U_0 = \frac{\partial \Psi}{\partial N} = s^\alpha f'(\zeta) \frac{\partial \zeta}{\partial N} = s^{\alpha-\beta} f'(\zeta), \tag{1.3.5}$$

$$V_0 = -\frac{\partial \Psi}{\partial s} = -\alpha s^{\alpha-1} f(\zeta) - s^\alpha f'(\zeta) \frac{\partial \zeta}{\partial s} = s^{\alpha-1} \big[\beta \zeta f'(\zeta) - \alpha f(\zeta)\big]. \tag{1.3.6}$$

We further need to know the derivatives of function U_0:

$$\frac{\partial U_0}{\partial s} = s^{\alpha-\beta-1}\big[(\alpha-\beta)f' - \beta\zeta f''\big], \qquad \frac{\partial U_0}{\partial N} = s^{\alpha-2\beta} f'', \qquad \frac{\partial^2 U_0}{\partial N^2} = s^{\alpha-3\beta} f'''. \tag{1.3.7}$$

Finally, it follows from (1.3.1) that

$$U_e \frac{dU_e}{ds} = mC^2 s^{2m-1}. \tag{1.3.8}$$

The substitution of (1.3.5)–(1.3.8) into (1.2.34a) yields

$$s^{2\alpha-2\beta-1} f'\big[(\alpha-\beta)f' - \beta\zeta f''\big] + s^{2\alpha-2\beta-1}\big(\beta\zeta f' - \alpha f\big)f''$$
$$= mC^2 s^{2m-1} + s^{\alpha-3\beta} f'''. \tag{1.3.9}$$

Since f is a function of ζ only, the above equation should not involve s, which is only possible if

$$2\alpha - 2\beta - 1 = 2m - 1, \qquad 2\alpha - 2\beta - 1 = \alpha - 3\beta.$$

Solving these equations for α and β, we have

$$\alpha = \frac{1+m}{2}, \qquad \beta = \frac{1-m}{2}. \tag{1.3.10}$$

By substituting (1.3.10) back into (1.3.9), we find that the equation for $f(\zeta)$ has the form

$$m\left(f'\right)^2 - \frac{m+1}{2} f f'' = mC^2 + f'''. \tag{1.3.11}$$

The boundary conditions for this equation are deduced by substituting (1.3.5) and (1.3.6) into (1.2.34c)–(1.2.34e). We start with the initial condition (1.2.34c). Using (1.3.10), we can express the longitudinal velocity component (1.3.5) in the form

$$U_0 = s^m f'(\zeta). \tag{1.3.12}$$

Since $m > 0$, we can see that the condition (1.2.34c) is satisfied automatically.

Now we turn to the no-slip conditions (1.2.34d). These have to be satisfied on the body surface ($N = 0$), where the similarity variable $\zeta = 0$. Therefore, we set $\zeta = 0$ in (1.3.12) and (1.3.6), and to make U_0 and V_0 zeros, we have to require

$$f(0) = f'(0) = 0. \tag{1.3.13}$$

Finally, we substitute (1.3.12) and (1.3.1) into the boundary condition (1.2.34e) at the outer edge of the boundary layer. We find that

$$f'(\infty) = C. \tag{1.3.14}$$

Equation (1.3.11) considered together with boundary conditions (1.3.12)–(1.3.14) constitute the *Falkner–Skan problem*, that should be solved to find function $f(\zeta)$. The affine transformations[11]

$$f = \sqrt{C}\,\varphi, \qquad \zeta = \frac{1}{\sqrt{C}}\eta, \tag{1.3.15}$$

allow us to cast this problem in the following canonical form:

$$\left.\begin{array}{l} \varphi''' + \dfrac{1+m}{2}\varphi\varphi'' = m\left(\varphi'^2 - 1\right), \\[2mm] \varphi(0) = \varphi'(0) = 0, \quad \varphi'(\infty) = 1. \end{array}\right\} \tag{1.3.16}$$

Problem (1.3.16) does not have an analytical solution, but it may be easily studied numerically. With $\varphi(0)$ and $\varphi'(0)$ known to be zeros, one can guess a value for $\varphi''(0)$ and calculate $\varphi(\eta)$ for $\eta \in [0, \eta_{\max}]$ using, for example, the Runge–Kutta routine. The solution has to satisfy the boundary condition, $\varphi'(\infty) = 1$. If the calculated value

---

[11]Here, it is assumed that the factor C in (1.3.1) is positive which, of course, is true for the flow considered; see Figure 1.11.

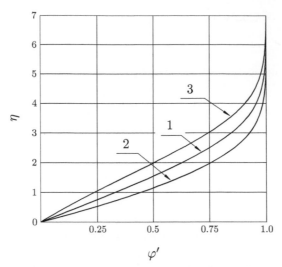

Fig. 1.12: The longitudinal velocity profile $\varphi'(\eta)$. Here, 1 denotes the Blasius solution $(m = 0)$, 2 is the solution with favourable pressure gradient $(m = 0.1)$, and 3 the solution with adverse pressure gradient $(m = -0.05)$.

of φ' at $\eta = \eta_{\max}$ differs from unity, then $\varphi''(0)$ should be iterated accordingly. The results shown in Figures 1.12–1.14 were obtained with the help of Newtonian iteration procedure.

In Figure 1.12, the Blasius solution $(m = 0)$ is compared with the solutions for the boundary layer exposed to a *favourable* (negative) pressure gradient $(m = 0.1)$ and *adverse* (positive) pressure gradient $(m = -0.05)$. For any positive value of the parameter m, the external velocity (1.3.1) increases with s and it follows from the Bernoulli equation, $P_e + \frac{1}{2}U_e^2 = const$, that the pressure $P_e(s)$ decreases in the downstream direction. This means that the fluid particles inside the boundary layer find themselves moving from a higher pressure region into a region with lower pressure. The fluid tends to accelerate under these conditions. Indeed, Figure 1.12 shows that with $m = 0.1$, the velocity φ' is larger than in the Blasius boundary layer. The adverse pressure gradient $(m = -0.05)$ causes a deceleration of the fluid.

To show the entire family of possible solutions, the graph of $\varphi''(0)$ versus m is plotted in Figure 1.13. Each point on the graph represents a solution of (1.3.16). In particular, point 1 corresponds to the Blasius solution. On the right-hand side of this point, $\varphi''(0)$ increases montonically with m, and the solution remains unique. To the left of the 'Blasius point' 1, the graph make a characteristic loop, which shows, first, that the solution exists only up to the critical value of $m_c \approx -0.0904$ and, second, that there are two solutions for all $m \in (m_c, 0)$. The velocity profiles at points 2 and 3 are shown in Figure 1.12. Figure 1.14 compares the velocity profiles for the upper branch solution (point 3 in Figure 1.13) and the lower branch solution (point 4) for the same value of parameter $m = -0.05$. The lower branch solution develops a negative velocity region $(\varphi' < 0)$ near the wall, where the fluid moves in the direction opposite to the main stream. The width of this region becomes progressively larger as the limit point

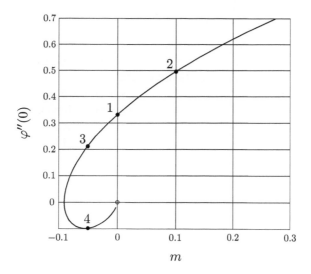

Fig. 1.13: The family of the solutions of the Falkner–Skan equation.

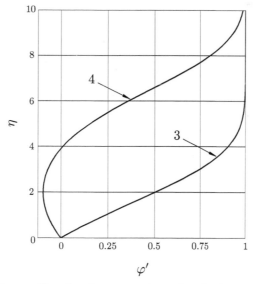

Fig. 1.14: The velocity profiles for the upper branch solution 3, and the lower branch solution 4, both calculated for $m = -0.05$.

(shown by an open circle in Figure 1.13) is approached.

Of course, the solutions with $m < 0$ have no relevance to the wedge flow shown in Figure 1.11; it follows from (1.3.2) that, for any $\theta \in (0, \pi)$, the exponent m in (1.3.1) is positive. Still, the Falkner–Skan solutions with negative m are helpful in revealing the reaction of the boundary layer on the action of an adverse pressure gradient.

Exercises 3

1. Assume that the body placed in an incompressible fluid flow has a rounded nose. Your task is to analyse the flow in a small vicinity of the front stagnation point, known as the *Hiemenz flow* (see Hiemenz, 1911).

   (a) First, perform your analysis based on the boundary-layer equations (1.2.34). Argue that the flow near the front stagnation point can be 'obtained' from the wedge flow (see Figure 1.11) by setting $\theta = \frac{1}{2}\pi$, and use equation (1.3.2) to show that in this flow $m = 1$. Hence, conclude that

$$U_e = Cs.$$

   Seek the solution to the equation (1.2.34a) in the form

$$\Psi = s\sqrt{C}\varphi(\eta), \qquad \eta = \sqrt{C}N,$$

   and show that function $\varphi(\eta)$ satisfies the equation

$$\varphi''' + \varphi\varphi'' - \varphi'^2 + 1 = 0. \tag{1.3.17}$$

   What are the boundary conditions for this equation?

   (b) Now use the full Navier–Stokes equations:

$$u\frac{\partial u}{\partial x} + v\frac{\partial u}{\partial y} = -\frac{\partial p}{\partial x} + \frac{1}{Re}\left(\frac{\partial^2 u}{\partial x^2} + \frac{\partial^2 u}{\partial y^2}\right), \tag{1.3.18a}$$

$$u\frac{\partial v}{\partial x} + v\frac{\partial v}{\partial y} = -\frac{\partial p}{\partial y} + \frac{1}{Re}\left(\frac{\partial^2 v}{\partial x^2} + \frac{\partial^2 v}{\partial y^2}\right), \tag{1.3.18b}$$

$$\frac{\partial u}{\partial x} + \frac{\partial v}{\partial y} = 0. \tag{1.3.18c}$$

   Seek their solution in the form

$$\psi(x, y) = xf(y), \qquad p = -\frac{1}{2}C^2 x^2 + g(y),$$

   where x is measured along the body surface from the stagnation point, and y in the perpendicular direction. Remember that the stream function $\psi(x, y)$ is related to the velocity components (u, v) by the equations

$$\frac{\partial \psi}{\partial x} = -v, \qquad \frac{\partial \psi}{\partial y} = u.$$

   Deduce an ordinary differential equation for $f(y)$, and show that the affine transformations

$$f = \sqrt{\frac{C}{Re}}\,\varphi, \qquad y = \frac{1}{\sqrt{CRe}}\,\eta$$

   reduce it to (1.3.17).

2. Consider a convergent channel which is built of two semi-infinite flat plates. The plates make an angle α to one another, and almost merge at point O, but still a small gap is left between them to serve as a sink centred at point O (see Figure 1.15). The fluid flux through the sink is $-\widehat{Q}$. You are advised to perform your analysis in dimensional variables.

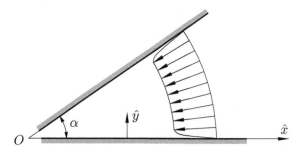

Fig. 1.15: The flow in the convergent channel.

Assume that the Reynolds number of the flow is large and perform the following tasks:

(a) Using the mass conservation in the 'inviscid core' of the flow, show that the velocity at the outer edge of the boundary layer

$$\widehat{U}_e = -\frac{\hat{q}}{\hat{x}},$$

where $\hat{q}$ is a positive constant, and $\hat{x}$ is the distance measured along the lower plate from the sink O.

(b) Write the boundary layer equations in the dimensional form,

$$\hat{u}\frac{\partial \hat{u}}{\partial \hat{x}} + \hat{v}\frac{\partial \hat{u}}{\partial \hat{y}} = \widehat{U}_e\frac{d\widehat{U}_e}{d\hat{x}} + \nu\frac{\partial^2 \hat{u}}{\partial \hat{y}^2},$$

$$\frac{\partial \hat{u}}{\partial \hat{x}} + \frac{\partial \hat{v}}{\partial \hat{y}} = 0,$$

and introduce the stream function $\hat{\psi}(\hat{x}, \hat{y})$ such that

$$\frac{\partial \hat{\psi}}{\partial \hat{x}} = -\hat{v}, \qquad \frac{\partial \hat{\psi}}{\partial \hat{y}} = \hat{u}.$$

Your task is to find the longitudinal velocity $\hat{u}$ in the boundary layer. To perform this task, seek the stream function $\hat{\psi}$ in the self-similar form

$$\hat{\psi} = -\sqrt{\nu\hat{q}}f(\eta), \qquad \eta = \frac{\hat{y}}{\hat{x}}\sqrt{\frac{\hat{q}}{\nu}},$$

and show that function $U = f'(\eta)$ satisfies the equation

$$U'' - U^2 + 1 = 0. \tag{1.3.19}$$

Argue that it has to be solved with the boundary conditions

$$U(0) = 0, \qquad U(\infty) = 1. \tag{1.3.20}$$

Multiply the three terms in (1.3.19) by U' and integrate the resulting equation with the second condition in (1.3.20). Show that

$$U' = -\sqrt{\frac{2}{3}} \, (U - 1)\sqrt{U + 2}. \tag{1.3.21}$$

Confirm (by direct substitution) that the solution of (1.3.21), satisfying the first condition in (1.3.20), is written as

$$U = 3 \tanh^2 \left(\frac{\eta}{\sqrt{2}} + \alpha \right) - 2, \qquad \alpha = \tanh^{-1} \sqrt{\frac{2}{3}}.$$

1.4   Shear Layer Flows

Shear layers are boundary layers that form inside a moving fluid. A typical example is shown in Figure 1.16. Here, the flow separates from a 'flat-faced body', which results in the formation of a separation region behind the body. The recirculation motion of the fluid in the separation region is rather slow, and if the fluid was inviscid then the tangential velocity would experience a jump across the *dividing streamline* that separates the recirculation region from the rest of the flow.[12] The viscosity, no matter how small, will act to smooth out the velocity jump in a thin region termed the *shear layer*. The characteristic thickness of the shear layer may be estimated through

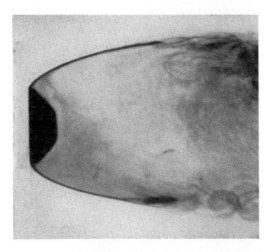

Fig. 1.16: Flow visualization by Flachsbart (1935). Copyright © 1935 WILEY-VCH Verlag GmbH & Co. KGaA, Weinheim.

---

[12]Remember that, in the Kirchhoff theory, the dividing streamline is termed the free streamline; see Section 3.8 in Part 1.

balancing the convective and viscous terms in the longitudinal momentum equation (1.2.8a). When analysing the flow in the shear layer, it is convenient to align the s-axis with the dividing streamline, and then we can write

$$\frac{V_\tau}{H_1}\frac{\partial V_\tau}{\partial s} \sim \frac{1}{Re}\frac{\partial^2 V_\tau}{\partial n^2}. \tag{1.4.1}$$

In the non-dimensional variables used in (1.4.1), the tangential velocity component, V_τ, is an order one quantity in the shear layer. Therefore, assuming $s = O(1)$, we immediately find that $n \sim Re^{-1/2}$. We see that the shear layer has the same characteristic thickness as the conventional boundary layer adjacent to a rigid body surface. This suggests that, at large values of the Reynolds number, the flow inside the shear layer is governed by Prandtl's boundary layer equations.

1.4.1 Chapman's problem

Here, we consider a rather simple example of a shear layer known as the *Chapman flow* (see Chapman, 1950). Let us assume that the shear layer forms as a result of the flow separation from a corner point of a rigid body contour (see Figure 1.17). We shall further assume that the front face of the body, AO, is flat and parallel to the velocity vector in the undisturbed free-stream flow before of the body. The modulus of the free-stream velocity is denoted, as usual, by V_∞. To describe this flow, we shall use Cartesian coordinates $(\hat{x}, \hat{y})$, with $\hat{x}$ measured from the corner point O parallel to the front face of the body AO.

If the fluid was inviscid, then the flow separation at point O would lead to a formation of the contact discontinuity surface along the $\hat{x}$-axis. In the separation region situated below the $\hat{x}$-axis, the fluid would remain at rest, while everywhere above the discontinuity the fluid would continue to flow with constant velocity V_∞. Of course, in reality, the internal viscosity will act to smooth out the jump in the velocity field, which leads to a formation of the shear layer. The velocity is expected to change smoothly across this layer, as sketched in Figure 1.17.

The viscosity will also lead to a formation of the boundary layer on the front face AO of the body. To simplify the problem, we shall assume that front face is 'covered' with a running belt whose speed equals the free-stream velocity V_∞. In this case, the flow will remain unperturbed up to the corner point O. Our task is to study the viscous

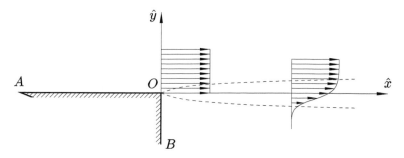

Fig. 1.17: Shear layer flow.

layer that forms downstream of this point. The Navier–Stokes equations describing the flow are written as

$$\hat{u}\frac{\partial \hat{u}}{\partial \hat{x}} + \hat{v}\frac{\partial \hat{u}}{\partial \hat{y}} = -\frac{1}{\rho}\frac{\partial \hat{p}}{\partial \hat{x}} + \nu\left(\frac{\partial^2 \hat{u}}{\partial \hat{x}^2} + \frac{\partial^2 \hat{u}}{\partial \hat{y}^2}\right), \tag{1.4.2a}$$

$$\hat{u}\frac{\partial \hat{v}}{\partial \hat{x}} + \hat{v}\frac{\partial \hat{v}}{\partial \hat{y}} = -\frac{1}{\rho}\frac{\partial \hat{p}}{\partial \hat{y}} + \nu\left(\frac{\partial^2 \hat{v}}{\partial \hat{x}^2} + \frac{\partial^2 \hat{v}}{\partial \hat{y}^2}\right), \tag{1.4.2b}$$

$$\frac{\partial \hat{u}}{\partial \hat{x}} + \frac{\partial \hat{v}}{\partial \hat{y}} = 0. \tag{1.4.2c}$$

Let the point of observation be situated a distance L downstream of the corner O. Then the Reynolds number may be defined as

$$Re = \frac{V_\infty L}{\nu}.$$

For large Re, the boundary-layer equations may be used to describe the flow in the shear layer. Keeping this in mind, we make the coordinates and velocity components dimensionless by means of the transformations

$$\hat{x} = Lx, \quad \hat{y} = LRe^{-1/2}Y, \quad \hat{u} = V_\infty U_0, \quad \hat{v} = V_\infty Re^{-1/2}V_0. \tag{1.4.3}$$

As far as the pressure $\hat{p}$ is concerned, we note that, according to the boundary-layer theory, it does not change across the viscous layer, and since the shear layer is 'in contact' with the separation region, where the fluid remains at rest in the leading-order approximation, we can conclude that the pressure is constant both in the separation region and in the shear layer.[13] We shall assume that it is equal to the free-stream pressure, $\hat{p} = p_\infty$.

By substituting (1.4.3) into (1.4.2a) and (1.4.2c), we find that in the leading-order approximation:

$$U_0\frac{\partial U_0}{\partial x} + V_0\frac{\partial U_0}{\partial Y} = \frac{\partial^2 U_0}{\partial Y^2}, \tag{1.4.4a}$$

$$\frac{\partial U_0}{\partial x} + \frac{\partial V_0}{\partial Y} = 0. \tag{1.4.4b}$$

Equations (1.4.4) should be solved with the initial condition

$$U_0\Big|_{x=0} = \begin{cases} 1 & \text{if } Y > 0, \\ 0 & \text{if } Y < 0, \end{cases} \tag{1.4.4c}$$

the matching condition with the unperturbed flow above the shear layer

$$U_0 = 1 \quad \text{at} \quad Y = \infty, \tag{1.4.4d}$$

---

[13]Later we will see that the entrainment effect of the shear layer causes a slow motion of the fluid in the separation region, which results in the pressure variations, $\hat{p} - p_\infty \sim \rho V_\infty^2 Re^{-1}$.

and matching condition with the stagnant fluid region below the shear layer,

$$U_0 = 0 \quad \text{at} \quad Y = -\infty. \tag{1.4.4e}$$

The boundary-value problem (1.4.4) also requires a boundary condition on V_0. We shall postpone the task of formulating this condition untill a more clear understanding of the physical properties of the flow is achieved.

Thanks to the continuity equation (1.4.4b), we can introduce the stream function $\Psi(x, Y)$, such that

$$\frac{\partial \Psi}{\partial x} = -V_0, \qquad \frac{\partial \Psi}{\partial Y} = U_0. \tag{1.4.5}$$

By analogy with the Blasius flow, we shall seek the solution for Ψ in the following self-similar form:

$$\Psi(x, Y) = \sqrt{x}\, \varphi(\eta), \tag{1.4.6a}$$

with the similarity variable η being

$$\eta = \frac{Y}{\sqrt{x}}. \tag{1.4.6b}$$

By substituting (1.4.6) into (1.4.5), we find that the velocity components

$$U_0 = \varphi'(\eta), \qquad V_0 = \frac{1}{2\sqrt{x}}(\eta\varphi' - \varphi). \tag{1.4.7}$$

The substitution of (1.4.7) into the momentum equation (1.4.4a) results in the Blasius equation for the function $\varphi(\eta)$:

$$\varphi''' + \frac{1}{2}\varphi\varphi'' = 0. \tag{1.4.8}$$

Two boundary conditions for this equation,

$$\varphi'(-\infty) = 0, \qquad \varphi'(\infty) = 1, \tag{1.4.9}$$

are obtained through substitution of the first of equations (1.4.5) into (1.4.4c), (1.4.4d), and (1.4.4e).

We can see that equation (1.4.8), as well as boundary conditions (1.4.9) are invariant with respect to an arbitrary shift d in η:

$$\eta = \tilde{\eta} + d. \tag{1.4.10}$$

Therefore, assuming that the function $\varphi(\eta)$ changes sign at some point $\eta = \eta_0$ inside the shear layer, we can choose the shift parameter d in (1.4.10) to be $d = \eta_0$, and then we will have

$$\varphi = 0 \quad \text{at} \quad \tilde{\eta} = 0. \tag{1.4.11}$$

Combining (1.4.11) with (1.4.8) and (1.4.9), we have the following boundary-value problem for $\varphi(\tilde{\eta})$:

$$\frac{d^3\varphi}{d\tilde{\eta}^3} + \frac{1}{2}\varphi\frac{d^2\varphi}{d\tilde{\eta}^2} = 0, \tag{1.4.12a}$$

$$\left.\frac{d\varphi}{d\tilde{\eta}}\right|_{\tilde{\eta}=-\infty} = 0, \quad \varphi\big|_{\tilde{\eta}=0} = 0, \quad \left.\frac{d\varphi}{d\tilde{\eta}}\right|_{\tilde{\eta}=\infty} = 1. \tag{1.4.12b}$$

The results of the numerical solution to this problem are displayed in Figure 1.18.

Interestingly enough, the shear-layer flow appears to belong to the Falkner–Skan family, and may be obtained as the limiting solution of the boundary-value problem (1.3.16) as $m \to 0^-$. The limit point is shown by an 'open circle' in Figure 1.13. The velocity profile at neighbouring point 4 is displayed in Figure 1.14. It has a characteristic reverse flow region where $\varphi' < 0$. The calculations show that the width of this region becomes progressively larger as $m \to 0^-$, and the fluid velocity $|\varphi'|$ tends to zero in the entire reverse flow region. Correspondingly, the shear layer moves away from the body surface, and the velocity profile inside the shear layer takes the form shown in Figure 1.18.

Note that the shift parameter d in (1.4.10) still remains undetermined. In order to find d, the process of interaction of the shear layer with the flow above it, as well and with the stagnation region below it, should be examined. The asymptotic behaviour of the solution of the Blasius equation (1.4.12a) at large values of $\tilde{\eta}$ was analysed in Section 1.1, and was found to be represented by equation (1.1.51). The results of this analysis remain applicable to the flow considered here. Hence, we can write

$$\varphi(\tilde{\eta}) = \tilde{\eta} + A + \cdots \quad \text{as} \quad \tilde{\eta} \to \infty. \tag{1.4.13}$$

The numerical solution of the problem (1.4.12) shows that

$$A = -0.5323.$$

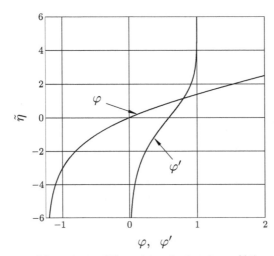

Fig. 1.18: Distribution of function $\varphi(\tilde{\eta})$ and its derivative $\varphi'(\tilde{\eta})$ across the shear layer.

Combining (1.4.13) with (1.4.10) we have

$$\varphi = \eta + (A - d) + \cdots .$$ (1.4.14)

If we substitute (1.4.14) into (1.4.7), then we find that at the upper edge of the shear layer

$$U_0\Big|_{\eta=\infty} = 1, \qquad V_0\Big|_{\eta=\infty} = -\frac{A - d}{2\sqrt{x}}.$$ (1.4.15)

The angle made by the velocity vector with the direction of unperturbed flow can be calculated with the help of (1.4.3) and (1.4.15):

$$\vartheta = \arctan\left(\frac{\hat{v}}{\hat{u}}\right) = Re^{-1/2}\frac{V_0}{U_0} = -Re^{-1/2}\frac{A - d}{2\sqrt{x}}.$$

Since ϑ is small, the displacement effect of the shear layer on the flow above it can be accounted for using the thin aerofoil theory.[14] It follows from this theory that, unless

$$A - d = 0,$$ (1.4.16)

the shear layer will cause $O(Re^{-1/2})$ perturbations of pressure in the inviscid flow above it, namely,

$$\hat{p} - p_\infty \sim \rho V_\infty^2 Re^{-1/2}.$$ (1.4.17)

We now need to see how these perturbations are transmitted across the shear layer. For this purpose, the $\hat{y}$-momentum equation (1.4.2b) should be used. Taking into account that in the shear layer the velocity components are represented by (1.4.3), we can express the convective terms on the left-hand side of (1.4.2b) as

$$\hat{u}\frac{\partial\hat{v}}{\partial\hat{x}} = \frac{V_\infty^2}{L}Re^{-1/2}U_0\frac{\partial V_0}{\partial x},$$ (1.4.18a)

$$\hat{v}\frac{\partial\hat{v}}{\partial\hat{y}} = \frac{V_\infty^2}{L}Re^{-1/2}V_0\frac{\partial V_0}{\partial Y}.$$ (1.4.18b)

Similarly, the viscous terms on the right-hand side of (1.4.2b) are

$$\nu\frac{\partial^2\hat{v}}{\partial\hat{x}^2} = \frac{V_\infty^2}{L}Re^{-3/2}U_0\frac{\partial^2 V_0}{\partial x^2},$$ (1.4.18c)

$$\nu\frac{\partial^2\hat{v}}{\partial\hat{y}^2} = \frac{V_\infty^2}{L}Re^{-1/2}U_0\frac{\partial^2 V_0}{\partial Y^2}.$$ (1.4.18d)

All that remains is to evaluate the pressure gradient term. Denoting the characteristic pressure variation across the shear layer by $\Delta\hat{p}$, and taking into account that the thickness of the shear layer $\Delta\hat{y} \sim LRe^{-1/2}$, we can write

$$\frac{1}{\rho}\frac{\partial\hat{p}}{\partial\hat{y}} \sim \frac{1}{\rho}\frac{\Delta\hat{p}}{\Delta\hat{y}} \sim \frac{Re^{1/2}}{\rho L}\Delta\hat{p}.$$ (1.4.19)

Clearly, the first viscous term (1.4.18c) is small as compared to the convective terms (1.4.18a), (1.4.18b), and the second viscous term (1.4.18d). The latter three terms are

---

[14]See Section 2.1 in Part 2 of this book series.

in balance with each other and should be in balance with the pressure gradient (1.4.19), that is

$$\frac{Re^{1/2}}{\rho L}\Delta\hat{p} \sim \frac{V_\infty^2}{L}Re^{-1/2}. \tag{1.4.20}$$

Solving equation (1.4.20) for $\Delta\hat{p}$, we find that the pressure variation across the shear layer,

$$\Delta\hat{p} \sim \rho V_\infty^2 Re^{-1},$$

is significantly smaller than the pressure perturbations (1.4.17) induced by the displacement effect of the shear layer at its upper edge. This means that the estimation (1.4.17) proves to be valid everywhere inside the shear layer and, in particular, at its lower edge.

Now, our task will be to investigate the flow in the separation region below the shear layer.

1.4.2 Entrainment effect

We start with the asymptotic analysis of the solution of the Blasius equation (1.4.12a) at the lower edge of the shear layer. It follows from the first boundary condition in (1.4.12b) and the numerical results displayed in Figure 1.18 that the leading-order term for $\varphi(\tilde{\eta})$ is

$$\varphi(\tilde{\eta}) = -\Lambda + \cdots \quad \text{as} \quad \tilde{\eta} \to -\infty, \tag{1.4.21}$$

where Λ is a positive constant. Using (1.4.21) instead of φ in the second term in the equation (1.4.12a) and denoting $d^2\varphi/d\tilde{\eta}^2$ as g, we have

$$\frac{dg}{d\tilde{\eta}} - \frac{\Lambda}{2}g = 0. \tag{1.4.22}$$

The general solution of equation (1.4.22) has the form

$$g = C_1 e^{\frac{1}{2}\Lambda\tilde{\eta}}, \tag{1.4.23}$$

where C_1 is an arbitrary constant. Integrating (1.4.23) twice with the boundary conditions

$$\varphi'(-\infty) = 0, \qquad \varphi(-\infty) = -\Lambda,$$

we can conclude that

$$\varphi = -\Lambda + \frac{4C_1}{\Lambda^2}e^{\frac{1}{2}\Lambda\tilde{\eta}} \cdots \quad \text{as} \quad \tilde{\eta} \to -\infty. \tag{1.4.24}$$

The numerical values for the two constants in (1.4.24) are

$$\Lambda = 1.2384, \qquad C_1 = 0.6237.$$

By substituting (1.4.10) into (1.4.24) and then into the second equation in (1.4.7), we find that at the lower edge of the shear layer

$$\hat{v} = V_\infty Re^{-1/2}V_0\Big|_{\eta=-\infty} = V_\infty Re^{-1/2}\frac{\Lambda}{2\sqrt{x}}. \tag{1.4.25}$$

Equation (1.4.25) shows that the shear layer makes the fluid particles in the separation region rise slowly. Once the fluid particles reach the shear layer, they experience a

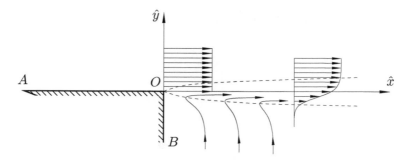

Fig. 1.19: Entrainment effect.

sharp acceleration due to the action of viscous forces, and are swept downstream with the rest of the fluid in the shear layer, as shown schematically in Figure 1.19. This phenomenon is referred to as the *entrainment effect*.

We shall now give a more detailed description of the flow in the separation region. The asymptotic analysis of the Navier–Stokes equations (1.4.2) in this region is based on the limit

$$\hat{x} = O(L), \quad \hat{y} = O(L), \quad Re \to \infty. \tag{1.4.26}$$

To predict the form of the asymptotic expansions of the velocity components $(\hat{u}, \hat{v})$ and the pressure $\hat{p}$ we note, first of all, that according to (1.4.25), the entrainment velocity, $\hat{v} \sim V_\infty Re^{-1/2}$. We further note that with $\hat{x} \sim \hat{y}$, a degeneration of the continuity equation (1.4.2c) can be avoided only if $\hat{u} \sim \hat{v}$. This means that in the region considered

$$\hat{u} \sim \hat{v} \sim V_\infty Re^{-1/2}. \tag{1.4.27}$$

Now, the pressure variations may be easily found through balancing the convective terms and pressure gradient in the $\hat{x}$- or $\hat{y}$-momentum equations, (1.4.2a), (1.4.2a). We have

$$\hat{p} - p_\infty \sim \hat{v}^2 \sim \rho V_\infty^2 Re^{-1}. \tag{1.4.28}$$

Clearly, this result comes in contradiction with the equation (1.4.17), which gives an estimation of the pressure perturbations in the flow above the shear layer. To resolve this contradiction, we need to 'suppress' the displacement effect of the shear, which is done by choosing the shift parameter d in (1.4.10) to satisfy condition (1.4.16), that is

$$d = A = -0.5323.$$

Now the solution in the shear layer is uniquely defined.

Being guided by (1.4.26)–(1.4.28), we represent the solution in the separation region in the form

$$\left. \begin{aligned} \hat{u} = V_\infty Re^{-1/2} u_0(x, y) + \cdots, \qquad \hat{v} = V_\infty Re^{-1/2} v_0(x, y) + \cdots, \\ \hat{p} = p_\infty + \rho V_\infty^2 Re^{-1} p_0(x, y) + \cdots, \end{aligned} \right\} \tag{1.4.29}$$

where

$$x = \frac{\hat{x}}{L}, \qquad y = \frac{\hat{y}}{L}.$$

The substitution of (1.4.29) into the Navier–Stokes equations (1.4.2) reduces them to the Euler equations:

$$u_0\frac{\partial u_0}{\partial x} + v_0\frac{\partial u_0}{\partial y} = -\frac{\partial p_0}{\partial x},\tag{1.4.30a}$$

$$u_0\frac{\partial v_0}{\partial x} + v_0\frac{\partial v_0}{\partial y} = -\frac{\partial p_0}{\partial y},\tag{1.4.30b}$$

$$\frac{\partial u_0}{\partial x} + \frac{\partial v_0}{\partial y} = 0.\tag{1.4.30c}$$

These can be simplified by cross-differentiation of (1.4.30a) and (1.4.30b) and elimination of p_0. We find that

$$u_0\frac{\partial \omega}{\partial x} + v_0\frac{\partial \omega}{\partial y} = 0,\tag{1.4.31}$$

where ω is the vorticity:

$$\omega = \frac{\partial v_0}{\partial x} - \frac{\partial u_0}{\partial y}.\tag{1.4.32}$$

Equation (1.4.31) shows that ω remains constant along each streamline. In what follows, we shall assume that far from the corner point O the fluid in the separation region (see Figure 1.19) is at rest, that is

$$u_0 = v_0 = 0 \quad \text{at} \quad x^2 + y^2 = \infty,\tag{1.4.33}$$

and, therefore,

$$\omega = 0 \quad \text{at} \quad x^2 + y^2 = \infty.\tag{1.4.34}$$

We shall further assume (subject to subsequent confirmation) that all the streamlines originate from a region where the condition (1.4.34) holds. Then, the vorticity appears to be zero everywhere in the flow:

$$\frac{\partial v_0}{\partial x} - \frac{\partial u_0}{\partial y} = 0.\tag{1.4.35}$$

Now, we use the continuity equation (1.4.30c) to introduce the stream function ψ such that

$$u_0 = \frac{\partial \psi}{\partial y}, \qquad v_0 = -\frac{\partial \psi}{\partial x}.\tag{1.4.36}$$

The substitution of (1.4.36) into (1.4.35) leads to the Laplace equation for ψ:

$$\frac{\partial^2 \psi}{\partial x^2} + \frac{\partial^2 \psi}{\partial y^2} = 0.\tag{1.4.37}$$

It has to be solved with the impermeability condition on the rear face OB of the body surface (see Figure 1.19),

$$\psi = 0 \quad \text{at} \quad x = 0, \ y \le 0,\tag{1.4.38}$$

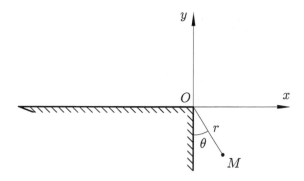

Fig. 1.20: Cylindrical polar coordinates.

and the matching condition with the solution in the shear layer. Comparing the asymptotic expansion of $\hat{v}$ in (1.4.29) with (1.4.25) we see that

$$v_0 = \frac{\Lambda}{2\sqrt{x}} \quad \text{at} \quad y = 0, \;\; x \geq 0. \tag{1.4.39}$$

This condition may be easily cast in terms of the stream function ψ. Combining (1.4.39) with the second equation in (1.4.36), we have

$$\frac{\partial \psi}{\partial x} = -\frac{\Lambda}{2\sqrt{x}},$$

which, being integrated with the initial condition (1.4.38), yields

$$\psi = -\Lambda\sqrt{x} \quad \text{at} \quad y = 0, \;\; x > 0. \tag{1.4.40}$$

When solving the boundary-value problem (1.4.37), (1.4.38), (1.4.40) it is convenient to use the cylindrical polar coordinates that are introduced as shown in Figure 1.20. The Laplace equation (1.4.37) and the boundary conditions (1.4.38), (1.4.40) are written in these coordinates as

$$\frac{\partial}{\partial r}\left(r\frac{\partial \psi}{\partial r}\right) + \frac{1}{r}\frac{\partial^2 \psi}{\partial \theta^2} = 0, \tag{1.4.41a}$$

$$\psi = 0 \qquad \text{at} \quad \theta = 0, \tag{1.4.41b}$$

$$\psi = -\Lambda\sqrt{r} \quad \text{at} \quad \theta = \frac{1}{2}\pi. \tag{1.4.41c}$$

Guided by (1.4.41c), we seek the solution for the stream function in the form

$$\psi = \sqrt{r}\,f(\theta). \tag{1.4.42}$$

The substitution of (1.4.42) into equation (1.4.41a) results in

$$f'' + \frac{1}{4}f = 0. \tag{1.4.43}$$

It further follows from the boundary conditions (1.4.41b), (1.4.41c) that

$$f(0) = 0, \qquad f\left(\tfrac{1}{2}\pi\right) = -\Lambda. \tag{1.4.44}$$

The solution of the boundary-value problem (1.4.43), (1.4.44) is written as

$$f = -\sqrt{2}\Lambda \sin\left(\tfrac{1}{2}\theta\right). \tag{1.4.45}$$

It remains to substitute (1.4.45) back into (1.4.42), and we can conclude that in the separation region

$$\psi = -\sqrt{2r}\Lambda \sin\left(\tfrac{1}{2}\theta\right). \tag{1.4.46}$$

Now, when the solution for the stream function is known, any other fluid-dynamic quantity can be easily found. In particular, the streamlines are obtained by setting $\psi = const$ (see Figure 1.21). The dimensional radial and circumferential velocity components are calculated as

$$\hat{V}_r = V_\infty Re^{-1/2} \frac{1}{r} \frac{\partial \psi}{\partial \theta} = -V_\infty Re^{-1/2} \frac{\Lambda}{\sqrt{2r}} \cos\left(\tfrac{1}{2}\theta\right),$$

$$\hat{V}_\theta = V_\infty Re^{-1/2} \left(-\frac{\partial \psi}{\partial r}\right) = V_\infty Re^{-1/2} \frac{\Lambda}{\sqrt{2r}} \sin\left(\tfrac{1}{2}\theta\right),$$

and, since the flow in the separation region is inviscid, the Bernoulli equation can be used to obtain the pressure:

$$\hat{p} = p_\infty - \rho V_\infty^2 Re^{-1} \frac{\Lambda^2}{4r}.$$

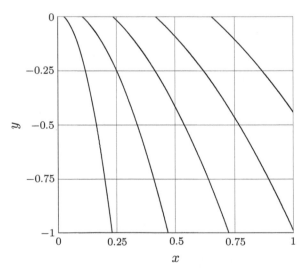

Fig. 1.21: Streamline pattern in the separation region; the streamlines shown are plotted for $\psi = -0.2, -0.4, -0.6, -0.8, -1$.

1.4.3   Prandtl's transposition theorem

When analysing the flow in the shear layer (see Figure 1.17), instead of using the Cartesian coordinates $(\hat{x}, \hat{y})$, it may be more convenient to use the body-fitted coordinates $(\hat{s}, \hat{n})$; see Figure 1.7. Indeed, let us align the $\hat{s}$-axis with the streamline that leaves the body surface at the separation point O, as shown in Figure 1.22. Remember that we call it the *dividing streamline*, since it separates the recirculating flow region that lies behind of the body base OB from the rest of the flow.

In the body-fitted coordinates, the position of any point M in the flow field is defined by two coordinates $(\hat{s}, \hat{n})$. The first one is the length of the arc $\hat{s}$ measured from the corner point O to the foot M' of the perpendicular dropped from M onto the dividing streamline. The second coordinate is the height $\hat{n}$ of the perpendicular MM'. The velocity vector $\widehat{\mathbf{V}}$ at point M is represented by the tangential and normal components $(\widehat{V}_\tau, \widehat{V}_n)$, that is

$$\widehat{\mathbf{V}} = \widehat{V}_\tau \mathbf{e}_1 + \widehat{V}_n \mathbf{e}_1. \tag{1.4.47}$$

Here, $\mathbf{e}_1$ and $\mathbf{e}_2$ are unit vectors, with $\mathbf{e}_1$ parallel to the tangent to the dividing streamline at point M', and $\mathbf{e}_2$ normal to $\mathbf{e}_1$.

We know that the dividing streamline lies inside the shear layer. Keeping this in mind, we write its equation as (see Figure 1.22):

$$\hat{y} = LRe^{-1/2}f(x), \qquad x = \frac{\hat{x}}{L}. \tag{1.4.48}$$

Since vector $\mathbf{e}_2$ is normal to the $\hat{s}$-axis, it can be found by writing (1.4.48) in the form

$$\Phi(\hat{x}, \hat{y}) = \hat{y} - LRe^{-1/2}f(x), \tag{1.4.49}$$

and using the well-known formula

$$\mathbf{e}_2 = \frac{1}{\sqrt{\left(\frac{\partial \Phi}{\partial \hat{x}}\right)^2 + \left(\frac{\partial \Phi}{\partial \hat{y}}\right)^2}} \left(\frac{\partial \Phi}{\partial \hat{x}}, \frac{\partial \Phi}{\partial \hat{y}}\right). \tag{1.4.50}$$

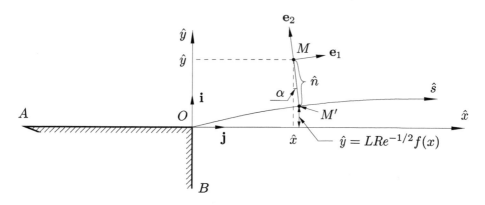

Fig. 1.22: Coordinate transformation.

By substituting (1.4.49) into (1.4.50) and disregarding $O(Re^{-1})$ terms, we find that

$$\mathbf{e}_2 = \left(- Re^{-1/2} f'(x) \, , \, 1 \right).$$

The tangential unit vector $\mathbf{e}_1$ is obtained by rotating $\mathbf{e}_2$ clockwise through the right angle. This gives

$$\mathbf{e}_1 = \left(1 \, , \, Re^{-1/2} f'(x) \right).$$

In Cartesian coordinates, the fluid velocity vector is written as

$$\widehat{\mathbf{V}} = \mathbf{i}\hat{u} + \mathbf{j}\hat{v}. \tag{1.4.51}$$

The tangential velocity component $\widehat{V}_\tau$ in the curvilinear coordinates can be found by taking the scalar product of (1.4.51) with $\mathbf{e}_1$. We have

$$\widehat{V}_\tau = \left(\widehat{\mathbf{V}} \cdot \mathbf{e}_1 \right) = \hat{u} + Re^{-1/2}\hat{v} f'(x). \tag{1.4.52}$$

Similarly, the normal velocity component is given by

$$\widehat{V}_n = \left(\widehat{\mathbf{V}} \cdot \mathbf{e}_2 \right) = \hat{v} - Re^{-1/2}\hat{u} f'(x). \tag{1.4.53}$$

Let us now deduce the corresponding formulae relating the curvilinear coordinates $(\hat{s}, \hat{n})$ of point M to the Cartesian coordinates $(\hat{x}, \hat{y})$ of this point. The length of the arc OM' is calculated as

$$\hat{s} = \int_0^{\hat{x}_{M'}} \sqrt{1 + \left(\frac{d\hat{y}}{d\hat{x}} \right)^2} \, d\hat{x}. \tag{1.4.54}$$

Here, $\hat{x}_{M'}$ denotes the $\hat{x}$-coordinate of point M'. Figure 1.22 shows that

$$\hat{x}_{M'} = \hat{x} + \hat{n}\sin\alpha,$$

where α is the angle between vectors $\mathbf{e}_2$ and $\mathbf{i}$. Keeping in mind that $\alpha = O(Re^{-1/2})$, and that inside the shear layer $\hat{n}$ is an order $Re^{-1/2}$ quantity, we can express the above equation in the form

$$\hat{x}_{M'} = \hat{x} + O(Re^{-1}). \tag{1.4.55}$$

Using (1.4.48) and (1.4.55) in (1.4.54), it is easy to see that

$$\hat{s} = \hat{x} + O(Re^{-1}). \tag{1.4.56}$$

Figure 1.22 further shows that the $\hat{y}$-coordinate of point M may be calculated as

$$\hat{y} = \hat{n}\cos\alpha + LRe^{-1/2}f(x).$$

Since $\cos\alpha = 1 + O(Re^{-1})$ and $\hat{n}$ is an order $Re^{-1/2}$ quantity, we can write

$$\hat{y} = \hat{n} + LRe^{-1/2}f(x) + O(Re^{-3/2}). \tag{1.4.57}$$

Let us now express the coordinate and velocity transformations in dimensionless form. In Cartesian coordinates, the non-dimensional variables have been introduced through scalings (1.4.3):

$$\hat{x} = Lx, \quad \hat{y} = LRe^{-1/2}Y, \quad \hat{u} = V_\infty U_0, \quad \hat{v} = V_\infty Re^{-1/2}V_0. \tag{1.4.58}$$

Similarly, in the curvilinear coordinates we shall write

$$\hat{s} = Ls, \quad \hat{n} = LRe^{-1/2}N, \quad \widehat{V}_\tau = V_\infty U_0', \quad \widehat{V}_n = V_\infty Re^{-1/2}V_0'. \tag{1.4.59}$$

Substituting (1.4.3) and (1.4.59) into (1.4.52), (1.4.53), (1.4.56), and (1.4.57), and working with the leading-order terms, we find that

$$U_0 = U_0', \quad V_0 = V_0' + U_0 f'(x), \quad s = x, \quad N = Y - f(x). \tag{1.4.60}$$

Of course, substitution of (1.4.59) into the Navier–Stokes equations (1.2.8) leads to the boundary-layer equations (1.2.34a), (1.2.34b), that are identical to the equations (1.4.4a), (1.4.4b) obtained by making use of (1.4.3). This observation suggests that the following statement, known as *Prandtl's transposition theorem*, is valid:

Theorem 1.1 *The boundary-layer equations*

$$U_0\frac{\partial U_0}{\partial x} + V_0\frac{\partial U_0}{\partial Y} = -\frac{\partial P_0}{\partial x} + \frac{\partial^2 U_0}{\partial Y^2}, \tag{1.4.61a}$$

$$\frac{\partial P_0}{\partial Y} = 0, \tag{1.4.61b}$$

$$\frac{\partial U_0}{\partial x} + \frac{\partial V_0}{\partial Y} = 0 \tag{1.4.61c}$$

are invariant with respect to the transformations

$$\left.\begin{array}{ccc} U_0 = U_0', & V_0 = V_0' + U_0 f'(x), & P_0 = P_0', \\ s = x, & N = Y - f(x). \end{array}\right\} \tag{1.4.62}$$

Proof To prove the theorem, we need to substitute (1.4.62) into equations (1.4.61), and confirm that in the new coordinate system, the equations take the same form as in the old one. We start with differentiation of pressure. Its transformation is written as

$$P_0(x, Y) = P_0'\big[s(x, Y), N(x, Y)\big], \tag{1.4.63}$$

where

$$s(x, Y) = x, \quad N(x, Y) = Y - f(x).$$

Applying the chain rule to (1.4.63), we have

$$\frac{\partial P_0}{\partial x} = \frac{\partial P_0'}{\partial s}\frac{\partial s}{\partial x} + \frac{\partial P_0'}{\partial N}\frac{\partial N}{\partial x} = \frac{\partial P_0'}{\partial s} - f'(x)\frac{\partial P_0'}{\partial N}, \tag{1.4.64a}$$

$$\frac{\partial P_0}{\partial Y} = \frac{\partial P_0'}{\partial s}\frac{\partial s}{\partial Y} + \frac{\partial P_0'}{\partial N}\frac{\partial N}{\partial Y} = \frac{\partial P_0'}{\partial N}. \tag{1.4.64b}$$

It follows from (1.4.64b) and (1.4.61b) that

$$\frac{\partial P_0'}{\partial N} = 0, \tag{1.4.65}$$

and then (1.4.64a) takes the form

$$\frac{\partial P_0}{\partial x} = \frac{\partial P_0'}{\partial x'}. \tag{1.4.66}$$

Similarly, differentiating the longitudinal velocity U_0, we have

$$\frac{\partial U_0}{\partial x} = \frac{\partial U_0'}{\partial s} - f'(x)\frac{\partial U_0'}{\partial N}, \qquad \frac{\partial U_0}{\partial Y} = \frac{\partial U_0'}{\partial N}, \qquad \frac{\partial^2 U_0}{\partial Y^2} = \frac{\partial^2 U_0'}{\partial N^2}. \tag{1.4.67}$$

By substituting (1.4.67) together with the transformation of the lateral velocity

$$V_0 = V_0' + U_0 f'(x) \tag{1.4.68}$$

into (1.4.61a), we find that

$$U_0'\frac{\partial U_0'}{\partial s} + V_0'\frac{\partial U_0'}{\partial N} = -\frac{\partial P_0'}{\partial s} + \frac{\partial^2 U_0'}{\partial N^2}. \tag{1.4.69}$$

It then remains to consider the continuity equation (1.4.61c). Differentiating (1.4.68) with respect to Y,

$$\frac{\partial V_0}{\partial Y} = \frac{\partial V_0'}{\partial Y'} + \frac{\partial U_0}{\partial Y}f'(x), \tag{1.4.70}$$

and substituting (1.4.70) together with the first of equations (1.4.67) into (1.4.61c), we find that

$$\frac{\partial U_0'}{\partial s} + \frac{\partial V_0'}{\partial N} = 0. \tag{1.4.71}$$

We see that the set of equations (1.4.61) is indeed invariant with respect to the transformations (1.4.62). ◻

Let us now return to the Chapman shear layer flow. Thanks to Theorem 1.1, the flow analysis in Section 1.4.1 leading to equation (1.4.8) and boundary conditions (1.4.9) remains unchanged if, instead of Cartesian coordinates $(\hat{x}, \hat{y})$, we use curvilinear coordinates $(\hat{s}, \hat{n})$; see Figure 1.22. However, if the $\hat{s}$-axis lies along the dividing streamline, then in addition to conditions (1.4.4c)–(1.4.4e), we will also have

$$V_0 = 0 \quad \text{at} \quad Y = 0. \tag{1.4.72}$$

The substitution of the second equation in (1.4.7) into (1.4.72) leads to the boundary condition

$$\varphi(0) = 0,$$

which closes the boundary-value problem (1.4.8), (1.4.9) for function $\varphi(\eta)$. Of course, the position of the dividing streamline has still to be found through analysis of the interaction of the shear layer with the flow above and below it.

1.5 Laminar Jet

Let us consider a steady flow produced by a jet that emerges through a narrow slit in a flat barrier OO', and penetrates into a semi-infinite region filled with a fluid at rest; see Figure 1.23.[15] Of course, due to the viscous forces, acting between the jet and surrounding fluid, the latter is also expected to start moving. When describing this motion, we shall assume for simplicity that the fluid in the jet is identical to that in the surrounding region.

We shall write the boundary-layer equations in dimensional variables:

$$\hat{u}\frac{\partial \hat{u}}{\partial \hat{x}} + \hat{v}\frac{\partial \hat{u}}{\partial \hat{y}} = -\frac{1}{\rho}\frac{\partial \hat{p}}{\partial \hat{x}} + \nu\frac{\partial^2 \hat{u}}{\partial \hat{y}^2}, \tag{1.5.1a}$$

$$\frac{\partial \hat{u}}{\partial \hat{x}} + \frac{\partial \hat{v}}{\partial \hat{y}} = 0, \tag{1.5.1b}$$

$$\frac{\partial \hat{p}}{\partial \hat{y}} = 0. \tag{1.5.1c}$$

Here, we use Cartesian coordinates $(\hat{x}, \hat{y})$ with $\hat{x}$ measured from the centre of the slit normal to the barrier OO'. The velocity components are denoted as usual by $\hat{u}$ and $\hat{v}$, and $\hat{p}$ denotes the pressure.

Since the fluid surrounding the jet is at rest, the pressure stays constant everywhere outside the jet. Denoting its value by $\hat{p}_0$, we arrive at the following boundary condition for equation (1.5.1c):

$$\hat{p} = \hat{p}_0 \quad \text{at} \quad \hat{y} = \pm\infty.$$

Integrating (1.5.1c) across the boundary layer, we can conclude that $\hat{p} = \hat{p}_0$ everywhere inside the boundary layer. This reduces the set of equations (1.5.1) to

$$\hat{u}\frac{\partial \hat{u}}{\partial \hat{x}} + \hat{v}\frac{\partial \hat{u}}{\partial \hat{y}} = \nu\frac{\partial^2 \hat{u}}{\partial \hat{y}^2}, \tag{1.5.2a}$$

$$\frac{\partial \hat{u}}{\partial \hat{x}} + \frac{\partial \hat{v}}{\partial \hat{y}} = 0. \tag{1.5.2b}$$

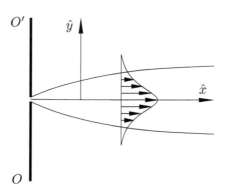

Fig. 1.23: Problem layout.

---

[15]This problem was first studied by Schlichting (1933).

We now need to formulate the boundary conditions for the equations (1.5.2). We note, first of all, that on both sides of the jet the longitudinal velocity should tend to zero,

$$\hat{u} = 0 \quad \text{at} \quad \hat{y} = \pm\infty, \quad \hat{x} > 0, \tag{1.5.3}$$

to satisfy the matching condition with the motionless fluid surrounding the jet. Secondly, due to the symmetry of the flow with respect to the $\hat{x}$-axis, we have

$$\hat{v} = 0 \quad \text{at} \quad \hat{y} = 0, \quad \hat{x} > 0. \tag{1.5.4}$$

Finally, we have to formulate an initial condition for $\hat{u}$. We expect the thickness of the jet to increase with distance $\hat{x}$ from the barrier. We shall restrict our attention to large enough values of $\hat{x}$ where the jet thickness becomes significantly larger than the slit width, and the initial condition may be written as

$$\hat{u} = 0 \quad \text{at} \quad \hat{x} = 0, \quad \hat{y} \neq 0.$$

Indeed, for any $\hat{y}$ comparable with the jet thickness, an 'observer' travelling upstream parallel to the $\hat{x}$-axis, encounters the barrier surface, that is impenetrable. The only exception is $\hat{y} = 0$, which corresponds to the opening in the barrier.

It is clear that, in order to support the motion of the fluid in the jet, the velocity through the slit should be made larger as the slit width decreases. A quantitative measure of the 'intensity' of the jet, emerging through the slit, may be given in terms of the total momentum flux through a plane perpendicular to the jet axis:

$$J = \rho \int\limits_{-\infty}^{\infty} \hat{u}^2 \, d\hat{y}.$$

It is easily shown that J remains constant along the jet. Indeed, if we multiply the continuity equation (1.5.2b) by $\hat{u}$ and add to the momentum equation (1.5.2a), then we will have

$$\frac{\partial}{\partial \hat{x}} \left(\hat{u}^2 \right) + \frac{\partial}{\partial \hat{y}} \left(\hat{u}\hat{v} \right) = \nu \frac{\partial^2 \hat{u}}{\partial \hat{y}^2}.$$

Integration of this equation across the jet leads to

$$\frac{d}{d\hat{x}} \int\limits_{-\infty}^{\infty} \hat{u}^2 \, d\hat{y} + \hat{u}\hat{v}\Big|_{-\infty}^{\infty} = \nu \frac{\partial \hat{u}}{\partial \hat{y}}\Big|_{-\infty}^{\infty}.$$

Since $\hat{u} \to 0$ as $\hat{y} \to \pm\infty$, it can be expected that $\partial\hat{u}/\partial\hat{y}$ also vanishes at $\hat{y} = \pm\infty$. Consequently,

$$\frac{d}{d\hat{x}} \int\limits_{-\infty}^{\infty} \hat{u}^2 \, d\hat{y} = 0.$$

This proves that the momentum flux J is indeed constant along the jet and coincides with its value J_0 at the slit. We shall therefore write

$$J_0 = \rho \int_{-\infty}^{\infty} \hat{u}^2 d\hat{y}. \tag{1.5.5}$$

Since the problem considered does not involve a characteristic length scale, we shall try to find its solution in a self-similar form. Based on the continuity equation (1.5.2b), we introduce the stream function $\hat{\psi}(\hat{x}, \hat{y})$ such that

$$\frac{\partial \hat{\psi}}{\partial \hat{x}} = -\hat{v}, \qquad \frac{\partial \hat{\psi}}{\partial \hat{y}} = \hat{u},$$

and seek the solution for the stream function in the form:

$$\hat{\psi}(\hat{x}, \hat{y}) = a\hat{x}^{\alpha} \varphi(\eta), \qquad \eta = b\frac{\hat{y}}{\hat{x}^{\beta}}. \tag{1.5.6}$$

We start with differentiation of the independent variable η:

$$\frac{\partial \eta}{\partial \hat{x}} = -\beta\frac{\eta}{\hat{x}}, \qquad \frac{\partial \eta}{\partial \hat{y}} = \frac{b}{\hat{x}^{\beta}}.$$

The velocity components are calculated as

$$\hat{u} = \frac{\partial \hat{\psi}}{\partial \hat{y}} = ab\hat{x}^{\alpha-\beta} \varphi'(\eta), \qquad \hat{v} = -\frac{\partial \hat{\psi}}{\partial \hat{x}} = -a\hat{x}^{\alpha-1}\big(\alpha\varphi - \beta\eta\varphi'\big). \tag{1.5.7}$$

We further have

$$\left.\begin{array}{c} \dfrac{\partial \hat{u}}{\partial \hat{x}} = ab\hat{x}^{\alpha-\beta-1}\big[(\alpha-\beta)\varphi' - \beta\eta\varphi''\big], \qquad \dfrac{\partial \hat{u}}{\partial \hat{y}} = ab^2\hat{x}^{\alpha-2\beta}\varphi'', \\[2mm] \dfrac{\partial^2 \hat{u}}{\partial \hat{y}^2} = ab^3\hat{x}^{\alpha-3\beta}\varphi'''. \end{array}\right\} \tag{1.5.8}$$

The substitution of (1.5.7) and (1.5.8) into the momentum equation (1.5.2a) yields

$$a^2b^2\hat{x}^{2\alpha-2\beta-1}\big[(\alpha-\beta)\varphi'^2 - \beta\eta\varphi'\varphi''\big]$$
$$- a^2b^2\hat{x}^{2\alpha-2\beta-1}\big(\alpha\varphi\varphi'' - \beta\eta\varphi'\varphi''\big) = \nu ab^3\hat{x}^{\alpha-3\beta}\varphi'''.$$

In order to eliminate $\hat{x}$ from this equation, we have to set

$$2\alpha - 2\beta - 1 = \alpha - 3\beta. \tag{1.5.9}$$

For convenience, we also choose

$$a^2b^2 = \nu ab^3. \tag{1.5.10}$$

As a result, the equation for φ takes the form

$$(\alpha - \beta)\varphi'^2 - \alpha\varphi\varphi'' = \varphi'''. \tag{1.5.11}$$

The substitution of (1.5.7) into the boundary conditions (1.5.3) and (1.5.4) yields

$$\varphi(0) = 0, \qquad \varphi'(-\infty) = \varphi'(\infty) = 0. \qquad (1.5.12)$$

It remains to make use of the integral condition (1.5.5). To express it in terms of the similarity variables $\eta = b\hat{y}/\hat{x}^{\beta}$, we note that the integration in (1.5.5) is performed across the jet with $\hat{x}$ kept constant. Therefore,

$$d\eta = \frac{b}{\hat{x}^{\beta}}d\hat{y}. \qquad (1.5.13)$$

Substituting (1.5.13) together with the first equation in (1.5.7) into (1.5.5), we have

$$J_0 = \rho a^2 b \hat{x}^{2\alpha-\beta} \int\limits_{-\infty}^{\infty} [\varphi'(\eta)]^2 d\eta. \qquad (1.5.14)$$

Since the momentum flux does not depend on $\hat{x}$, we have to set

$$2\alpha - \beta = 0. \qquad (1.5.15)$$

If we further choose

$$J_0 = \rho a^2 b, \qquad (1.5.16)$$

then equation (1.5.14) assumes the form

$$\int\limits_{\infty}^{-\infty} [\varphi'(\eta)]^2 d\eta = 1. \qquad (1.5.17)$$

Solving (1.5.15) and (1.5.9) for α and β, we have

$$\alpha = \frac{1}{3}, \qquad \beta = \frac{2}{3}.$$

From (1.5.10) and (1.5.16), it follows that

$$a = \sqrt[3]{\frac{J_0 \nu}{\rho}}, \qquad b = \sqrt[3]{\frac{J_0}{\rho \nu^2}}.$$

Consequently, the sought solution for the stream function (1.5.6) may be written as

$$\hat{\psi}(\hat{x}, \hat{y}) = \sqrt[3]{\frac{J_0 \nu}{\rho}}\hat{x}^{1/3}\varphi(\eta), \qquad \eta = \sqrt[3]{\frac{J_0}{\rho\nu^2}}\frac{\hat{y}}{\hat{x}^{2/3}},$$

and equation (1.5.11) for $\varphi(\eta)$ assumes the form

$$\varphi''' + \frac{1}{3}\varphi'^2 + \frac{1}{3}\varphi\varphi'' = 0.$$

This equation is easily integrated to give

$$\varphi'' + \frac{1}{3}\varphi\varphi' = C. \tag{1.5.18}$$

The constant of integration C may be found from the boundary condition (1.5.12) at the outer edge of the jet where $\varphi'(\infty) = 0$. Assuming, subject to subsequent confirmation, that φ'' also tends to zero as $\eta \to \infty$, we have to set $C = 0$, which turns equation (1.5.18) into

$$\varphi'' + \frac{1}{3}\varphi\varphi' = 0.$$

It can be integrated again to yield

$$\varphi' + \frac{1}{6}\varphi^2 = C_1.$$

Consequently, we can write

$$\frac{d\varphi}{d\eta} = \frac{1}{6}(6C_1 - \varphi^2). \tag{1.5.19}$$

We expect the longitudinal velocity $\hat{u}$ to be positive in the jet. It then follows from the first equation in (1.5.7) that φ' is positive. The latter is only possible if $C_1 > 0$ and

$$-\sqrt{6C_1} < \varphi < \sqrt{6C_1}. \tag{1.5.20}$$

Separating the variables in (1.5.19), we have

$$d\eta = \frac{6d\varphi}{6C_1 - \varphi^2}.$$

In order to integrate this equation, we represent its right-hand side in the form of simple fractions

$$d\eta = \frac{3}{\sqrt{6C_1}}\left(\frac{d\varphi}{\varphi + \sqrt{6C_1}} - \frac{d\varphi}{\varphi - \sqrt{6C_1}}\right),$$

and we see that

$$\eta = \frac{3}{\sqrt{6C_1}} \ln\left|\frac{\varphi + \sqrt{6C_1}}{\varphi - \sqrt{6C_1}}\right| + C_2. \tag{1.5.21}$$

The constant of integration, C_2, may be found from the first condition in (1.5.12) to be

$$C_2 = 0.$$

Taking further into account the range (1.5.20) of variation of $\varphi(\eta)$, we can write (1.5.21) in the form

$$\eta = \frac{3}{\sqrt{6C_1}} \ln \frac{\sqrt{6C_1} + \varphi}{\sqrt{6C_1} - \varphi}. \tag{1.5.22}$$

Equation (1.5.22) is easily solved for $\varphi(\eta)$ to give

$$\varphi(\eta) = \sqrt{6C_1} \frac{1 - \exp\left(-\sqrt{\frac{2}{3}C_1}\,\eta\right)}{1 + \exp\left(-\sqrt{\frac{2}{3}C_1}\,\eta\right)}. \tag{1.5.23}$$

It remains to find constant C_1. This is done by substituting (1.5.23) into the integral condition (1.5.17). It appears that

$$C_1 = \frac{1}{2}\sqrt[3]{\frac{3}{4}}.$$

Exercises 4

1. Consider the flow in the separation region below the shear layer (see Figure 1.19), where the stream function is given by the equation (1.4.46). Using Figure 1.20, deduce that

$$\sin\left(\tfrac{1}{2}\theta\right) = \sqrt{\tfrac{1}{2}(1 - \cos\theta)} = \sqrt{\frac{1}{2}\left(1 + \frac{y}{r}\right)},$$

and show that, in this flow, the streamlines ($\psi = const$) are parabolas:

$$y = \frac{\psi^2}{2\Lambda^2} - \frac{\Lambda^2}{2\psi^2}x^2.$$

2. Consider the laminar jet shown in Figure 1.23. You can use without proof the fact that the solution of the boundary-layer equations for the flow inside the jet is written as

$$\hat{\psi}(\hat{x}, \hat{y}) = \sqrt[3]{\frac{J_0\nu}{\rho}}\,\hat{x}^{1/3}\varphi(\eta), \qquad \eta = \sqrt[3]{\frac{J_0}{\rho\nu^2}}\,\frac{\hat{y}}{\hat{x}^{2/3}}, \qquad (1.5.24)$$

where $(\hat{x}, \hat{y})$ are Cartesian coordinates with $\hat{x}$ measured along the axis of symmetry of the flow, $\hat{\psi}$ is the stream function, ρ and ν are the fluid density and viscosity, and J_0 is the momentum of the fluid in the jet.

You can further use without proof the fact that function $\varphi(\eta)$ in (1.5.24) is given by

$$\varphi(\eta) = \sqrt{6C_1}\,\frac{1 - \exp\left(-\sqrt{\tfrac{2}{3}C_1}\,\eta\right)}{1 + \exp\left(-\sqrt{\tfrac{2}{3}C_1}\,\eta\right)}, \qquad C_1 = \frac{1}{2}\sqrt[3]{\frac{3}{4}}.$$

Your task is to analyse the entrainment effect of the jet, and to describe the slow fluid motion above and below the jet. You can perform this task in the following steps:

(a) Concentrate your attention on the flow above the jet, and show that at the upper edge of the jet

$$\psi = \Lambda x^{1/3},$$

where

$$\Lambda = \sqrt{6C_1}\,\sqrt[3]{\frac{J_0\nu}{\rho}}.$$

(b) Now analyse the flow above the jet. When performing this task you may use without proof the fact that this flow is inviscid (in the leading-order approximation) and potential. Hence, introduce the complex potential

$$w(z) = \varphi(x, y) + i\psi(x, y), \qquad z = \hat{x} + i\hat{y}$$

and formulate two boundary conditions for the imaginary part of $w(z)$:
   (i) impermeability condition on the barrier OO',
   (ii) matching condition with the solution in the jet.
(c) Seek the solution for the complex potential $w(z)$ in the form

$$w(z) = Cz^{1/3}.$$

Find the real and imaginary parts of constant C, and show that the pressure above the jet is given by

$$p = p_\infty - \frac{2}{9}\rho\Lambda^2\hat{r}^{-4/3},$$

where $\hat{r} = |z|$ is the distance from the observation point to the slit in the wall from which the jet emerges.

1.6   Viscous Wake Behind a Rigid Body

The viscous wake is an integral part of any real flow past a solid body. To explain how it forms, let us consider, for example, the flow past an aerofoil; see Figure 1.24. When analysing this flow, one needs to examine two boundary layers, one on the upper side of the aerofoil, the other on the lower side. In either boundary layer the fluid velocity decreases from its inviscid value at the outer edge of the boundary layer to zero on the aerofoil surface. This creates a deficit of the fluid momentum in the boundary layers, that cannot disappear at once as the fluid leaves the aerofoil at the trailing edge. Instead, it persists in the flow behind the aerofoil, where the two boundary layers merge to form the *viscous wake*. The thickness of the wake is, obviously, an order $Re^{-1/2}$ quantity, which means that it still obeys the boundary-layer equations. The viscous forces acting in the wake are expected to perform a function of gradual transformation of the velocity profile, as sketched in Figure 1.24, into a uniform flow further downstream. Our task here will be to describe this process through analysis of the far-field behaviour of the flow.[16]

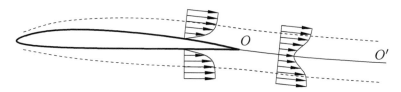

Fig. 1.24: Merger of the two boundary layers into viscous wake.

[16]Tollmien (1931) was the first to study this flow.

1.6.1 Inviscid flow

We start, as usual, with the inviscid region. The far-field behaviour of the inviscid flow past a solid body is given by equation (3.4.47) on page 173 in Part 1 of this book series:

$$\frac{dw}{dz} = V_\infty e^{-i\alpha} + \frac{\Gamma}{2\pi i z} + O\left(\frac{1}{z^2}\right) \quad \text{as} \quad z \to \infty. \tag{1.6.1}$$

Here, V_∞ is the free-stream velocity, Γ is the circulation of the velocity vector around the aerofoil, and $w(z) = \hat{\varphi} + i\hat{\psi}$ is the complex potential. The latter is known to be an analytic function of $z = \hat{x} + i\hat{y}$. All the variables in (1.6.1) and in the analysis that follows will be assumed dimensional.

If we choose the Cartesian coordinates $(\hat{x}, \hat{y})$ such that the $\hat{x}$-axis is parallel to the free-stream velocity (see Figure 1.25), then the angle of attack α is zero, and we can write

$$\frac{dw}{dz} = V_\infty + \frac{\Gamma}{2\pi i z} + O\left(\frac{1}{z^2}\right) \quad \text{as} \quad z \to \infty. \tag{1.6.2}$$

Integration of (1.6.2) yields

$$w = V_\infty z + \frac{\Gamma}{2\pi i} \ln z + O\left(\frac{1}{z}\right). \tag{1.6.3}$$

The constant of integration in (1.6.3) has been disregarded in view of the fact that the complex potential $w(z)$ is defined to within an arbitrary constant.

The wake develops along the streamline OO' (see Figure 1.25) that leaves the aerofoil surface at that trailing edge O. Our first task will be to find the shape of this streamline and the distribution of the inviscid velocity $U_e(s)$ along it. To perform this task, we write (1.6.3) as

$$\hat{\varphi} + i\hat{\psi} = V_\infty(\hat{x} + i\hat{y}) + \frac{\Gamma}{2\pi i}\left(\ln\sqrt{\hat{x}^2 + \hat{y}^2} + i\arg z\right) + \cdots,$$

and we see that

$$\hat{\psi} = V_\infty \hat{y} - \frac{\Gamma}{2\pi} \ln \sqrt{\hat{x}^2 + \hat{y}^2} + \cdots.$$

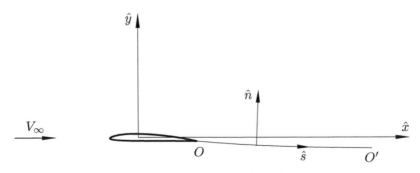

Fig. 1.25: Coordinate systems used in the flow analysis.

We know that the stream function $\hat{\psi}$ is constant along any streamline, including streamline OO'. Hence, we write

$$V_\infty \hat{y} - \frac{\Gamma}{2\pi} \ln \sqrt{\hat{x}^2 + \hat{y}^2} + \cdots = C. \tag{1.6.4}$$

Equation (1.6.4) is easily solved for $\hat{y}$, taking into account that $\hat{y}/\hat{x} \ll 1$ for large values of $\hat{x}$. We have

$$\hat{y} = \frac{\Gamma}{2\pi V_\infty} \ln \hat{x} + O(1) \quad \text{as} \quad \hat{x} \to \infty. \tag{1.6.5}$$

Let us now determine the velocity distribution along OO'. Remember that the derivative of $w(z)$ equals the complex conjugate velocity $\overline{V} = \hat{u} - i\hat{v}$.[17] This means that equation (1.6.2) may be written as

$$\hat{u} - i\hat{v} = V_\infty + \frac{\Gamma}{2\pi i(\hat{x} + i\hat{y})} + \cdots . \tag{1.6.6}$$

Separating the real and imaginary parts in (1.6.6), we find that

$$\hat{u} = V_\infty - \frac{\Gamma}{2\pi} \frac{\hat{y}}{\hat{x}^2 + \hat{y}^2} + \cdots , \qquad \hat{v} = \frac{\Gamma}{2\pi} \frac{\hat{x}}{\hat{x}^2 + \hat{y}^2} + \cdots . \tag{1.6.7}$$

To determine $\hat{u}$ and $\hat{v}$ on the streamline OO', one needs to substitute (1.6.5) into (1.6.7), and expand the resulting equations assuming that $\hat{x} \to \infty$. This yields

$$\hat{u} = V_\infty - \frac{\Gamma^2}{4\pi^2 V_\infty} \frac{\ln \hat{x}}{\hat{x}^2} + \cdots , \qquad \hat{v} = \frac{\Gamma}{2\pi} \frac{1}{\hat{x}} + \cdots .$$

Finally, the modulus of the velocity is calculated as

$$\widehat{U}_e = \sqrt{\hat{u}^2 + \hat{v}^2} = V_\infty - \frac{\Gamma^2}{4\pi^2 V_\infty} \frac{\ln \hat{x}}{\hat{x}^2} + \cdots \quad \text{as} \quad \hat{x} \to \infty. \tag{1.6.8}$$

1.6.2 Flow in the wake

We are ready now to turn to the viscous flow in the wake. As already mentioned, the fluid motion in the wake is described by the boundary-layer equations

$$\widehat{V}_\tau \frac{\partial \widehat{V}_\tau}{\partial \hat{s}} + \widehat{V}_n \frac{\partial \widehat{V}_\tau}{\partial \hat{n}} = \widehat{U}_e \frac{d\widehat{U}_e}{d\hat{s}} + \nu \frac{\partial^2 \widehat{V}_\tau}{\partial \hat{n}^2}, \tag{1.6.9a}$$

$$\frac{\partial \widehat{V}_\tau}{\partial \hat{s}} + \frac{\partial \widehat{V}_n}{\partial \hat{n}} = 0. \tag{1.6.9b}$$

Here we use curvilinear orthogonal coordinates $(\hat{s}, \hat{n})$, with $\hat{s}$ measured from the trailing edge of the aerofoil along the streamline OO'; see Figure 1.25. The velocity components in these coordinates are denoted as $(\widehat{V}_\tau, \widehat{V}_n)$.

---

[17]See equation (3.4.6) on page 160 in Part 1 of this book series.

Equations (1.6.9) require two boundary conditions for $\widehat{V}_\tau$. These are obtained by matching with the solution in the inviscid region:

$$\widehat{V}_\tau = \widehat{U}_e(\hat{s}) \quad \text{at} \quad \hat{n} = \pm\infty. \tag{1.6.10}$$

We also need a boundary condition for $\widehat{V}_n$. Since the velocity vector is tangent to the streamline OO', we can write

$$\widehat{V}_n = 0 \quad \text{at} \quad \hat{n} = 0. \tag{1.6.11}$$

Strictly speaking, we also need to formulate an initial condition for $\widehat{V}_\tau$, say, at the trailing edge of the aerofoil. However, we will see that the asymptotic solution of equations (1.6.9) does not require detailed information about the initial velocity profile; it depends only on integral drag of the aerofoil.

We seek the solution of (1.6.9)–(1.6.11) in the form

$$\left. \begin{aligned} \widehat{V}_\tau &= V_\infty + \hat{s}^{-\alpha} f(\eta) + \cdots, \\ \widehat{V}_n &= \hat{s}^{-\gamma} g(\eta) + \cdots \end{aligned} \right\} \quad \text{as} \quad \hat{s} \to \infty, \tag{1.6.12}$$

where

$$\eta = b \frac{\hat{n}}{\hat{s}^\beta}. \tag{1.6.13}$$

Since all the perturbations are expected to decay far downstream of the aerofoil, we have to restrict our attention to positive values of α and γ.

We start with the continuity equation (1.6.9b). The substitution of (1.6.12), (1.6.13) into (1.6.9b) yields

$$-\hat{s}^{-\alpha-1}\big[\alpha f(\eta) + \beta\eta f'(\eta)\big] + \hat{s}^{-\gamma-\beta} bg'(\eta) = 0.$$

To avoid degeneration in this equation, we have to set

$$\alpha + 1 = \gamma + \beta, \tag{1.6.14}$$

and then we will have

$$bg' = \alpha f + \beta\eta f'. \tag{1.6.15}$$

Now, we turn to the momentum equation (1.6.9a). The convective terms on the left-hand side of this equation are calculated as

$$\widehat{V}_\tau \frac{\partial \widehat{V}_\tau}{\partial \hat{s}} = -\hat{s}^{-\alpha-1} V_\infty \big[\alpha f(\eta) + \beta\eta f'(\eta)\big] + \cdots, \tag{1.6.16}$$

$$\widehat{V}_n \frac{\partial \widehat{V}_\tau}{\partial \hat{n}} = \hat{s}^{-\gamma-\alpha-\beta} bg(\eta) f'(\eta) + \cdots. \tag{1.6.17}$$

Before calculating the pressure gradient on the right-hand side of (1.6.9a), we need to cast equation (1.6.8) for $\widehat{U}_e$ in terms of the arc length $\hat{s}$ along the streamline OO'.

Denoting the position of the trailing edge of the aerofoil in Cartesian coordinates by $\hat{x}_0$, we have

$$\hat{s} = \int_{\hat{x}_0}^{\hat{x}} \sqrt{1 + [\hat{y}'(\hat{x})]^2}\, d\hat{x}. \qquad (1.6.18)$$

Using (1.6.5) in (1.6.18), one can see that

$$\hat{s} = \hat{x} + O(1) \quad \text{as} \quad \hat{x} \to \infty,$$

which, being substituted into (1.6.8), yields

$$\widehat{U}_e(\hat{s}) = V_\infty - \frac{\Gamma^2}{4\pi^2 V_\infty} \frac{\ln \hat{s}}{\hat{s}^2} + \cdots \quad \text{as} \quad \hat{s} \to \infty. \qquad (1.6.19)$$

Thus, the two terms on the right-hand side of (1.6.9a) are calculated as

$$\widehat{U}_e \frac{d\widehat{U}_e}{d\hat{s}} = \frac{\Gamma^2}{2\pi^2} \frac{\ln \hat{s}}{\hat{s}^3} + \cdots, \qquad \nu \frac{\partial^2 \widehat{V}_\tau}{\partial \hat{n}^2} = \hat{s}^{\alpha-2\beta} \nu b^2 f''(\eta) + \cdots. \qquad (1.6.20)$$

Substituting (1.6.16), (1.6.17), and (1.6.20) into (1.6.9a), we have

$$-\hat{s}^{-\alpha-1} V_\infty (\alpha f + \beta \eta f') + \hat{s}^{-\gamma-\alpha-\beta} bg f' = \frac{\Gamma^2}{2\pi^2} \frac{\ln \hat{s}}{\hat{s}^3} + \hat{s}^{\alpha-2\beta} \nu b^2 f''. \qquad (1.6.21)$$

It follows from (1.6.14) that the second term on the left-hand side of this equation is $\hat{s}^\alpha$ times smaller than the first one, and should be disregarded in the limit $\hat{s} \to \infty$ for all $\alpha > 0$. If we assume (subject to subsequent confirmation) that

$$\alpha < 2, \qquad (1.6.22)$$

then the first term on the right-hand side of (1.6.21), which represents the pressure gradient, appears to be small compared to the first term on the left-hand side, which reduces (1.6.21) to

$$-\hat{s}^{-\alpha-1} V_\infty (\alpha f + \beta \eta f') = \hat{s}^{\alpha-2\beta} \nu b^2 f''. \qquad (1.6.23)$$

To avoid further degeneration in this equation, we have to set

$$\beta = \frac{1}{2}. \qquad (1.6.24)$$

Constant b in the definition of the similarity variable (1.6.13) still remains arbitrary, and may be used to exclude dimensional constants V_∞ and ν from equation (1.6.23). Choosing

$$b = \sqrt{\frac{V_\infty}{\nu}}, \qquad (1.6.25)$$

we have equation (1.6.23) in the following form:

$$f'' + \frac{1}{2}\eta f' + \alpha f = 0. \qquad (1.6.26)$$

Two boundary conditions for this equation are obtained by substituting (1.6.19) together with the asymptotic expansion for $\widehat{V}_\tau$ in (1.6.12) into (1.6.10). We find that for any α satisfying condition (1.6.22),

$$f(-\infty) = f(\infty) = 0. \tag{1.6.27}$$

Equation (1.6.26) considered together with boundary conditions (1.6.27) admits a trivial solution $f(\eta) \equiv 0$. Of course, we are interested in a non-trivial solution. Hence, we have to treat the boundary-value problem (1.6.26), (1.6.27) as an eigenvalue problem with parameter α playing the role of an eigenvalue to be found.

The general solution of equation (1.6.26) may be written in the form

$$f(\eta) = C_1 f_1(\eta) + C_2 f_2(\eta), \tag{1.6.28}$$

where $f_1(\eta)$ and $f_2(\eta)$ are two complementary solutions of (1.6.26) defined by the initial conditions

$$f_1(0) = 1, \qquad f_1'(0) = 0, \tag{1.6.29}$$

$$f_2(0) = 0, \qquad f_2'(0) = 1. \tag{1.6.30}$$

We start with function $f_1(\eta)$. It satisfies equation (1.6.26), which may be transformed into Kummer's equation

$$z\frac{d^2 f_1}{dz^2} + \left(\frac{1}{2} - z\right)\frac{d^2 f_1}{dz^2} - \alpha f_1 = 0 \tag{1.6.31}$$

by introducing instead of η a new independent variable

$$z = -\frac{1}{4}\eta^2. \tag{1.6.32}$$

The properties of Kummer's equation are well known.[18] In general, it is written as

$$z\frac{d^2 f_1}{dz^2} + (b - z)\frac{d^2 f_1}{dz^2} - a f_1 = 0, \tag{1.6.33}$$

where a and b are constants. Equation (1.6.33) has two complementary solutions: the confluent hypergeometric function of the first kind $M(a, b, z)$, known as Kummer's function, and confluent hypergeometric function of the second kind $U(a, b, z)$, the Tricomi function. Kummer's function $M(a, b, z)$ is regular at $z = 0$, and may be represented by the Taylor series

$$M(a, b, z) = 1 + \frac{az}{b} + \frac{(a)_2 z^2}{(b)_2 2!} + \cdots + \frac{(a)_n z^n}{(b)_n n!} + \cdots , \tag{1.6.34}$$

convergent for all finite z, real or complex. The coefficients of the series (1.6.34) are calculated according to the rule

$$(a)_0 = 1, \quad (a)_n = a(a + 1)(a + 2) \cdots (a + n - 1).$$

---

[18]See, for example, Abramowitz and Stegun (1965).

The Tricomi function, $U(a, b, z)$, is defined as

$$U(a, b, z) = \frac{\Gamma(1-b)}{\Gamma(1+a-b)} M(a, b, z) + \frac{\Gamma(b-1)}{\Gamma(a)} z^{1-b} M(1+a-b, 2-b, z), \quad (1.6.35)$$

where Γ is Euler's gamma function.

Comparing (1.6.33) with (1.6.31), we can see that in our case

$$b = \frac{1}{2}, \qquad a = \alpha,$$

and it follows from (1.6.34) and (1.6.35) that

$$\left.\begin{array}{l}
M\left(\alpha, \dfrac{1}{2}, -\dfrac{\eta^2}{4}\right) = 1 - \dfrac{\alpha}{2}\eta^2 + \cdots, \\[3mm]
U\left(\alpha, \dfrac{1}{2}, -\dfrac{\eta^2}{4}\right) = \dfrac{\Gamma(1/2)}{\Gamma(\alpha+1/2)} + i\dfrac{\Gamma(-1/2)}{2\Gamma(\alpha)}\eta + \cdots
\end{array}\right\} \quad \text{as} \quad \eta \to 0,$$

which means that the solution for $f_1(\eta)$, satisfying conditions (1.6.29), is given by

$$f_1(\eta) = M\left(\alpha, \frac{1}{2}, -\frac{\eta^2}{4}\right). \qquad (1.6.36)$$

Notice that $f_1(\eta)$ is symmetric with respect to the streamline OO' (see Figure 1.24). Consequently, when dealing with the boundary conditions (1.6.27) we only need to investigate the behaviour of $f_1(\eta)$ at large positive values of η. The asymptotic behaviour of Kummer's function depends on the route the point of observation takes in the complex z-plane. If z tends to infinity along a ray lying in the left half-plane ($\Re\{z\} < 0$), then

$$M(a, b, z) = \frac{\Gamma(b)}{\Gamma(b-a)}(-z)^{-a} + \cdots \quad \text{as} \quad z \to \infty. \qquad (1.6.37)$$

Using (1.6.37) in the solution (1.6.36) for $f_1(\eta)$, yields

$$f_1(\eta) = \frac{4^\alpha\, \Gamma(1/2)}{\Gamma(1/2 - \alpha)} \frac{1}{\eta^{2\alpha}} + \cdots \quad \text{as} \quad \eta \to \infty. \qquad (1.6.38)$$

We see that $f_1(\eta)$ satisfies boundary condition (1.6.27) with any positive value of α. Hence, in order to determine α, we need to make the condition (1.6.27) more restrictive. We shall require that $f_1(\eta)$ not just tends to zero as $\eta \to \infty$ but does so faster than any power of η. This requirement is justified as follows.

The substitution of (1.6.38) into (1.6.28), and then into the asymptotic expansion (1.6.12) for $\widehat{V}_\tau$, results in

$$\widehat{V}_\tau = V_\infty + C_1\frac{4^\alpha\, \Gamma(1/2)}{\Gamma(1/2 - \alpha)} \frac{1}{(\eta^2 \hat{s})^\alpha} + \cdots. \qquad (1.6.39)$$

It further follows from (1.6.13), (1.6.24), and (1.6.25) that

$$\eta^2 \hat{s} = \frac{V_\infty}{\nu} \hat{n}^2,$$

which, being substituted into (1.6.39), leads to a conclusion that at large values of $\hat{s}$, the asymptotic behaviour of $\widehat{V}_\tau$ at the outer edge of the wake is given by

$$\widehat{V}_\tau = V_\infty + C_1 \left(\frac{4\nu}{V_\infty} \right)^\alpha \frac{\Gamma(1/2)}{\Gamma(1/2 - \alpha)} \frac{1}{\hat{n}^{2\alpha}} + \cdots . \tag{1.6.40}$$

Now, our task will be to study the behaviour $\widehat{V}_\tau$ at large values of $\hat{n}$ in the boundary layer on the aerofoil surface and in the wake. To perform this task, we need to assume that $\hat{s}$ is finite. Being guided by (1.6.40), we seek the solution at the outer edge of the boundary layer in the form

$$\widehat{V}_\tau = \widehat{U}_e(\hat{s}) + F(\hat{s}) \frac{1}{\hat{n}^{2\alpha}} + \cdots \quad \text{as} \quad \hat{n} \to \infty, \quad \hat{s} = O(L). \tag{1.6.41}$$

To find function $F(\hat{s})$, we need to use the boundary-layer equations (1.6.9). It follows from the continuity equation (1.6.9b) that

$$\widehat{V}_n = -\widehat{U}'_e(\hat{s})\hat{n} + \cdots \quad \text{as} \quad \hat{n} \to \infty. \tag{1.6.42}$$

Substituting (1.6.41) together with (1.6.42) into the momentum equation (1.6.9a), and working with $O(\hat{n}^{-2\alpha})$ terms, we arrive at the following equation for function $F(\hat{s})$:

$$\widehat{U}_e F' + (1 + 2\alpha) F \widehat{U}'_e = 0.$$

It is easily integrated to yield

$$F = \frac{\widetilde{C}}{\widehat{U}_e^{1+2\alpha}}. \tag{1.6.43}$$

The constant of integration $\widetilde{C}$ may be found from the solution for the boundary layer on the aerofoil surface. For example, when the aerofoil is represented by a flat plate, the longitudinal velocity near the outer edge of the boundary layer is obtained by substituting (1.1.53) into the first equation in (1.1.48). This gives

$$U_0 = 1 - \frac{C}{2(\eta - A_-)} e^{-(\eta - A_-)^2/4} + \cdots .$$

Clearly, no algebraic term is present in this solution, which means that $\widetilde{C} = 0$. It then follows from (1.6.43) that the algebraic term $F(\hat{s})/\hat{n}^{2\alpha}$ can be present in (1.6.40) neither the boundary layer, nor in the solution for the wake.

When dealing with the flow past an aerofoil with a rounded nose, one should start with a small vicinity of the front stagnation point where Hiemenz's solution holds (see Problem 1 in Exercises 3). It may be deduced from equation (1.3.17) that

$$\varphi = \eta + A + Be^{-\eta^2/2} + \cdots \quad \text{as} \quad \eta \to \infty,$$

which again precludes the algebraic term from appearing in (1.6.41).

Now, returning back to the asymptotic formula (1.6.38), we see that function $f_1(\eta)$ can only exhibit the desired behaviour if

$$\Gamma(1/2 - \alpha) = \infty.$$

The gamma function is known to be infinite at zero and all negative interger values of the argument. Consequently, the sought eigenvalues are

$$\alpha = \frac{1}{2} + m, \qquad m = 0,\ 1,\ 2,\ \dots . \tag{1.6.44}$$

We shall refer to (1.6.44) as the *symmetric solution spectrum*. The first of (1.6.44),

$$\alpha = \frac{1}{2}, \tag{1.6.45}$$

satisfies the assumption (1.6.22) and turns (1.6.36) into[19]

$$f_1(\eta) = M\left(\frac{1}{2}, \frac{1}{2}, -\frac{\eta^2}{4}\right) = e^{-\eta^2/4}. \tag{1.6.46}$$

The second complementary solution, $f_2(\eta)$, in (1.6.28) can be analysed in the same way.[20] We find that function $f_2(\eta)$ represents an anti-symmetric part of the solution. Its spectrum is

$$\alpha = 1 + m, \qquad m = 0,\ 1,\ 2,\ \dots . \tag{1.6.47}$$

The first of these, $\alpha = 1$, satisfies condition (1.6.22), and the correspondent eigen-function is given by

$$f_2(\eta) = \eta e^{-\eta^2/4}.$$

Summarizing the results of the above analysis, we can conclude that the down-stream asymptotic expansion (1.6.12) of $\widehat{V}_\tau$ involves two families of eigenfunctions, symmetric with the spectrum (1.6.44) and anti-symmetric with the spectrum (1.6.47). At large distance downstream of the aerofoil, the term with smallest value of α dominates the velocity field. It is given by (1.6.45), with the corresponding eigenfunction (1.6.46). Substituting (1.6.46) into (1.6.28) and then into (1.6.12), we have

$$\widehat{V}_\tau = V_\infty + C_1 \hat{s}^{-1/2} e^{-\eta^2/4} + \cdots \qquad \text{as} \quad s \to \infty. \tag{1.6.48a}$$

The similarity variable η is defined by (1.6.13), which may be written, using (1.6.24) and (1.6.25), as

$$\eta = \sqrt{\frac{V_\infty}{\nu}}\,\frac{\hat{n}}{\hat{s}^{1/2}}. \tag{1.6.48b}$$

---

[19]It is known that with $b = a$ the Kummer function $M(a, a, z) = e^z$.
[20]See Problem 1 in Exercises 5

1.6.3 Integral momentum equation

We shall now show that factor C_1 in (1.6.48a) is related to the aerofoil drag $\widehat{D}$. For this purpose, the integral momentum equation will be used,[21]

$$\mathbf{\widehat{F}} = \iint\limits_{S_c} \mathbf{\hat{p}}_n dS - \iint\limits_{S_c} \rho(\mathbf{\widehat{V}} \cdot \mathbf{n})\mathbf{\widehat{V}} \, dS. \tag{1.6.49}$$

Here $\mathbf{\widehat{F}}$ denotes the aerodynamic force acting on the aerofoil, S_c is the surface of the control volume, $\mathbf{n}$ is the unit vector normal to S_c, and $\mathbf{\hat{p}}_n$ is the stress vector. Its components may be calculated with the help of equations (1.2.11) on page 13 in Part 1 of this book series.

Projecting (1.6.49) upon the $\hat{x}$-axis, we find that the aerofoil drag

$$\widehat{D} = \iint\limits_{S_c} \hat{p}_{nx} dS - \iint\limits_{S_c} \rho(\mathbf{\widehat{V}} \cdot \mathbf{n})\hat{u} \, dS. \tag{1.6.50}$$

The $\hat{x}$-component $\hat{p}_{nx}$, of the stress vector $\mathbf{p}_n$ is given by[22]

$$\hat{p}_{nx} = n_x \hat{p}_{xx} + n_y \hat{p}_{yx} + n_z \hat{p}_{zx}.$$

Hence, (1.6.50) can be written as

$$\widehat{D} = \iint\limits_{S_c} (n_x \hat{p}_{xx} + n_y \hat{p}_{yx}) \, ds - \iint\limits_{S_c} \rho(\hat{u}n_x + \hat{v}n_y)\hat{u} \, ds. \tag{1.6.51}$$

Here, it is taken into account that the flow considered is two-dimensional. Correspondingly, we shall assume that the surface S_c of the control volume has cylindrical shape with the generatrix perpendicular the $(\hat{x}, \hat{y})$-plane.

For our purposes, it is convenient to choose the control volume with a circular cross-section, as shown in Figure 1.26. We shall assume that its radius R is large as compared to the aerofoil chord. Of course, if the flow were fully inviscid, then the aerofoil drag would be zero.[23] Hence, when calculating the integrals on the right-hand side of (1.6.51) we must pay attention to the differences between the flow considered and its inviscid counterpart. The first obvious place to look is at the wake. Here, a deficit of the velocity produces a loss in the momentum flux, which is represented by the second integral in (1.6.51). Since the thickness of the wake is an order $Re^{-1/2}$ quantity, we can expect the wake to produce an $O(Re^{-1/2})$ contribution to the drag $\widehat{D}$. On the rest of the integration surface the solution for the inviscid region should be used. As mentioned, the leading-order solution produces no drag. However, the $O(Re^{-1/2})$ correction terms in the asymptotic expansions (1.2.40) for the inviscid part of the flow are expected to give a contribution to $\widehat{D}$ comparable to that of the wake.

---

[21]See equation (1.7.29) on page 69 in Part 1 of this book series.
[22]See the first of equations (1.2.11) in Part 1.
[23]Remember that this result is known as d'Alambert's paradox.

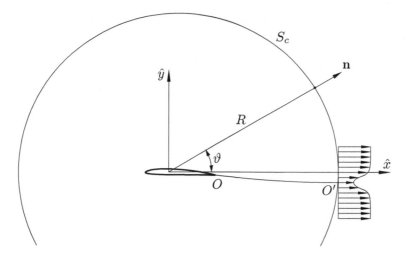

Fig. 1.26: Control volume.

Since the equations (1.2.41) for the $O(Re^{-1/2})$ terms in (1.2.40) are inviscid, we can use the Bernoulli equation

$$\frac{\hat{p}}{\rho} + \frac{\hat{u}^2 + \hat{v}^2}{2} = \frac{p_\infty}{\rho} + \frac{V_\infty^2}{2}. \tag{1.6.52}$$

We can also represent the complex conjugate velocity in the far field by the Laurent series:[24]

$$\hat{u} - i\hat{v} = a_0 + \frac{a_1}{z} + O\left(\frac{1}{z^2}\right) \quad \text{as} \quad z \to \infty. \tag{1.6.53}$$

Setting $z = \infty$ on the right-hand side of (1.6.53), we have to recover the free-stream velocity, which means that $a_0 = V_\infty$. The coefficient a_1 in the next order term is defined by equation (3.4.46) in Part 1,

$$\Delta\varphi + i\Delta\psi = a_1 2\pi i,$$

where the increment of the velocity potential $\Delta\varphi$ coincides with the circulation Γ around the aerofoil, and the increment of the stream function $\Delta\psi$ equals the fluid volume flux q through surface S_c. In purely inviscid flow $q = 0$, which gives $a_1 = \Gamma/2\pi i$. However, the wake creates a deficit of the fluid flux

$$q = \int\limits_{-\infty}^{\infty} (V_\infty - \hat{V}_\tau) \, d\hat{n} \tag{1.6.54}$$

when crossing the control surface S_c. It should be compensated by an equal fluid flux in the inviscid part of the flow. Consequently, we have to set $\Delta\psi = q$, and we will have

---

[24]See equation (3.4.45) on page 173 in Part 1 of this book series.

$$a_1 = \frac{\Gamma + iq}{2\pi i}.$$ (1.6.55)

On surface S_c of the control volume (see Figure 1.26),

$$z = Re^{i\vartheta}.$$ (1.6.56)

Substituting (1.6.55) and (1.6.56) into (1.6.53), and separating the real and imaginary parts, we find that outside the wake

$$\hat{u} = V_\infty - \frac{\Gamma \sin \vartheta - q \cos \vartheta}{2\pi R} + O\Big(\frac{1}{R^2}\Big), \quad \hat{v} = \frac{\Gamma \cos \vartheta + q \sin \vartheta}{2\pi R} + O\Big(\frac{1}{R^2}\Big).$$ (1.6.57)

Now, we can determine the pressure by substituting (1.6.57) into the Bernoulli equation (1.6.52). We find that

$$\hat{p} = p_\infty + \frac{\rho V_\infty}{2\pi} \frac{\Gamma \sin \vartheta - q \cos \vartheta}{R} + O\Big(\frac{1}{R^2}\Big).$$ (1.6.58)

When performing the task of calculating the integrals on the right-hand side of (1.6.51), it is convenient to introduce a function

$$\Phi = n_x \hat{p}_{xx} + n_y \hat{p}_{yx} - \rho \left(\hat{u} n_x + \hat{v} n_y \right) \hat{u},$$ (1.6.59)

and write equation (1.6.51) as

$$\hat{D} = \iint\limits_{S_c} \Phi_{\text{inv}} \, dS + \iint\limits_{S_c} (\Phi - \Phi_{\text{inv}}) \, dS.$$ (1.6.60)

Here, Φ_{inv} stands for function Φ, calculated based on the inviscid solution. In the inviscid flow the stress is represented by the pressure only, which means that[25]

$$\hat{p}_{xx} = -\hat{p}, \qquad \hat{p}_{xy} = 0.$$

Keeping further in mind that on S_c (see Figure 1.26)

$$n_x = \cos \vartheta, \qquad n_y = \sin \vartheta,$$

and using (1.6.57) and (1.6.58) in (1.6.59), we find that

$$\Phi_{\text{inv}} = -(p_\infty + \rho V_\infty^2) \cos \vartheta - \frac{\rho V_\infty q}{2\pi R} + O\Big(\frac{1}{R^2}\Big).$$

We shall assume that the control volume has a unit hight (in the direction perpendicular to the sketch in Figure 1.26). Then the area dS of the surface element of S_c is calculated as $dS = R \, d\vartheta$. Hence,

$$\iint\limits_{S_c} \Phi_{\text{inv}} \, dS = \int\limits_0^{2\pi} \Phi_{\text{inv}} R \, d\vartheta = -\rho V_\infty q.$$ (1.6.61)

---

[25]See equation (1.2.14) on page 14 in Part 1 of this book series.

Turning to the second integral in (1.6.60), we note that $\Phi - \Phi_{inv}$ differs from zero only inside the wake, where n_x and n_y may be approximated as

$$n_x = 1, \qquad n_y = 0.$$

Therefore,

$$\Phi = \hat{p}_{xx} - \rho \hat{u}^2. \tag{1.6.62}$$

The components of the stress tensor are given by equation (1.7.1) on page 61 in Part 1. In particular, $\hat{p}_{xx}$ is calculated as

$$\hat{p}_{xx} = -\hat{p} + 2\mu \frac{\partial \hat{u}}{\partial \hat{x}}.$$

This allows us to express (1.6.62) in the form

$$\Phi = -\hat{p} - \rho \hat{u}^2 + \mu \frac{\partial \hat{u}}{\partial \hat{x}}. \tag{1.6.63}$$

The inviscid counterpart of (1.6.62) is obtained by setting $\mu = 0$ and taking into account that inviscid velocity immediately outside the wake is given by (1.6.8). This leads to

$$\Phi_{inv} = -p - \rho V_\infty^2 + O\left(\frac{\ln \hat{x}}{\hat{x}^2}\right). \tag{1.6.64}$$

With (1.6.64) and (1.6.63), the second integral in (1.6.60) may be expressed as

$$\iint\limits_{S_c} (\Phi - \Phi_{inv}) \, dS = \int\limits_{-\infty}^{\infty} \left[\rho\left(V_\infty^2 - \hat{u}^2\right) + \underbrace{\mu \frac{\partial \hat{u}}{\partial \hat{x}} + O\left(\frac{\ln \hat{x}}{\hat{x}^2}\right)} \right] dS. \tag{1.6.65}$$

It follows from the solution (1.6.48) in the wake that $\partial \hat{u}/\partial \hat{x} \sim \hat{x}^{-3/2}$, while the width of the wake is estimated as $\hat{n} \sim \hat{x}^{1/2}$. This means that, at large distances from the aerofoil, the two underbraced terms in (1.6.65) can be disregarded, and we will have

$$\iint\limits_{S_c} (\Phi - \Phi_{inv}) \, dS = \int\limits_{-\infty}^{\infty} \rho\left(V_\infty^2 - \hat{u}^2\right) dS = \rho \int\limits_{-\infty}^{\infty} (V_\infty + \hat{u})(V_\infty - \hat{u}) \, dS. \tag{1.6.66}$$

The last integral in (1.6.66) can be simplified by taking into account that in the wake $\hat{u} \approx \widehat{V}_\tau$. We have

$$\iint\limits_{S_c} (\Phi - \Phi_{inv}) \, dS = \rho \int\limits_{-\infty}^{\infty} 2V_\infty(V_\infty - \widehat{V}_\tau) \, dS = 2\rho\, V_\infty q. \tag{1.6.67}$$

Adding (1.6.61) and (1.6.67) together, we find that the aerofoil drag is given by

$$\widehat{D} = \rho\, V_\infty q. \tag{1.6.68}$$

It remains to calculate the integral (1.6.54) for fluid flux deficit q in the wake. This is done by substituting (1.6.48) into (1.6.54). We have

$$q = -C_1\sqrt{\frac{\nu}{V_\infty}} \int_{-\infty}^{\infty} e^{-\eta^2/4} d\eta = -2C_1\sqrt{\frac{\pi\nu}{V_\infty}}. \tag{1.6.69}$$

Finally, we can substitute (1.6.69) into (1.6.68). Solving the resulting equation for C_1, we find

$$C_1 = -\frac{\widehat{D}}{2\rho\sqrt{\pi\nu V_\infty}}.$$

Exercises 5

1. Using the results of the solution of Problem 3 in Exercises 1, find the value of constant C_1 in the solution (1.6.48) for the viscous wake downstream of the flat plate placed at zero incidence to the oncoming flow.

2. The longitudinal velocity $\widehat{V}_\tau$ in the wake is given by (1.6.48). Deduce the corresponding equation for the normal velocity component $\widehat{V}_n$.

3. By direct substitution, show that the second symmetrical solution of the eigenvalue problem (1.6.26), (1.6.27), corresponding to $m = 1$ in (1.6.44), is

$$f = \left(1 - \frac{1}{2}\eta^2\right)e^{-\eta^2/4}.$$

4. The second complementary solution in (1.6.28), the function $f_2(\eta)$, is the solution of equation (1.6.26) which satisfies initial conditions (1.6.30). To find $f_2(\eta)$, you need to perform the following tasks:

   (a) Seek the solution for $f_2(\eta)$ in the form

$$f_2(\eta) = \eta\, h(z), \qquad z = -\frac{1}{4}\eta^2,$$

   and show that $h(z)$ satisfies Kummer's equation

$$z\frac{d^2h}{dz^2} + (b - z)\frac{dh}{dz} - ah = 0 \tag{1.6.70}$$

   with parameters

$$b = \frac{3}{2}, \qquad a = \alpha + \frac{1}{2}.$$

   (b) Taking into account that two complementary solutions of (1.6.70) are given by (1.6.34) and (1.6.35), argue that the solution suitable for $f_2(\eta)$ is given by

$$h(z) = M\left(\alpha + \frac{1}{2}, \frac{3}{2}, z\right),$$

   and, therefore,

$$f_2(\eta) = \eta M\left(\alpha + \frac{1}{2}, \frac{3}{2}, -\frac{\eta^2}{4}\right). \tag{1.6.71}$$

(c) Use (1.6.37) to show that

$$f_2(\eta) = \frac{4^{\alpha+1/2}\Gamma(3/2)}{\Gamma(1-\alpha)} \frac{1}{\eta^{2\alpha}} + \cdots \quad \text{as} \quad \eta \to \infty,$$

and explain why α should belong to the set

$$\alpha = 1 + m, \qquad m = 0, 1, 2, \ldots.$$

Consider the first eigenvalue, $\alpha = 1$, and using the fact that $M(a, a, z) = e^z$, deduce that

$$f_2(\eta) = \eta e^{-\eta^2/4}.$$

1.7 Von Mises Variables

When dealing with a two-dimensional flow it might be convenient to use the von Mises variables. While they are applicable to the full set of the Navier–Stokes equations, here we shall restrict our attention to the boundary-layer equations.

1.7.1 Boundary-layer equations in von Mises variables

Let us consider, as an example, the boundary layer on the surface of a blunt body as shown in Figure 1.7. For convenience, we reproduce the flow layout in Figure 1.27 below. To describe the flow, we use the body-fitted coordinate system where the position of point M is defined by two coordinates, $\hat{s}$ and $\hat{n}$. The first of these is the arc length measured from the front stagnation point O to the point M' that lies in the foot of the perpendicular dropped from the point M to the body surface. The second coordinate, $\hat{n}$, is the distance between the point M and body surface. The velocity vector is decomposed onto tangential component $\widehat{V}_\tau$ and normal component $\widehat{V}_n$. We denote the pressure by $\hat{p}$, density by ρ, and kinematic viscosity coefficient by ν.

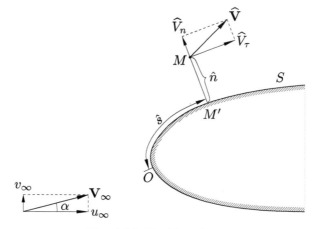

Fig. 1.27: Problem layout.

In the boundary layer, we scale the independent variables and the fluid-dynamic functions as

$$\hat{s} = Ls, \qquad \hat{n} = LRe^{-1/2}N,$$
$$\hat{V}_\tau = V_\infty U_0, \qquad \hat{V}_n = V_\infty Re^{-1/2} V_0, \qquad \hat{p} = p_\infty + \rho V_\infty^2 P_0.$$

This allows us to write the boundary-layer equations in the following dimensionless form:

$$U_0 \frac{\partial U_0}{\partial s} + V_0 \frac{\partial U_0}{\partial N} = -\frac{\partial P_0}{\partial s} + \frac{\partial^2 U_0}{\partial N^2}, \tag{1.7.1a}$$

$$\frac{\partial P_0}{\partial N} = 0, \tag{1.7.1b}$$

$$\frac{\partial U_0}{\partial s} + \frac{\partial V_0}{\partial N} = 0. \tag{1.7.1c}$$

We shall now write these equations using, instead of (s, N), new independent variables (x, Ψ). The latter are formally introduced through the equations

$$x = x(s, N), \qquad \Psi = \Psi(s, N). \tag{1.7.2}$$

We shall assume that function $x(s, N)$ simply coincides with s, that is

$$x(s, N) = s, \tag{1.7.3}$$

and $\Psi(s, N)$ is a stream function defined by the equations,

$$\frac{\partial \Psi}{\partial s} = -V_0, \qquad \frac{\partial \Psi}{\partial N} = U_0. \tag{1.7.4}$$

Correspondingly, instead of $U_0(s, N)$, $V_0(s, N)$, and $P_0(s, N)$, we will now deal with functions $U(x, \Psi)$, $V(x, \Psi)$, and $P(x, \Psi)$ defined as

$$U_0(s, N) = U\big[x(s, N), \Psi(s, N)\big], \tag{1.7.5a}$$
$$V_0(s, N) = V\big[x(s, N), \Psi(s, N)\big], \tag{1.7.5b}$$
$$P_0(s, N) = P\big[x(s, N), \Psi(s, N)\big]. \tag{1.7.5c}$$

Let us start with equation (1.7.1c). Differentiating (1.7.5c) with respect to N, we have

$$\frac{\partial P_0}{\partial N} = \frac{\partial P}{\partial x} \frac{\partial x}{\partial N} + \frac{\partial P}{\partial \Psi} \frac{\partial \Psi}{\partial N}.$$

Taking into account that

$$\frac{\partial x}{\partial N} = 0, \qquad \frac{\partial \Psi}{\partial N} = U_0 = U, \tag{1.7.6}$$

we find

$$\frac{\partial P_0}{\partial N} = U \frac{\partial P}{\partial \Psi}. \tag{1.7.7}$$

The substitution of (1.7.7) into (1.7.1b) results in

$$U\frac{\partial P}{\partial \Psi} = 0.$$

In the boundary layer, U can only become zero on isolated lines. This, of course, includes the body surface where the no-slip condition has to be satisfied. Additionally, when the flow separates from the body surface, a line of zero longitudinal velocity forms is observed in the reversed flow region. It separates the near-wall region where $U < 0$ from the rest of the flow. However, since everywhere else $U \neq 0$, we have to conclude that

$$\frac{\partial P}{\partial \Psi} = 0. \tag{1.7.8}$$

Let us now turn to the longitudinal momentum equation (1.7.1a). Differentiating (1.7.5a) with respect to s, we have

$$\frac{\partial U_0}{\partial s} = \frac{\partial U}{\partial x}\frac{\partial x}{\partial s} + \frac{\partial U}{\partial \Psi}\frac{\partial \Psi}{\partial s}. \tag{1.7.9}$$

It follows from (1.7.3), (1.7.4), and (1.7.5b) that

$$\frac{\partial x}{\partial s} = 1, \qquad \frac{\partial \Psi}{\partial s} = -V_0 = -V. \tag{1.7.10}$$

Therefore,

$$\frac{\partial U_0}{\partial s} = \frac{\partial U}{\partial x} - V\frac{\partial U}{\partial \Psi}. \tag{1.7.11}$$

Similarly, the first and second derivatives of U_0 with respect to N are found to be

$$\frac{\partial U_0}{\partial N} = U\frac{\partial U}{\partial \Psi}, \qquad \frac{\partial^2 U_0}{\partial N^2} = U\frac{\partial}{\partial \Psi}\left(U\frac{\partial U}{\partial \Psi}\right). \tag{1.7.12}$$

It remains to calculate the longitudinal pressure gradient, $\partial P_0/\partial s$. Differentiating (1.7.5c) with respect to s,

$$\frac{\partial P_0}{\partial s} = \frac{\partial P}{\partial x}\frac{\partial x}{\partial s} + \frac{\partial P}{\partial \Psi}\frac{\partial \Psi}{\partial s},$$

and using (1.7.8), we find that

$$\frac{\partial P_0}{\partial s} = \frac{\partial P}{\partial x}. \tag{1.7.13}$$

The substitution of (1.7.11), (1.7.12), and (1.7.13) into (1.7.1a) yields

$$U\frac{\partial U}{\partial x} = -\frac{dP_e}{dx} + U\frac{\partial}{\partial \Psi}\left(U\frac{\partial U}{\partial \Psi}\right). \tag{1.7.14}$$

Here, $P_e(x)$ is the pressure in the inviscid flow immediately outside the boundary layer.

Notice that in the von Mises variables, the momentum equation (1.7.14) does not involve V. With given pressure distribution, $P_e(x)$, it serves to calculate the longitudinal velocity $U(x, \Psi)$. In order to find the lateral velocity component V, we need to use the continuity equation (1.7.1c). The first term in the continuity equation is given by (1.7.11). To express the second term in the von Mises variables, we differentiate (1.7.5b) with respect to N. We find

$$\frac{\partial V_0}{\partial N} = U \frac{\partial V}{\partial \Psi}.$$

Consequently, the continuity equation assumes the form

$$\frac{\partial U}{\partial x} - V \frac{\partial U}{\partial \Psi} + U \frac{\partial V}{\partial \Psi} = 0,$$

or, equivalently,

$$\frac{\partial}{\partial x} \left(\frac{1}{U} \right) - \frac{\partial}{\partial \Psi} \left(\frac{V}{U} \right) = 0. \tag{1.7.15}$$

It follows from (1.7.15) that there exists a function $\mathcal{N}(x, \Psi)$ such that

$$\frac{\partial \mathcal{N}}{\partial x} = \frac{V}{U}, \qquad \frac{\partial \mathcal{N}}{\partial \Psi} = \frac{1}{U}. \tag{1.7.16}$$

To clarify the physical meaning of this function, let us consider a pair of functions

$$s = s(x, \Psi), \qquad N = N(x, \Psi), \tag{1.7.17}$$

that are inverse to the functions (1.7.2). The substitution of (1.7.2) into (1.7.17) leads to the identities

$$s \equiv s\big[x(s, N), \Psi(s, N)\big], \qquad N \equiv N\big[x(s, N), \Psi(s, N)\big]. \tag{1.7.18}$$

Differentiation of the second of these with respect to s and N results in

$$0 = \frac{\partial N}{\partial x} \frac{\partial x}{\partial s} + \frac{\partial N}{\partial \Psi} \frac{\partial \Psi}{\partial s}, \qquad 1 = \frac{\partial N}{\partial x} \frac{\partial x}{\partial N} + \frac{\partial N}{\partial \Psi} \frac{\partial \Psi}{\partial N}.$$

Using (1.7.6) and (1.7.10), the above equations may be expressed in the form

$$0 = \frac{\partial N}{\partial x} + \frac{\partial N}{\partial \Psi} (-V), \qquad 1 = \frac{\partial N}{\partial \Psi} U.$$

Solving these equations for $\partial N / \partial x$ and $\partial N / \partial \Psi$, we find that

$$\frac{\partial N}{\partial x} = \frac{V}{U}, \qquad \frac{\partial N}{\partial \Psi} = \frac{1}{U}. \tag{1.7.19}$$

Comparing (1.7.19) with (1.7.16), we see that $\mathcal{N}(s, \psi)$ is simply the distance N from an observation point in the flow field to the body surface.

The stream function Ψ may be chosen such that it is zero on the body surface, that is

$$\mathcal{N}\Big|_{\Psi=0} = 0. \tag{1.7.20}$$

Integrating the second equation in (1.7.16) with the boundary condition (1.7.20), we can write

$$\mathcal{N}(x, \Psi) = \int_0^\Psi \frac{1}{U(x, \psi)} \, d\psi. \tag{1.7.21}$$

It remains to substitute (1.7.21) into the first of equations (1.7.16), and we arrive at a conclusion that the lateral velocity component may be calculated as

$$V = U \frac{\partial}{\partial x} \int_0^\Psi \frac{1}{U} \, d\psi.$$

1.7.2 Batchelor problem

Here we shall use the von Mises variables to solve the following problem suggested by Batchelor (1956). We shall consider the flow of an incompressible viscous fluid inside a circular cylinder of radius R rotating around its centre with angular velocity Ω_0. A part of the cylinder surface is shielded by an inner sleeve of length αR, which is held motionless (see Figure 1.28). If the Reynolds number is large, then the flow may be subdivided into two parts, the inviscid 'core' of the flow, and a thin boundary layer that forms on the cylinder surface. According to the *Prandtl–Batchelor theorem* (see Problem 1 in Exercises 6), in the 'core' the fluid rotates as a solid body. We shall show that the angular velocity Ω of the 'core' rotation may be determined by means of the flow analysis in the boundary layer.

Using Bernoulli's equation

$$\frac{dP_e}{dx} = -U_e \frac{dU_e}{dx},$$

we can write the boundary-layer momentum equation (1.7.14) as

$$U \frac{\partial U}{\partial x} = U_e \frac{dU_e}{dx} + U \frac{\partial}{\partial \Psi} \left(U \frac{\partial U}{\partial \Psi} \right). \tag{1.7.22}$$

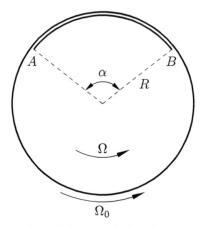

Fig. 1.28: Batchelor flow.

To be definite, we choose the arc length x to be measured along the cylinder surface in the counter-clockwise direction from the left-hand side edge A of the sleeve. The velocity at the outer edge of the boundary layer $U_e(x)$ is found by analysing the fluid motion in the 'core' of the flow. Since the 'core' rotates as a solid body, we have

$$U_e = \Omega R.$$

It does not depend on x, and, therefore, equation (1.7.22) may be written as

$$U \frac{\partial U}{\partial x} = U \frac{\partial}{\partial \Psi} \left(U \frac{\partial U}{\partial \Psi} \right).$$

Cancelling U on both sides of this equation and rearranging the viscous term, we have

$$\frac{\partial U}{\partial x} = \frac{\partial^2}{\partial \Psi^2} \left(\frac{U^2}{2} \right). \tag{1.7.23}$$

Let us now integrate equation (1.7.23) with respect to x along a streamline.[26] We start at an arbitrary point inside the boundary layer and, after making a full circle along the chosen streamline ($\Psi = \Psi_0$), we return back to the same point. As a result, U returns to its original value and the left-hand side of (1.7.23) disappears, leading to

$$\frac{d^2}{d\Psi^2} \int\limits_0^{2\pi R} U^2 dx = 0. \tag{1.7.24}$$

If we integrate (1.7.24) twice with respect to Ψ, then we will have

$$\int\limits_0^{2\pi R} U^2 dx = C_1 \Psi + C_2. \tag{1.7.25}$$

Here, C_1 and C_2 are constants of integration. They may be found from the boundary conditions on the wall ($\Psi = 0$) and at the outer edge of the boundary layer ($\Psi = \infty$). In order to formulate the first of these conditions, we shall assume that the thickness of the sleeve is much smaller than the thickness of the boundary layer. Then we will have

$$U \Big|_{\Psi=0} = \begin{cases} \Omega_0 R & \text{if } x \in \big[0 \,,\, (2\pi - \alpha)R\big], \\ 0 & \text{if } x \in \big[(2\pi - \alpha)R \,,\, 2\pi R\big]. \end{cases} \tag{1.7.26}$$

Setting $\Psi = 0$ in (1.7.25) and using (1.7.26), we find that

$$C_2 = (\Omega_0 R)^2 (2\pi - \alpha) R. \tag{1.7.27}$$

The boundary condition at the outer edge of the boundary layer is written as

$$U \Big|_{\Psi=\infty} = U_e = \Omega R. \tag{1.7.28}$$

---

[26] Remember that the stream function Ψ stays constant on each streamline.

Calculating the integral on the left-hand side of (1.7.25) with the help of (1.7.28), we have

$$U_e^2 2\pi R = C_1 \Psi + C_2 \quad \text{as} \quad \Psi \to \infty.$$

Since the left-hand side in this equation remains finite as $\Psi \to \infty$, we have to set $C_1 = 0$, and we see that

$$C_2 = (\Omega R)^2 2\pi R. \tag{1.7.29}$$

Eliminating C_2 from (1.7.29) and (1.7.27), and solving the resulting equation for Ω, we arrive at a conclusion that the angular velocity of the 'core' rotation is given by

$$\Omega = \Omega_0 \sqrt{1 - \frac{\alpha}{2\pi}}.$$

Exercises 6

1. Suppose that a steady two-dimensional flow of an incompressible fluid has a recirculation region $\mathcal{D}$ with closed streamlines. A typical example is the separation eddy behind a circular cylinder; see Figure 1.29.

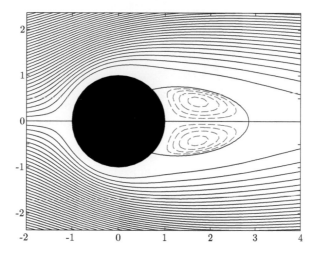

Fig. 1.29: Viscous flow past a circular cylinder at Reynolds number $Re = 20$. The dashed lines show the streamlines in the recirculation eddies.

Perform the following tasks:

(a) Consider one of the closed streamlines and denote it by C. Assume that the body force $\mathbf{f}$ has potential U, such that $\mathbf{f} = -\nabla U$, and write the Navier–Stokes equations in the following vector form:[27]

$$[\boldsymbol{\omega} \times \mathbf{V}] + \nabla \left(\frac{V^2}{2} + U + \frac{1}{\rho} p \right) = -\nu \operatorname{curl} \boldsymbol{\omega}. \tag{1.7.30}$$

---

[27]See equation (1.8.33) on page 82 in Part 1 of this book series.

Denote an increment of the position vector along the streamline C by $d\mathbf{r}$, and use equation (1.7.30) to show that for any $\nu \neq 0$

$$\oint_C (\text{curl}\,\boldsymbol{\omega} \cdot d\mathbf{r}) = 0. \tag{1.7.31}$$

(b) Show that in a two-dimensional flow equation (1.7.31) may be written as

$$\oint_C \left(\frac{\partial \omega}{\partial x} dy - \frac{\partial \omega}{\partial y} dx \right) = 0, \tag{1.7.32}$$

where ω is the only non-zero component of the vorticity vector, $\boldsymbol{\omega} = \mathbf{k}\,\omega$; with $\mathbf{k}$ being the unit vector perpendicular to the flow plane.

(c) Now consider the inviscid flow limit, $\nu \to 0$, and recall that, in this case, the vorticity ω remains constant along each streamline, and, therefore, depends on the streamfunction ψ only,

$$\omega = \omega(\psi). \tag{1.7.33}$$

Show that with (1.7.33) the left-hand side of equation (1.7.32) assumes the form

$$\oint_C \left(\frac{\partial \omega}{\partial x} dy - \frac{\partial \omega}{\partial y} dx \right) = -\frac{d\omega}{d\psi} \Gamma,$$

with Γ being the circulation of the velocity vector along contour C. Hence, conclude that the vorticity remains constant over the entire recirculation region $\mathcal{D}$.

2. Consider a device made of two concentric circular cylinders. The outer cylinder is hollow and has a segment of angle α removed as shown in Figure 1.30. The skin of the outer cylinder is very thin, such that the radii of the two cylinders may be assumed almost the same; a small gap is left between the cylinders to allow them rotate with respect to one another.

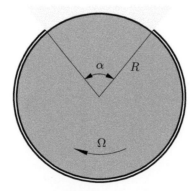

Fig. 1.30: The flow round a circular cylinder with a partially moving surface.

The device is placed into an incompressible fluid in a tank whose size is large compared to the cylinder radius. The outer cylinder is kept motionless, while the inner is brought into rotation with angular velocity Ω. Due to viscosity, the fluid particles in the tank start to go in circles around the device, and after sufficient time the motion of the fluid takes the form of the 'potential vortex', with the velocity components and pressure given by[28]

$$V_r = 0, \qquad V_\phi = \frac{\Gamma}{2\pi r}, \qquad V_z = 0, \qquad p = p_\infty - \frac{\rho\Gamma^2}{8\pi^2 r^2}. \qquad (1.7.34)$$

You need to perform the following tasks:

(a) Confirm the validity of this statement by direct substitution of (1.7.34) into the Navier–Stokes equations written in the cylindrical coordinates.[29]

(b) Assume the Reynolds number is large and show that

$$\Gamma = \sqrt{2\pi\alpha}\,\Omega R^2,$$

where R is the common radius of the cylinders.

Suggestion: At large values of the Reynolds number, the potential vortex solution remains valid everywhere except in a thin boundary layer forming near the surface of the device. Write the boundary-layer equations in the von Mises variables,

$$U\frac{\partial U}{\partial x} = U_e\frac{dU_e}{dx} + U\frac{\partial}{\partial\Psi}\left(U\frac{\partial U}{\partial\Psi}\right),$$

and take into account that the velocity U returns to its original value when a full circle is made along a streamline inside the boundary layer.

1.8   Flow Past a Rotating Cylinder

Let a circular cylinder of radius a be placed in a uniform flow of incompressible fluid with density ρ and free-stream velocity V_∞. We shall assume that the cylinder rotates around its axis with the angular velocity Ω. It is well known from experimental observations that the flow past a stationary cylinder ($\Omega = 0$) develops a separation; see Figure I.1(b). However, rotation of the cylinder with large enough Ω suppresses the separation completely, as Figure 1.31(b) shows, and then the solution for the inviscid part of the flow may be constructed as described in Section 3.4.2 in Part 1 of this book series. In particular, the velocity on the cylinder surface is given by equation (3.4.34) on page 169 in Part 1:

$$\overline{V} = ie^{-i\theta}\left(2V_\infty\sin\vartheta - \frac{\Gamma}{2\pi a}\right). \qquad (1.8.1)$$

Here, $\overline{V}$ in the complex conjugate velocity, Γ is the circulation of the velocity vector around the cylinder, and ϑ is the angle that defines the position of the observation point on the cylinder surface, as shown in Figure 1.32.

---

[28]See equation (3.4.18) on page 164 in Part 1 of this book series.
[29]These are equations (1.8.45) on page 84 in Part 1.

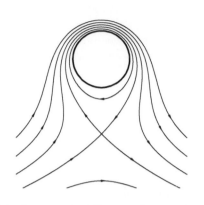 

(a) Theoretical predictions; $\Gamma = -6\pi a V_\infty$.   (b) Experimental observations; $a\Omega/V_\infty = 6$.

Fig. 1.31: Comparison of theoretical predictions for the flow past a circular cylinder with experimental observations by Prandtl and Tietjens (1934).

Remember that the circulation Γ cannot be found from the inviscid flow analysis. At the same time, experimental observations clearly show that Γ depends on the angular velocity of the cylinder rotation, Ω. Our task here will be to show that this dependence may be established through analysis of the boundary-layer flow. We start by writing the boundary-layer equations in dimensional form:

$$\left.\begin{aligned} \widehat{V}_\tau \frac{\partial \widehat{V}_\tau}{\partial \hat{s}} + \widehat{V}_n \frac{\partial \widehat{V}_\tau}{\partial \hat{n}} &= \widehat{U}_e \frac{d\widehat{U}_e}{d\hat{s}} + \nu \frac{\partial^2 \widehat{V}_\tau}{\partial \hat{n}^2}, \\ \frac{\partial \widehat{V}_\tau}{\partial \hat{s}} + \frac{\partial \widehat{V}_n}{\partial \hat{n}} &= 0. \end{aligned}\right\} \tag{1.8.2}$$

Here $(\hat{s}, \hat{n})$ are the body-fitted coordinates, with $\hat{s}$ measured from point O that lies at the intersection of the cylinder contour with a straight line drawn throught the cylinder centre parallel to the free-stream velocity vector; see Figure 1.32.

Assuming Ω positive when the cylinder rotates in the clockwise direction, we can write the no-slip conditions on the cylinder surface as

$$\widehat{V}_\tau = a\Omega, \quad \widehat{V}_n = 0 \quad \text{at} \quad \hat{n} = 0. \tag{1.8.3}$$

We also need to impose the following boundary condition at the outer edge of the boundary layer:

$$\widehat{V}_\tau = \widehat{U}_e(\hat{s}) \quad \text{at} \quad \hat{n} = \infty. \tag{1.8.4}$$

The function $\widehat{U}_e(\hat{x})$ is obtained by taking modulus on both sides of (1.8.1):

$$\widehat{U}_e = 2V_\infty \sin \vartheta - \frac{\Gamma}{2\pi a}. \tag{1.8.5}$$

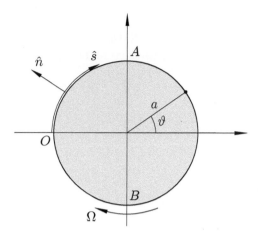

Fig. 1.32: Problem layout.

From Figure 1.32 we can see that

$$\hat{s} = a(\pi - \vartheta). \tag{1.8.6}$$

Let us now cast equations (1.8.2) and boundary conditions (1.8.3), (1.8.4) in dimensionless form. For this purpose, we use the transformations:

$$\left.\begin{array}{ll} \hat{s} = as, & \hat{n} = aRe^{-1/2}N, \\ \widehat{V}_\tau = \Omega aU, & \widehat{V}_n = \Omega aRe^{-1/2}V, \quad \widehat{U}_e = \Omega aU_e, \end{array}\right\} \tag{1.8.7}$$

with the Reynolds number defined as

$$Re = \frac{\Omega a^2}{\nu}.$$

The substitution of (1.8.7) into the boundary-layer equations (1.8.2) turns them into

$$U\frac{\partial U}{\partial s} + V\frac{\partial U}{\partial N} = U_e\frac{dU_e}{ds} + \frac{\partial^2 U}{\partial N^2}, \tag{1.8.8a}$$

$$\frac{\partial U}{\partial s} + \frac{\partial V}{\partial N} = 0. \tag{1.8.8b}$$

The boundary conditions (1.8.3), (1.8.4) are written in the non-dimensional form as

$$\left.\begin{array}{lll} U = 1, & V = 0 & \text{at} \quad N = 0, \\ U = U_e(s) & & \text{at} \quad N = \infty. \end{array}\right\} \tag{1.8.9}$$

The transformations (1.8.7) should also be applied to the equations (1.8.5) and (1.8.6). Thus, we find that

$$U_e(s) = \gamma + 2\varepsilon\sin s, \tag{1.8.10}$$

where constant γ is given by

$$\gamma = -\frac{\Gamma}{2\pi\Omega a^2}, \tag{1.8.11}$$

and ε is the ratio of the free-stream velocity to the circumferential velocity of the cylinder surface:

$$\varepsilon = \frac{V_\infty}{\Omega a}. \tag{1.8.12}$$

In what follows, we shall assume that ε is small.

For our purposes, it is convenient to express the boundary-value problem (1.8.8), (1.8.9) in terms of the stream function Ψ. The latter is related to the velocity components (U, V) as follows:

$$U = \frac{\partial\Psi}{\partial N}, \qquad V = \frac{\partial\Psi}{\partial s}. \tag{1.8.13}$$

With (1.8.13), the continuity equation (1.8.8b) is satisfied automatically. The substitution of (1.8.13) together with (1.8.10) into the momentum equation (1.8.8a) results in

$$\frac{\partial\Psi}{\partial N}\frac{\partial^2\Psi}{\partial s\partial N} - \frac{\partial\Psi}{\partial s}\frac{\partial^2\Psi}{\partial N^2} = 2\varepsilon\gamma\cos s + 2\varepsilon^2\sin 2s + \frac{\partial^3\Psi}{\partial N^3}. \tag{1.8.14}$$

The boundary conditions (1.8.9) are written in terms of the stream function as

$$\left.\begin{aligned}
\Psi = 0, \quad \frac{\partial\Psi}{\partial N} &= 1 \quad \text{at} \quad N = 0, \\
\frac{\partial\Psi}{\partial N} &= \gamma + 2\varepsilon\sin s \quad \text{at} \quad N = \infty.
\end{aligned}\right\} \tag{1.8.15}$$

Finally, for obvious reasons, we shall require that Ψ is periodic in s with period 2π:

$$\Psi(s, N) = \Psi(s + 2\pi, N). \tag{1.8.16}$$

Notice that when solving the boundary-value problem (1.8.14)–(1.8.16), both the stream Ψ and the scaled circulation γ should be treated as unknowns. We represent the solution in the form of asymptotic expansions

$$\left.\begin{aligned}
\Psi &= \Psi_0(s, N) + \varepsilon\Psi_1(s, N) + \varepsilon^2\Psi_2(s, N) + \cdots, \\
\gamma &= \gamma_0 + \varepsilon\gamma_1 + \varepsilon^2\gamma_2 + \cdots
\end{aligned}\right\} \quad \text{as} \quad \varepsilon \to 0. \tag{1.8.17}$$

By substituting (1.8.17) into (1.8.14) and (1.8.15), we find that the leading-order problem is written as

$$\frac{\partial\Psi_0}{\partial N}\frac{\partial^2\Psi_0}{\partial s\partial N} - \frac{\partial\Psi_0}{\partial s}\frac{\partial^2\Psi_0}{\partial N^2} = \frac{\partial^3\Psi_0}{\partial N^3}, \tag{1.8.18a}$$

$$\Psi_0 = 0, \quad \frac{\partial\Psi_0}{\partial N} = 1 \quad \text{at} \quad N = 0, \tag{1.8.18b}$$

$$\frac{\partial\Psi_0}{\partial N} = \gamma_0 \quad \text{at} \quad N = \infty. \tag{1.8.18c}$$

We can see that equation (1.8.18a) and boundary conditions (1.8.18b) and (1.8.18c), are invariant with respect to an arbitrary shift in s. This is due to the fact that with

$\varepsilon = 0$, the free-stream velocity turns zero, and the flow we are dealing with reduces to the one where the fluid far from the rotating cylinder is at rest. The latter, of course, is axisymmetric. Keeping this in mind, we shall seek the solution to (1.8.18) in the form

$$\Psi_0(s, N) = h_0(N). \tag{1.8.19}$$

The substitution of (1.8.19) into (1.8.18) results in

$$h_0''' = 0, \tag{1.8.20a}$$
$$h_0(0) = 0, \quad h_0'(0) = 1, \tag{1.8.20b}$$
$$h_0'(\infty) = \gamma_0. \tag{1.8.20c}$$

The general solution of the equation (1.8.20a) is written as

$$h_0 = C_1 + C_2 N + C_3 N^2. \tag{1.8.21}$$

The first two constants in (1.8.21) are found from the boundary conditions (1.8.20b) to be $C_1 = 0$, $C_2 = 1$. It further follows from (1.8.20c) that h_0' should remain finite as $N \to \infty$, which is only possible if $C_3 = 0$. Consequently,

$$h_0 = N,$$

and it follows from (1.8.20c) that

$$\gamma_0 = 1.$$

Taking this into account, we can now write asymptotic expansions (1.8.17) as

$$\left.\begin{array}{l} \Psi = N + \varepsilon\Psi_1(s, N) + \varepsilon^2\Psi_2(s, N) + \cdots, \\ \gamma = 1 + \varepsilon\gamma_1 + \varepsilon^2\gamma_2 + \cdots \end{array}\right\} \quad \text{as} \quad \varepsilon \to 0. \tag{1.8.22}$$

The boundary-value problem for $\Psi_1(s, N)$ is formulated by substituting (1.8.22) into (1.8.14) and (1.8.15), and working with the $O(\varepsilon)$ terms. We find that

$$\frac{\partial^2 \Psi_1}{\partial s \partial N} = 2 \cos s + \frac{\partial^3 \Psi_1}{\partial N^3}, \tag{1.8.23a}$$

$$\Psi_1 = \frac{\partial \Psi_1}{\partial N} = 0 \quad \text{at} \quad N = 0, \tag{1.8.23b}$$

$$\frac{\partial \Psi_1}{\partial N} = \gamma_1 + 2 \sin s \quad \text{at} \quad N = \infty. \tag{1.8.23c}$$

We shall seek the periodic solution to (1.8.23) in the form

$$\Psi_1(x, N) = h_1(N) + f_1(N)e^{is} + \overline{f}_1(N)e^{-is}, \tag{1.8.24}$$

where $\overline{f}_1$ is the complex conjugate of f_1. By substituting (1.8.24) into (1.8.23), we find that function $h_1(N)$ has to be determined by solving the boundary-value problem

$$h_1''' = 0, \tag{1.8.25a}$$
$$h_1(0) = h_1'(0) = 0, \tag{1.8.25b}$$
$$h_1'(\infty) = \gamma_1. \tag{1.8.25c}$$

We also find that the boundary-value problem for function $f_1(Y)$ is written as

$$f_1''' - if_1' + 1 = 0, \tag{1.8.26a}$$
$$f_1(0) = f_1'(0) = 0, \tag{1.8.26b}$$
$$f_1'(\infty) = -i. \tag{1.8.26c}$$

The solution of problem (1.8.25) may be constructed in the same way as it was done when solving problem (1.8.20). We find that

$$h_1 = 0, \tag{1.8.27a}$$
$$\gamma_1 = 0. \tag{1.8.27b}$$

Now, turning to the boundary-value problem (1.8.26), it is convenient to introduce a new function $g = f_1'$, and then equation (1.8.26a) may be written as

$$g'' - ig + 1 = 0. \tag{1.8.28}$$

The general solution of this equation is easily seen to be

$$g = -i + C_1 e^{\varkappa N} + C_2 e^{-\varkappa N}, \tag{1.8.29}$$

where $\varkappa = (1+i)/\sqrt{2}$.

It follows from (1.8.26c) that $g(N)$ should remain finite as $N \to \infty$, which is only possible if $C_1 = 0$. Using further the second condition in (1.8.26b), we find that $C_2 = i$, which reduces (1.8.29) to

$$g = -i + i e^{-\varkappa N}. \tag{1.8.30}$$

Now we need to integrate (1.8.30) with respect to N, which results in

$$f_1 = C_3 - iN - \frac{1+i}{\sqrt{2}} e^{-\varkappa N}.$$

The constant of integration C_3 may be found from the first condition in (1.8.26b) to be $C_3 = (1+i)/\sqrt{2}$. Hence, we can conclude that the solution of the problem (1.8.26) is written as

$$f_1 = -iN + \frac{1+i}{\sqrt{2}} \left(1 - e^{-\varkappa N} \right), \qquad \varkappa = \frac{1+i}{\sqrt{2}}. \tag{1.8.31}$$

Finally, we shall consider the boundary-value problem for Ψ_2:

$$\frac{\partial^2 \Psi_2}{\partial s \partial N} + \frac{\partial \Psi_1}{\partial N} \frac{\partial^2 \Psi_1}{\partial s \partial N} - \frac{\partial \Psi_1}{\partial s} \frac{\partial^2 \Psi_1}{\partial N^2} = 2\sin 2s + \frac{\partial^3 \Psi_2}{\partial N^3}, \tag{1.8.32a}$$

$$\Psi_2 = \frac{\partial \Psi_2}{\partial N} = 0 \quad \text{at} \quad N = 0, \tag{1.8.32b}$$

$$\frac{\partial \Psi_2}{\partial N} = \gamma_2 \qquad \text{at} \quad N = \infty. \tag{1.8.32c}$$

The forcing terms in (1.8.32a) may be calculated using (1.8.24), (1.8.27a), and (1.8.31). We have

$$\frac{\partial \Psi_1}{\partial N}\frac{\partial^2 \Psi_1}{\partial s \partial N} - \frac{\partial \Psi_1}{\partial s}\frac{\partial^2 \Psi_1}{\partial N^2} = i\Big[(f_1')^2 - f_1 f_1''\Big]e^{i2s}$$
$$+ i\Big[f_1''\overline{f}_1 - f_1\overline{f}_1''\Big] - i\Big[(\overline{f}_1')^2 - \overline{f}_1\overline{f}_1''\Big]e^{-i2s}.$$

This suggests that the solution of the boundary-value problem (1.8.32) should be sought in the form

$$\Psi_2 = h_2(N) + f_2(N)e^{i2s} + \overline{f}_2(N)e^{-i2s}.$$

Our primary goal here is to find the correction γ_2 to the circulation of the velocity vector around the cylinder. This is done by analysing function $h_2(N)$. The boundary-value problem for this function is written as

$$h_2''' = i\big(f_1''\overline{f}_1 - f_1\overline{f}_1''\big), \tag{1.8.33a}$$
$$h_2(0) = h_2'(0) = 0, \tag{1.8.33b}$$
$$h_2'(\infty) = \gamma_2. \tag{1.8.33c}$$

Calculating the right-hand side in equation (1.8.33a) with the help of (1.8.31), we have

$$h_2''' = 2Ne^{-N/\sqrt{2}}\sin\left(\frac{N}{\sqrt{2}} - \frac{\pi}{4}\right) + 2e^{-N/\sqrt{2}}\cos\frac{N}{\sqrt{2}} - 2e^{-\sqrt{2}N},$$

which, when integrated twice, gives

$$h_2' = C_1 + C_2 N - 2e^{-N/\sqrt{2}}\sin\frac{N}{\sqrt{2}} - e^{-\sqrt{2}N}$$
$$+ 2e^{-N/\sqrt{2}}\left[(N + \sqrt{2})\cos\left(\frac{N}{\sqrt{2}} - \frac{\pi}{4}\right) - \sqrt{2}\sin\left(\frac{N}{\sqrt{2}} - \frac{\pi}{4}\right)\right].$$

Since h_2' has to remain bounded as $N \to \infty$, we set $C_2 = 0$. It further follows from the second condition in (1.8.33b) that $C_1 = -3$. Finally, using condition (1.8.33c), we can conclude that

$$\gamma_2 = -3. \tag{1.8.34}$$

Substituting (1.8.34) and (1.8.27b) back into the asymptotic expansion for γ in (1.8.22), we have

$$\gamma = 1 - 3\varepsilon^2 + \cdots. \tag{1.8.35}$$

It remains to combine (1.8.35) with (1.8.11) and (1.8.12), and we can conclude that the circulation around the cylinder

$$\Gamma = -2\pi\Omega a^2\left(1 - \frac{3V_\infty^2}{\Omega^2 a^2} + \cdots\right).$$

The theory presented was developed by Glauert (1957). It relies on the assumption that $\varepsilon = V_\infty/\Omega a$ is small. If we want to extend the solution to finite values of ε, then the problem (1.8.14), (1.8.15) has to be solved numerically. The results of the calculations obtained with the help of Crank–Nicolson method (see Section 1.9) are displayed in Figure 1.33. Figure 1.33(a) shows how $\gamma = -\Gamma/2\pi\Omega a^2$ depends on ε. It is interesting to notice that the asymptotic theory gives a rather accurate prediction of the flow behaviour.

Another important result is that the solution in the boundary layer develops a singularity when parameter ε reaches a critical value $\varepsilon_* = 0.25525$. What happens is that between points O and A (see Figure 1.32), the boundary layer is exposed to a favourable pressure gradient, which causes the fluid in the boundary layer to accelerate. Then between points A and B, the pressure grows monotonically, and the fluid particles in the boundary layer experience a deceleration. As a result, the velocity profile develops a characteristic minimum. The latter deepens as ε becomes larger, and at the critical value of parameter $\varepsilon = \varepsilon_*$ the minimal velocity becomes zero, as shown in Figure 1.33(b). Interestingly enough, this first happens at point B $(s = \frac{3}{2}\pi)$, where the pressure is maximal. The solution to the boundary-layer equations (1.8.14), (1.8.15) develops a singularity at this point, signifying that the boundary layer can no longer remain attached to the cylinder surface. The reader interested in the process incipience of the separation on the rotating cylinder is referred to studies of Sychev (1986), Negoda and Sychev (1986), and Timoshin (1996).

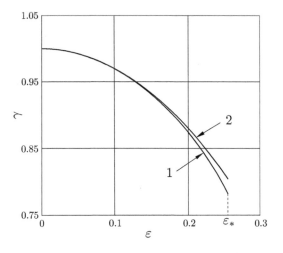
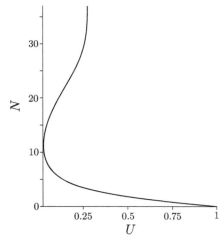

(a) Comparison of the asymptotic formula (1.8.35) with numerical solution of the problem (1.8.14), (1.8.15); 1: numerical solution, 2: asymptotic theory.

(b) Velocity profile at $s = \frac{3}{2}\pi$ at the critical value, ε_*, of the parameter (1.8.12).

Fig. 1.33: Results of the numerical solution of the boundary-value problem (1.8.14), (1.8.15).

1.9 Numerical Solution of the Boundary-Layer Equations

The flow examples considered so far in this chapter are concerned with situations when the boundary-layer equations allow for significant simplifications. We started with the flow past a flat plate, and demonstrated that, for this flow, the solution of the boundary-layer equations had a self-similar form, which allowed us to reduces the boundary-layer equations to an ordinary differential equation, known as the Blasius equation. We then considered Chapman's shear layer, Schlichting's laminar jet, and the boundary layer in the convergent channel (see Problem 2 in Exercises 3). These flows are also described by self-similar solutions valid in the entire flow field. In this respect, the Falkner–Skan solutions and Tollmien's wake are different; they represent asymptotic solutions. The first one is valid in a small vicinity of the wedge tip, and the second in the flow far downstream of the aerofoil. We have seen that, for some of these flows, the resulting ordinary differential equation may be solved analytically. For others, it requires numerical solution. Still, all the flows with self-similar solutions are traditionally assigned to the category of *exact solutions* of the boundary-layer equations. This is to highlight their importance in uncovering fundamental the properties of the boundary-layer flows.

More often, however, the flow in the boundary layer has to be studied by means of numerical solution of the boundary-layer equations. One of the techniques used for this purpose is the *Crank–Nicolson method* (see Crank and Nicolson, 1947). We shall describe this method using as an example symmetric flow past a circular cylinder.

1.9.1 Problem formulation

Let a circular cylinder of radius a be placed in the uniform flow of an incompressible fluid with the free-stream velocity V_∞. We shall define the Reynolds number as

$$Re = \frac{V_\infty a}{\nu},$$

where ν is the kinematic viscosity coefficient, and perform the flow analysis under the assumption that Re is large. Following Prandtl's hierarchical strategy, we first need to consider the inviscid part of the flow. A detailed discussion of this flow is given in Section 3.4.2 in Part 1 of this book series. In particular, we found that the tangential velocity on the cylinder surface is given by[30]

$$\widehat{V}_\tau = 2V_\infty \sin \vartheta. \tag{1.9.1}$$

Here, ϑ is the angle that defines the position of the observation point M on the cylinder surface; see Figure 1.34.

When dealing with the flow in the boundary layer we use the body-fitted coordinates $(\hat{s}, \hat{n})$ with $\hat{s}$ measured along the cylinder surface from the front stagnation point O as shown in Figure 1.34. The velocity components in these coordinates are denoted

---

[30]See equation (3.4.30) on Page 167 in Part 1.

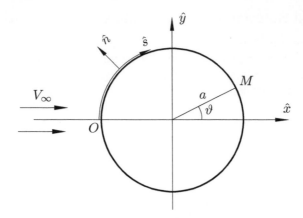

Fig. 1.34: Flow past a circular cylinder.

as $(\widehat{V}_\tau, \widehat{V}_n)$. The 'hat' is used, as before, to indicate that the corresponding variables are dimensional. If we introduce the non-dimensional variables through the scalings

$$\hat{s} = a\,s, \qquad \hat{n} = aRe^{-1/2}N, \qquad \widehat{V}_\tau = V_\infty U, \qquad \widehat{V}_n = V_\infty Re^{-1/2}V,$$

then the boundary-layer equations assume the form

$$U\frac{\partial U}{\partial s} + V\frac{\partial U}{\partial N} = U_e\frac{dU_e}{ds} + \frac{\partial^2 U}{\partial N^2}, \qquad (1.9.2a)$$

$$\frac{\partial U}{\partial s} + \frac{\partial V}{\partial N} = 0. \qquad (1.9.2b)$$

They have to be solved with the boundary conditions

$$U = 0 \qquad \text{at} \quad s = 0, \ \ N \in [0, \infty), \qquad (1.9.2c)$$

$$U = V = 0 \quad \text{at} \quad N = 0, \ \ s \ge 0, \qquad (1.9.2d)$$

$$U = U_e(s) \quad \text{at} \quad N = \infty, \ \ s \ge 0. \qquad (1.9.2e)$$

Here, the initial condition (1.9.2c) states that at the front stagnation point O the fluid velocity is zero. The velocity at the outer edge of the boundary layer,

$$U_e(s) = 2\sin s,$$

is obtained by expressing (1.9.1) in the non-dimensional variables.

1.9.2   Crank–Nicolson method

We start by introducing a mesh $\{s_k, N_j\}$ with uniform spacing Δs, ΔN in the s- and N-directions, respectively, such that

$$s_k = k\,\Delta s, \qquad k = 0,\ 1,\ 2,\ \dots,\ K,$$
$$N_j = j\,\Delta N, \qquad j = 0,\ 1,\ 2,\ \dots,\ J.$$

Thanks to the fact that the boundary-layer equations are parabolic, their solution may be constructed using a marching procedure where the calculations progress from one

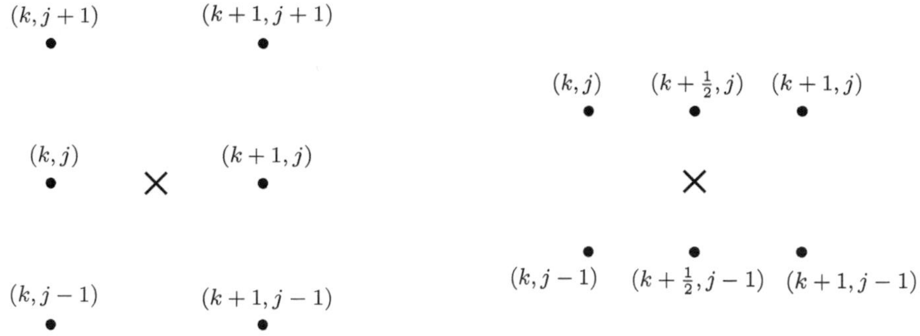

$(k, j+1)$ $(k+1, j+1)$

(k, j) $(k+\frac{1}{2}, j)$ $(k+1, j)$

(k, j) × $(k+1, j)$

×

$(k, j-1)$ $(k+1, j-1)$

$(k, j-1)$ $(k+\frac{1}{2}, j-1)$ $(k+1, j-1)$

(a) Finite-differencing stencil for the momentum equation (1.9.2a).

(b) Finite-differencing stencil for the continuity equation (1.9.2b).

Fig. 1.35: Finite-difference approximation of equations (1.9.2).

mesh line, $s = s_k$, to the next one, $s = s_{k+1}$, situated a step further downstream. Figure 1.35(a) shows the stencil employed for finite-differencing the momentum equation (1.9.2a). The derivatives in the equation are represented by the formulae

$$\left.\begin{aligned}
\frac{\partial U}{\partial s} &= \frac{U_{k+1,j} - U_{k,j}}{\Delta s}, \\
\frac{\partial U}{\partial N} &= \frac{1}{2}\left(\frac{U_{k,j+1} - U_{k,j-1}}{2\Delta N} + \frac{U_{k+1,j+1} - U_{k+1,j-1}}{2\Delta N}\right), \\
\frac{\partial^2 U}{\partial N^2} &= \frac{1}{2}\left(\frac{U_{k,j+1} - 2U_{k,j} + U_{k,j-1}}{(\Delta N)^2} + \frac{U_{k+1,j+1} - 2U_{k+1,j} + U_{k+1,j-1}}{(\Delta N)^2}\right),
\end{aligned}\right\} \quad (1.9.3)$$

all being second-order accurate with respect to the central point in the stencil; the latter is shown by the cross in Figure 1.35(a).

The substitution of (1.9.3) into (1.9.2a) results in the following set of algebraic equations:

$$a_j U_{k+1,j+1} + b_j U_{k+1,j} + c_j U_{k+1,j-1} = d_j, \qquad j = 1, 2, \ldots, J-1, \qquad (1.9.4)$$

where

$$a_j = \frac{V_j^{(c)}}{4\Delta N} - \frac{1}{2(\Delta N)^2}, \qquad b_j = \frac{U_j^{(c)}}{\Delta s} + \frac{1}{(\Delta N)^2}, \qquad c_j = -\frac{V_j^{(c)}}{4\Delta N} - \frac{1}{2(\Delta N)^2},$$

$$d_j = U_e \frac{dU_e}{ds}\bigg|_{k+1/2} + \frac{U_{k,j+1} - 2U_{k,j} + U_{k,j-1}}{2(\Delta N)^2} + U_j^{(c)}\frac{U_{k,j}}{\Delta s} - V_j^{(c)}\frac{U_{k,j+1} - U_{k,j-1}}{4\Delta N},$$

with $U_j^{(c)}$ and $V_j^{(c)}$ being the values of the tangential and normal velocity components at the central point of the stencil.

The set of equations (1.9.4) may be solved in various ways. One can use, for example, the Thomas method. This is an elimination technique with the solution sought in the form

$$U_{k+1,j} = R_j U_{k+1,j-1} + Q_j, \qquad j = 1, 2, \ldots, J. \qquad (1.9.5)$$

The calculations are performed in two steps. First, the Thomas coefficients, R_j, Q_j, are calculated using the recurrent equations

$$R_j = -\frac{c_j}{b_j + a_j R_{j+1}}, \qquad Q_j = \frac{d_j - a_j Q_{j+1}}{b_j + a_j R_{j+1}}, \qquad j = J - 1, \ldots, 1. \qquad (1.9.6)$$

These are deduced as follows. If we apply equation (1.9.5) to $U_{k+1,j+1}$, then we will have

$$U_{k+1,j+1} = R_{j+1} U_{k+1,j} + Q_{j+1} = R_{j+1}(R_j U_{k+1,j-1} + Q_j) + Q_{j+1}$$
$$= R_{j+1} R_j U_{k+1,j-1} + R_{j+1} Q_j + Q_{j+1}. \qquad (1.9.7)$$

Substituting (1.9.7) together with (1.9.5) into (1.9.4) results in

$$(a_j R_{j+1} R_j + b_j R_j + c_j) U_{k+1,j-1} + a_j R_{j+1} Q_j + b_j Q_j + a_j Q_{j+1} = d_j.$$

If we want this equation to be satisfied independent of what value $U_{k+1,j-1}$ takes, then we need to set

$$a_j R_{j+1} R_j + b_j R_j + c_j = 0, \qquad a_j R_{j+1} Q_j + b_j Q_j + a_j Q_{j+1} = d_j. \qquad (1.9.8)$$

Solving equations (1.9.8) for R_j and Q_j respectively, leads to (1.9.6).

To start the calculations in (1.9.6), we need to know R_M and Q_M. These are provided by the boundary condition at the outer edge of the boundary layer (1.9.2e),

$$U_{k+1,J} = U_e(s_{k+1}). \qquad (1.9.9)$$

Setting $j = J$ in (1.9.5) and substituting the result into (1.9.9), we have

$$R_J U_{k+1,J-1} + Q_J = U_e(s_{k+1}).$$

This equation is satisfied independently of the value of $U_{k+1,M-1}$ provided that

$$R_J = 0, \qquad Q_J = U_e(s_{k+1}).$$

Once the Thomas coefficients are found, the recurrent equation (1.9.5) may be used to determine the distribution of the tangential velocity component U along the mesh line s_{k+1}. The calculations begin by setting

$$U_{k+1,0} = 0,$$

to ensure that the no-slip condition on the body surface (1.9.2d) is satisfied.

It should be noted that the coefficients of equation (1.9.4) depend not only on the values of U on the previous mesh line s_k, which are known,[31] but also on the velocity

---

[31] For $k = 0$, these are defined by the initial condition (1.9.2c) to be
$$U_{0,j} = 0, \qquad j = 1, 2, \ldots J.$$

components $U_j^{(c)}$, $V_j^{(c)}$ at the stencil centre, which are not known in advance. For this reason, an iteration process is used on each mesh line s_{k+1}. For the first iteration one can simply choose

$$U_j^{(c)} = U_{k,j}, \qquad V_j^{(c)} = V_{k,j},$$

but after the distribution of U on the mesh line s_{k+1} is calculated at least once, the value of $U_j^{(c)}$ can be updated using the equation

$$U_j^{(c)} = \frac{1}{2}\Big(U_{k+1,j} + U_{k,j}\Big),$$

which is second-order accurate at the stencil centre; see Figure 1.35(a). To update $V_j^{(c)}$, we use the continuity equation (1.9.2b). The derivatives in (1.9.2b) are finite-differenced as

$$\left.\begin{aligned}
\frac{\partial U}{\partial s} &= \frac{1}{2}\left(\frac{U_{k+1,j} - U_{k,j}}{\Delta s} + \frac{U_{k+1,j-1} - U_{k,j-1}}{\Delta s}\right), \\
\frac{\partial V}{\partial N} &= \frac{V_j^{(c)} - V_{j-1}^{(c)}}{\Delta N}.
\end{aligned}\right\} \tag{1.9.10}$$

Both equations in (1.9.10) are second-order accurate with respect to the central point of the stencil in Figure 1.35(b). By substituting (1.9.10) into (1.9.2b), and solving the resulting equation for $V_j^{(c)}$, we find that

$$V_j^{(c)} = V_{j-1}^{(c)} - \frac{\Delta N}{2}\left(\frac{U_{k+1,j} - U_{k,j}}{\Delta s} + \frac{U_{k+1,j-1} - U_{k,j-1}}{\Delta s}\right), \tag{1.9.11}$$

for $j = 1, 2, \ldots, J$. Here, it is taken into account that the point $(k+\frac{1}{2}, j)$ of the stencil in Figure 1.35(b) coincides with the central point of the stencil in Figure 1.35(a). The calculations in (1.9.11) are initiated by setting

$$V_0^{(c)} = 0,$$

as required by the second condition in (1.9.2d). The solution has to be iterated on the mesh line s_{k+1} until the convergence is achieved, and then one can progress to the next mesh line.

1.9.3 Goldstein singularity

The results of the calculations for the circular cylinder flow are displayed in Figure 1.36, where we show the behaviour of two functions: the dimensionless skin friction τ_w, and the displacement function δ^*. The former is defined as

$$\tau_w = \left.\frac{\partial U}{\partial N}\right|_{N=0},$$

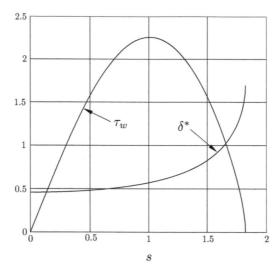

Fig. 1.36: Skin friction and displacement thickness distributions in the boundary layer on the circular cylinder surface.

and is related to dimensional skin friction (1.1.54) by means of the equation

$$\frac{\hat{\tau}_w}{\rho V_\infty^2} = \frac{1}{\sqrt{Re}}\tau_w.$$

Numerically it was calculated using the second-order accurate formula

$$\tau_w\bigg|_{s=s_k} = \frac{4U_{k,1} - 3U_{k,0} - U_{k,2}}{2\Delta N}.$$

We see that at the front stagnation point, $\tau_w = 0$, which, of course, is a consequence of the initial condition (1.9.2c). As soon as the boundary layer leaves the front stagnation point, the flow finds itself under the action of favourable pressure gradient $(dP_e/ds < 0)$, and the skin friction starts to grow. It should be noted that near the cylinder surface the fluid velocity

$$U = \tau_w(s)N + \cdots, \qquad (1.9.12)$$

which makes the skin friction τ_w a convenient function to reveal the flow behaviour. The skin friction reaches its maximum at $s = 1.009$, long before the pressure gradient becomes zero; this is due to the dissipative effect of the viscosity. The pressure gradient becomes adverse $(dP_e/ds > 0)$ after the point $s = \frac{1}{2}\pi$ that lies on the top of the cylinder, and this leads to a sharp drop in τ_w. The skin friction becomes zero at $s_0 = 1.82214$, which corresponds to the angle of 104.5° between the radii drawn from the cylinder centre to the point $s = s_0$, and to the front stagnation point $s = 0$.

One could expect that behind the point $s = s_0$ the skin friction would become negative which, in view of (1.9.12), would suggest that a reverse flow region $(U < 0)$ forms near the cylinder surface. This is why the point of zero skin friction is thought to be the point of the flow separation. However, the calculations show that the solution of the boundary-layer equations develops a singularity as $s \to s_0 - 0$. It manifests itself through square-root behaviour of the skin friction and displacement function

$$\left.\begin{aligned}\tau_w &= a_0(s_0 - s)^{1/2} + \cdots, \\ \delta^* &= \delta_0^* - d_0(s_0 - s)^{1/2} + \cdots\end{aligned}\right\} \quad \text{as} \quad s \to s_0 - 0. \qquad (1.9.13)$$

For the flow past a circular cylinder the constants a_0, δ_0^*, and d_0 are found to be

$$a_0 = 1.904, \qquad \delta_0^* = 1.692, \qquad d_0 = 1.693.$$

The formulae (1.9.13) have been deduced by Landau and Lifshitz (1944), and then by Goldstein (1948). Goldstein also proved that this singularity prevents the solution of the boundary-layer equation to exist behind $s = s_0$. We will discuss this issue in more detail in Section 5.3.2.

1.10 Compressible Boundary Layers

Here, our task will be to study the effects of compressibility on flow behaviour in the boundary layer. We shall start with the boundary layer on the surface of a flat plate that is aligned with the velocity vector in the free stream; see Figure 1.37. When dealing with this flow it is convenient to use Cartesian coordinates $(\hat{x}, \hat{y})$, with $\hat{x}$ measured along the flat plate surface from its leading edge O, and $\hat{y}$ in the perpendicular direction. In what follows we shall assume that the flow is steady and two-dimensional. We shall also assume that the body force $\mathbf{f}$ is negligibly small. Then the Navier–Stokes equations may be written in the form[32]

$$\hat{\rho}\left(\hat{u}\frac{\partial \hat{u}}{\partial \hat{x}} + \hat{v}\frac{\partial \hat{u}}{\partial \hat{y}}\right) = -\frac{\partial \hat{p}}{\partial \hat{x}} + \frac{\partial}{\partial \hat{x}}\left[\hat{\mu}\left(\frac{4}{3}\frac{\partial \hat{u}}{\partial \hat{x}} - \frac{2}{3}\frac{\partial \hat{v}}{\partial \hat{y}}\right)\right] + \frac{\partial}{\partial \hat{y}}\left[\hat{\mu}\left(\frac{\partial \hat{u}}{\partial \hat{y}} + \frac{\partial \hat{v}}{\partial \hat{x}}\right)\right], \quad (1.10.1\text{a})$$

$$\hat{\rho}\left(\hat{u}\frac{\partial \hat{v}}{\partial \hat{x}} + \hat{v}\frac{\partial \hat{v}}{\partial \hat{y}}\right) = -\frac{\partial \hat{p}}{\partial \hat{y}} + \frac{\partial}{\partial \hat{y}}\left[\hat{\mu}\left(\frac{4}{3}\frac{\partial \hat{v}}{\partial \hat{y}} - \frac{2}{3}\frac{\partial \hat{u}}{\partial \hat{x}}\right)\right] + \frac{\partial}{\partial \hat{x}}\left[\hat{\mu}\left(\frac{\partial \hat{u}}{\partial \hat{y}} + \frac{\partial \hat{v}}{\partial \hat{x}}\right)\right], \quad (1.10.1\text{b})$$

$$\hat{\rho}\left(\hat{u}\frac{\partial \hat{h}}{\partial \hat{x}} + \hat{v}\frac{\partial \hat{h}}{\partial \hat{y}}\right) = \hat{u}\frac{\partial \hat{p}}{\partial \hat{x}} + \hat{v}\frac{\partial \hat{p}}{\partial \hat{y}} + \frac{1}{Pr}\left[\frac{\partial}{\partial \hat{x}}\left(\hat{\mu}\frac{\partial \hat{h}}{\partial \hat{x}}\right) + \frac{\partial}{\partial \hat{y}}\left(\hat{\mu}\frac{\partial \hat{h}}{\partial \hat{y}}\right)\right]$$
$$+ \hat{\mu}\left(\frac{4}{3}\frac{\partial \hat{u}}{\partial \hat{x}} - \frac{2}{3}\frac{\partial \hat{v}}{\partial \hat{y}}\right)\frac{\partial \hat{u}}{\partial \hat{x}} + \hat{\mu}\left(\frac{4}{3}\frac{\partial \hat{v}}{\partial \hat{y}} - \frac{2}{3}\frac{\partial \hat{u}}{\partial \hat{x}}\right)\frac{\partial \hat{v}}{\partial \hat{y}} + \hat{\mu}\left(\frac{\partial \hat{u}}{\partial \hat{y}} + \frac{\partial \hat{v}}{\partial \hat{x}}\right)^2, \quad (1.10.1\text{c})$$

$$\frac{\partial \hat{\rho}\hat{u}}{\partial \hat{x}} + \frac{\partial \hat{\rho}\hat{v}}{\partial \hat{y}} = 0, \qquad (1.10.1\text{d})$$

$$\hat{h} = \frac{\gamma}{\gamma - 1}\frac{\hat{p}}{\hat{\rho}}. \qquad (1.10.1\text{e})$$

These are in turn the $\hat{x}$- and $\hat{y}$-momentum equations, the energy equation, the continuity equation, and the state equation. The latter is applicable for a perfect gas, like air, with parameter γ being the ratio of specific heats.[33] The equations are written in

---

[32]See equations (1.7.21) on page 66 in Part 1 of this book series.

[33]For details see Section 1.3 in Part 1.

Fig. 1.37: Problem layout.

usual notations: $(\hat{u}, \hat{v})$ are the velocity components, $\hat{\rho}$ the gas density, $\hat{p}$ the pressure, $\hat{h}$ the enthalpy, and $\hat{\mu}$ the dynamic viscosity coefficient. Parameter Pr in the energy equation (1.10.1c) is the Prandtl number. For air, $Pr_{\mathrm{air}} = 0.713$ and $\gamma_{\mathrm{air}} = 1.4$.

We denote the length of the plate by L; the velocity, density, viscosity, and pressure in the unperturbed flow upstream of the plate are denoted as V_∞, ρ_∞, μ_∞, and p_∞, respectively. Using these quantities, we introduce the non-dimensional variables as follows:

$$\begin{aligned}
\hat{x} &= Lx, & \hat{y} &= Ly, & \hat{u} &= V_\infty u, & \hat{v} &= V_\infty v, \\
\hat{\rho} &= \rho_\infty \rho, & \hat{p} &= p_\infty + \rho_\infty V_\infty^2 p, & \hat{h} &= V_\infty^2 h, & \hat{\mu} &= \mu_\infty \mu.
\end{aligned} \tag{1.10.2}$$

The substitution of (1.10.2) into the Navier–Stokes equations (1.10.1) turns them into the non-dimensional form:

$$\rho\left(u\frac{\partial u}{\partial x} + v\frac{\partial u}{\partial y}\right) = -\frac{\partial p}{\partial x} + \frac{1}{Re}\left\{\frac{\partial}{\partial x}\left[\mu\left(\frac{4}{3}\frac{\partial u}{\partial x} - \frac{2}{3}\frac{\partial v}{\partial y}\right)\right]\right.$$
$$\left. + \frac{\partial}{\partial y}\left[\mu\left(\frac{\partial u}{\partial y} + \frac{\partial v}{\partial x}\right)\right]\right\}, \tag{1.10.3a}$$

$$\rho\left(u\frac{\partial v}{\partial x} + v\frac{\partial v}{\partial y}\right) = -\frac{\partial p}{\partial y} + \frac{1}{Re}\left\{\frac{\partial}{\partial y}\left[\mu\left(\frac{4}{3}\frac{\partial v}{\partial y} - \frac{2}{3}\frac{\partial u}{\partial x}\right)\right]\right.$$
$$\left. + \frac{\partial}{\partial x}\left[\mu\left(\frac{\partial u}{\partial y} + \frac{\partial v}{\partial x}\right)\right]\right\}, \tag{1.10.3b}$$

$$\rho\left(u\frac{\partial h}{\partial x} + v\frac{\partial h}{\partial y}\right) = u\frac{\partial p}{\partial x} + v\frac{\partial p}{\partial y} + \frac{1}{Re}\left\{\frac{1}{Pr}\left[\frac{\partial}{\partial x}\left(\mu\frac{\partial h}{\partial x}\right) + \frac{\partial}{\partial y}\left(\mu\frac{\partial h}{\partial y}\right)\right]\right.$$
$$\left. + \mu\left(\frac{4}{3}\frac{\partial u}{\partial x} - \frac{2}{3}\frac{\partial v}{\partial y}\right)\frac{\partial u}{\partial x} + \mu\left(\frac{4}{3}\frac{\partial v}{\partial y} - \frac{2}{3}\frac{\partial u}{\partial x}\right)\frac{\partial v}{\partial y} + \mu\left(\frac{\partial u}{\partial y} + \frac{\partial v}{\partial x}\right)^2\right\}, \tag{1.10.3c}$$

$$\frac{\partial \rho u}{\partial x} + \frac{\partial \rho v}{\partial y} = 0, \tag{1.10.3d}$$

$$h = \frac{1}{(\gamma - 1)M_\infty^2}\frac{1}{\rho} + \frac{\gamma}{\gamma - 1}\frac{p}{\rho}. \tag{1.10.3e}$$

Here, M_∞ is the free-stream Mach number, $M_\infty = V_\infty/a_\infty$, with a_∞ being the speed of sound in the unperturbed flow upstream of the plate. It is calculated as

$$a_\infty = \sqrt{\gamma \frac{p_\infty}{\rho_\infty}}.$$

1.10.1 Boundary-layer equations

We shall assume that the Mach number M_∞ is an order one quantity. If $M_\infty < 1$, then the flow outside the boundary layer is subsonic; if $M_\infty > 1$, then it is supersonic. The Reynolds number

$$Re = \frac{\rho_\infty V_\infty L}{\mu_\infty}$$

will be assumed large. The flow analysis in the boundary layer is based on the limit

$$x = O(1), \quad Y = Re^{1/2}y = O(1), \quad Re \to \infty,$$

with the solution of the Navier–Stokes equations (1.10.3) sought in the form of the asymptotic expansions

$$\left.\begin{aligned}
u(x,y;Re) &= U_0(x,Y) + \cdots, & v(x,y;Re) &= Re^{-1/2}V_0(x,Y) + \cdots, \\
\rho(x,y;Re) &= \rho_0(x,Y) + \cdots, & p(x,y;Re) &= Re^{-1/2}P_1(x,Y) + \cdots, \\
h(x,y;Re) &= h_0(x,Y) + \cdots, & \mu(x,y;Re) &= \mu_0(x,Y) + \cdots.
\end{aligned}\right\} \quad (1.10.4)$$

These are analogous to the corresponding asymptotic expansions (1.1.28), (1.2.47) for the incompressible flow past a flat plate.

By substituting (1.10.4) into the Navier–Stokes equations (1.10.3) and setting $Re \to \infty$, we arrive at the following set of the boundary-layer equations:

$$\rho_0 U_0 \frac{\partial U_0}{\partial x} + \rho_0 V_0 \frac{\partial U_0}{\partial Y} = \frac{\partial}{\partial Y}\left(\mu_0 \frac{\partial U_0}{\partial Y}\right), \tag{1.10.5a}$$

$$\rho_0 U_0 \frac{\partial h_0}{\partial x} + \rho_0 V_0 \frac{\partial h_0}{\partial Y} = \frac{1}{Pr}\frac{\partial}{\partial Y}\left(\mu_0 \frac{\partial h_0}{\partial Y}\right) + \mu_0\left(\frac{\partial U_0}{\partial Y}\right)^2, \tag{1.10.5b}$$

$$\frac{\partial(\rho_0 U_0)}{\partial x} + \frac{\partial(\rho_0 V_0)}{\partial Y} = 0, \tag{1.10.5c}$$

$$h_0 = \frac{1}{(\gamma-1)M_\infty^2}\frac{1}{\rho_0}. \tag{1.10.5d}$$

Similar to the momentum equation (1.10.5a), the energy equation (1.10.5b) is parabolic.[34] It therefore requires an initial condition for h_0 at the leading edge of the plate, as well as boundary conditions at the outer edge of the boundary layer and

---

[34]This issue is discussed in detail at the end of Section 1.1.1.

on the plate surface. From (1.10.2) and (1.10.4) we can see that in the unperturbed flow $\rho_0 = 1$, which being substituted into (1.10.5d) yields

$$h_0 = \frac{1}{(\gamma - 1)M_\infty^2}.$$

We also know that in the unperturbed flow $U_0 = 1$. Consequently, at the leading edge of the plate we have to impose the following conditions:

$$U_0 = 1, \quad h_0 = \frac{1}{(\gamma - 1)M_\infty^2} \quad \text{at} \quad x = 0, \ Y \in [0, \infty). \qquad (1.10.6)$$

Similarly, the boundary conditions at the outer edge of the boundary layer are written as

$$U_0 = 1, \quad h_0 = \frac{1}{(\gamma - 1)M_\infty^2} \quad \text{at} \quad Y = \infty, \ x \in [0, 1]. \qquad (1.10.7)$$

On the plate surface, the velocity components satisfy the no-slip conditions

$$U_0 = V_0 = 0 \quad \text{at} \quad Y = 0, \ x \in [0, 1]. \qquad (1.10.8)$$

They should be supplemented with a thermal condition for h_0. Remember that the enthalpy $\hat{h}$ is proportional to the gas temperature $\widehat{T}$, namely,[35]

$$\hat{h} = c_p \widehat{T},$$

with c_p being the specific heat at constant pressure. If the temperature $\widehat{T}$ varies in the flow field, it then leads to a heat transfer from one part of the flow to another, which is described by the heat conduction vector $\mathbf{q}$ being proportional to the temperature gradient:[36]

$$\hat{\mathbf{q}} = -\hat{\kappa} \nabla \widehat{T}. \qquad (1.10.9)$$

Here, $\hat{\kappa}$ is the heat conductivity coefficient.

A typical physical situation may be described as follows. When a plate is placed in the flow, a heat starts to exchange between the plate and the moving gas. But after initial adjustment, an equilibrium temperature distribution will be reached, which prevents further heat transfer through the plate surface. This is expressed mathematically by setting the normal component of the heat conduction vector (1.10.9) to zero on the plate surface, namely,

$$\frac{\partial h_0}{\partial Y} = 0 \quad \text{at} \quad Y = 0, \ x \in [0, 1]. \qquad (1.10.10a)$$

An alternative formulation is a given distribution of the temperature along the plate:

$$h_0 = F(x) \quad \text{at} \quad Y = 0, \ x \in [0, 1]. \qquad (1.10.10b)$$

The latter may be achieved in practice when the plate is artificially cooled or heated.

---

[35]See equation (1.3.28) on page 23 in Part 1 of this book series.
[36]See Section 1.6.3 in Part 1.

1.10.2 Self-similar solution

The boundary-value problem (1.10.5)–(1.10.10) admits a self-similar solution for a thermally isolated wall when condition (1.10.10b) holds, as well as in the case when the wall temperature is constant, that is, function $F(x)$ in (1.10.10b) does not depend on x. In the latter case, the boundary condition (1.10.10b) can be written as

$$h_0 = h_w = \frac{c_p T_w}{V_\infty^2} \quad \text{at} \quad Y = 0, \quad x \in [0, 1], \tag{1.10.11}$$

where T_w is the wall temperature. Since the flow considered here represents a 'compressible analogue' of the Blasius flow (see Section 1.1.2), we seek the self-similar solution in the form

$$\left. \begin{aligned} U_0(x, Y) &= U(\eta), \quad V_0(x, Y) = \frac{1}{\sqrt{x}} V(\eta), \quad \rho_0(x, Y) = \rho(\eta), \\ h_0(x, Y) &= h(\eta), \qquad \mu_0(x, Y) = \mu(\eta), \end{aligned} \right\} \tag{1.10.12}$$

where

$$\eta = \frac{Y}{\sqrt{x}}.$$

The substitution of (1.10.12) into the momentum equation (1.10.5a) results in

$$-\frac{1}{2}\eta \rho\, U \frac{dU}{d\eta} + \rho V \frac{dU}{d\eta} = \frac{d}{d\eta}\left(\mu \frac{dU}{d\eta}\right). \tag{1.10.13}$$

The energy equation (1.10.5b) turns into

$$-\frac{1}{2}\eta \rho\, U \frac{dh}{d\eta} + \rho V \frac{dh}{d\eta} = \frac{1}{Pr}\frac{d}{d\eta}\left(\mu \frac{dh}{d\eta}\right) + \mu \left(\frac{dU}{d\eta}\right)^2, \tag{1.10.14}$$

and the continuity equation takes the form

$$-\frac{1}{2}\eta \frac{d}{d\eta}\left(\rho U\right) + \frac{d}{d\eta}\left(\rho V\right) = 0. \tag{1.10.15}$$

Let us assume, to be definite, that the plate is thermally isolated. Then, using (1.10.12) in (1.10.6)–(1.10.8) and (1.10.10a), we can see that the boundary conditions for (1.10.13)–(1.10.15) are

$$U = 1, \quad h = \frac{1}{(\gamma - 1)M_\infty^2} \quad \text{at} \quad \eta = \infty, \tag{1.10.16a}$$

$$U = V = 0, \quad \frac{dh}{d\eta} = 0 \quad \text{at} \quad \eta = 0. \tag{1.10.16b}$$

Before solving equations (1.10.13)–(1.10.15), it is convenient to rearrange them as follows. Note that the continuity equation (1.10.15) may be written as

$$-\frac{1}{2}\frac{d}{d\eta}\left(\eta \rho U\right) + \frac{1}{2}\rho U + \frac{d}{d\eta}\left(\rho V\right) = 0.$$

Therefore, if we introduce function $f(\eta)$ such that

$$\frac{df}{d\eta} = \rho U, \tag{1.10.17a}$$

$$f(0) = 0, \tag{1.10.17b}$$

then we will have

$$-\frac{1}{2}\frac{d}{d\eta}(\eta\rho U) + \frac{1}{2}\frac{df}{d\eta} + \frac{d}{d\eta}(\rho V) = 0. \tag{1.10.18}$$

Equation (1.10.18) can be integrated to yield

$$-\frac{1}{2}\eta\rho U + \frac{1}{2}f + \rho V = const. \tag{1.10.19}$$

Using the conditions on V in (1.10.16b) and the fact that $f(0) = 0$, we can see that the constant of integration in (1.10.19) is zero. Hence,

$$-\frac{1}{2}\eta\rho U + \rho V = -\frac{1}{2}f. \tag{1.10.20}$$

Substituting (1.10.20) into (1.10.13) and (1.10.14), we have

$$-\frac{1}{2}f\frac{dU}{d\eta} = \frac{d}{d\eta}\left(\mu\frac{dU}{d\eta}\right), \tag{1.10.21}$$

$$-\frac{1}{2}f\frac{dh}{d\eta} = \frac{1}{Pr}\frac{d}{d\eta}\left(\mu\frac{dh}{d\eta}\right) + \mu\left(\frac{dU}{d\eta}\right)^2. \tag{1.10.22}$$

Now we introduce a new independent variable ξ such that

$$\xi = \int_0^\eta \rho(\eta')\,d\eta'. \tag{1.10.23}$$

Equation (1.10.23) is known as the *Dorodnitsyn–Howarth transformation*. It converts equations (1.10.21), (1.10.22), and (1.10.17a) into

$$-\frac{1}{2}f\frac{dU}{d\xi} = \frac{d}{d\xi}\left(\mu\rho\frac{dU}{d\xi}\right), \tag{1.10.24a}$$

$$-\frac{1}{2}f\frac{dh}{d\xi} = \frac{1}{Pr}\frac{d}{d\xi}\left(\mu\rho\frac{dh}{d\xi}\right) + \mu\rho\left(\frac{dU}{d\xi}\right)^2, \tag{1.10.24b}$$

$$\frac{df}{d\xi} = U. \tag{1.10.24c}$$

The boundary conditions for (1.10.24)

$$U = 1, \quad h = \frac{1}{(\gamma - 1)M_\infty^2} \quad \text{at} \quad \xi = \infty, \tag{1.10.25a}$$

$$U = f = 0, \quad \frac{dh}{d\xi} = 0 \quad \text{at} \quad \xi = 0 \tag{1.10.25b}$$

are obtained by combining (1.10.16) with (1.10.17b).

To close the set of equations (1.10.24), we use the fact that the enthalpy h and the density ρ are related to one another through the state equation (1.10.5d). It is written in terms of the self-similar variables (1.10.12) as

$$h = \frac{1}{(\gamma - 1)M_\infty^2} \frac{1}{\rho}. \tag{1.10.26}$$

We also need to specify the dependence of the viscosity coefficient $\hat{\mu}$ on the temperature. We shall adopt the *Sutherland law*:

$$\frac{\hat{\mu}}{\mu_0} = \left(\frac{\hat{T}}{T_0}\right)^{3/2} \frac{T_0 + S}{\hat{T} + S}. \tag{1.10.27}$$

Here $\hat{T}$ is gas temperature measured on the Kelvin scale; for air $S = 110.4\,\mathrm{K}$, $T_0 = 273.1\,\mathrm{K}$, and μ_0 is the value of the viscosity coefficient at $0°\mathrm{C}$. Applying equation (1.10.27) to the unperturbed flow upstream of the plate, where $\hat{T} = T_\infty$, we have

$$\frac{\mu_\infty}{\mu_0} = \left(\frac{T_\infty}{T_0}\right)^{3/2} \frac{T_0 + S}{T_\infty + S}.$$

It then follows that the non-dimensional viscosity coefficient is calculated as

$$\mu = \frac{\hat{\mu}}{\mu_\infty} = \left(\frac{\hat{T}}{T_\infty}\right)^{3/2} \frac{T_\infty + S}{\hat{T} + S}. \tag{1.10.28}$$

We know that[37]

$$\hat{h} = c_p \hat{T}, \qquad h_\infty = c_p T_\infty,$$

which allows us to express (1.10.28) in the form

$$\mu = \left(\frac{\hat{h}}{h_\infty}\right)^{3/2} \frac{1 + S/T_\infty}{\hat{h}/h_\infty + S/T_\infty}. \tag{1.10.29}$$

We further know that[38]

$$h_\infty = \frac{a_\infty^2}{\gamma - 1}. \tag{1.10.30}$$

Consequently, taking into account that the non-dimensional enthalpy is introduced through the scaling $h = \hat{h}/V_\infty^2$, we can write

$$\frac{\hat{h}}{h_\infty} = \frac{V_\infty^2}{h_\infty} h = (\gamma - 1)\frac{V_\infty^2}{a_\infty^2} h = (\gamma - 1)M_\infty^2 h. \tag{1.10.31}$$

It remains to substitute (1.10.31) into (1.10.29), and we will have

$$\mu = \left[(\gamma - 1)M_\infty^2 h\right]^{3/2} \frac{1 + S/T_\infty}{(\gamma - 1)M_\infty^2 h + S/T_\infty}. \tag{1.10.32}$$

---

[37] See equation (1.3.28) on page 23 in Part 1 of this book series.
[38] See equation (4.3.3) on page 250 in Part 1.

The results of numerical solution of the boundary-value problem (1.10.24), (1.10.25) with (1.10.26) and (1.10.32) are displayed in Figure 1.38. The problem involves four non-dimensional parameters: the Prandtl number Pr, the specific heats ratio γ, the temperature ratio S/T_∞, and the free-stream Mach number M_∞. We chose $Pr = 0.713$, $\gamma = 1.4$, and $T_\infty = 250\,\mathrm{K}$. The latter is the temperature at the altitude of 5,000 meters in the 'standard atmosphere'. The calculations were performed for different values of M_∞. It appeared that for $M_\infty = 0.1$, the velocity distribution across the boundary layer (see curve 1 in Figure 1.38a) is almost indistinguishable from that of incompressible flow (see Figure 1.5). For all values of the Mach number, the density ρ equals unity at the outer edge of the boundary layer, and it decreases monotonically as the plate is approached. However, for $M_\infty = 0.1$ it only changes by less than 0.2%, which is why the corresponding curve has been omitted in Figure 1.38(b). Interestingly enough, the velocity profile in the boundary layer changes very little in the entire subsonic range of the Mach number, as curves 1 and 2 in Figure 1.38(a) clearly show. As the Mach number increases beyond $M_\infty = 1$, the gas density inside the boundary layer becomes progressively smaller. This is because a deceleration of fluid particles in the boundary layer leads to the conversion of the kinetic energy into heat. As a result, the gas temperature increases, and, according to (1.10.26), the density becomes smaller.

Another important conclusion that may be drawn from the results in Figure (1.38) is that the thickness of the boundary layer increases monotonically with M_∞. This

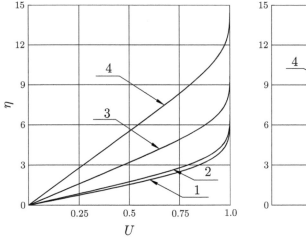
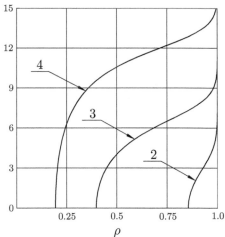

(a) Velocity distribution across the boundary layer.

(b) Density distribution.

Fig. 1.38: Results of numerical solution of the boundary-value problem (1.10.24), (1.10.25) with $Pr = 0.713$, $\gamma = 1.4$, and $T_\infty = 250\,\mathrm{K}$. The velocity and density distributions are shown for the following values of the free-stream Mach number: graph 1 corresponds to $M_\infty = 0.1$; graphs 2, 3, and 4 correspond to $M_\infty = 1$, $M_\infty = 3$, and $M_\infty = 5$, respectively.

effect may be explained by balancing the first convective term on the left-hand side of equation (1.10.5a) with the viscous term:

$$\rho_0 U_0 \frac{\partial U_0}{\partial x} \sim \frac{\partial}{\partial Y}\left(\mu_0 \frac{\partial U_0}{\partial Y}\right).$$

In the boundary layer, the non-dimensional longitudinal velocity U_0 and the coordinate x are order one quantities, which reduces the above order-of-magnitude equation to

$$\rho_o \sim \frac{\mu_0}{Y^2},$$

and we see that

$$Y \sim \sqrt{\frac{\mu_0}{\rho_0}}. \qquad (1.10.33)$$

Since the viscosity coefficient μ_0 increases with the temperature, and the density ρ_0 decreases, it follows from (1.10.33) that the thickness of the boundary layer does indeed grow with M_∞.

1.10.3 Crocco's integral

Crocco's integral is an integral of the energy equation. It is applicable to the compressible boundary layers on various body shapes. As an example, consider a body with a rounded nose; see Figure 1.39. When analysing the flow in the boundary layer that forms on the surface of such a body, we shall use again the body-fitted coordinates. Remember that in these coordinates the position of a point M is defined by coordinates $\hat{s}$ and $\hat{n}$. The first of these is the distance measured along the body contour S from the front stagnation point O to the point M' that is obtained by dropping a perpendicular from M to the body surface S. The second coordinate $\hat{n}$ is the distance between M

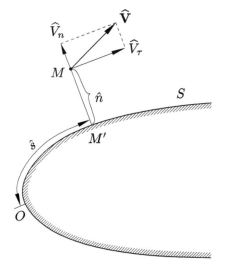

Fig. 1.39: Body with a rounded nose.

and the body surface. The velocity vector is represented in these coordinates by the tangential and normal components $(\widehat{V}_\tau, \widehat{V}_n)$. As before, we use the 'hat' to indicate that the correspondent variables are dimensional.

In the boundary layer, the fluid-dynamic functions are written as

$$
\left.
\begin{aligned}
\widehat{V}_\tau &= V_\infty U_0(s, N) + \cdots, & \widehat{V}_n &= V_\infty Re^{-1/2} V_0(s, N) + \cdots, \\
\hat{\rho} &= \rho_\infty \rho_0(s, N) + \cdots, & \hat{p} &= p_\infty + \rho_\infty V_\infty^2 P_0(s, N) + \cdots, \\
\hat{h} &= V_\infty^2 h_0(s, N) + \cdots, & \hat{\mu} &= \mu_\infty \mu_0(s, N) + \cdots,
\end{aligned}
\right\}
\tag{1.10.34}
$$

where

$$
\hat{s} = Ls, \qquad \hat{n} = LRe^{-1/2} N,
$$

with L being a characteristic length scale, and

$$
Re = \frac{\rho_\infty V_\infty L}{\mu_\infty}.
$$

The boundary-layer equations have the form

$$
\rho_0 U_0 \frac{\partial U_0}{\partial s} + \rho_0 V_0 \frac{\partial U_0}{\partial N} = -\frac{\partial P_0}{\partial s} + \frac{\partial}{\partial N}\left(\mu_0 \frac{\partial U_0}{\partial N}\right),
\tag{1.10.35a}
$$

$$
\frac{\partial P_0}{\partial N} = 0,
\tag{1.10.35b}
$$

$$
\rho_0 U_0 \frac{\partial h_0}{\partial s} + \rho_0 V_0 \frac{\partial h_0}{\partial N} = U_0 \frac{\partial P_0}{\partial s} + \frac{1}{Pr}\frac{\partial}{\partial N}\left(\mu_0 \frac{\partial h_0}{\partial N}\right) + \mu_0 \left(\frac{\partial U_0}{\partial N}\right)^2,
\tag{1.10.35c}
$$

$$
\frac{\partial(\rho_0 U_0)}{\partial s} + \frac{\partial(\rho_0 V_0)}{\partial N} = 0,
\tag{1.10.35d}
$$

$$
h_0 = \frac{1}{(\gamma - 1)M_\infty^2 \rho_0} + \frac{\gamma}{\gamma - 1}\frac{P_0}{\rho_0}.
\tag{1.10.35e}
$$

Now we shall formulate the boundary conditions for these equations. We start with the conditions at the outer edge of the boundary layer. The inviscid flow outside the boundary layer is described by the Euler equations.[39] Their solution represents the first step in Prandtl's hierarchical strategy when the existence of the boundary layer can be disregarded. As a part of the solution, the tangential velocity $U_e(s)$ and the enthalpy $h_e(s)$ on the body surface are found. Using these functions, the matching conditions with the corresponding functions in the boundary layer are written as

$$
U_0 = U_e(s), \quad h_0 = h_e(s) \quad \text{at} \quad N = \infty.
\tag{1.10.36}
$$

It should be noted that functions $U_e(s)$ and $h_e(s)$ are not independent of one another. Indeed, we know that in any inviscid flow (with or without shock waves), the Bernoulli equation holds:[40]

$$
\hat{h} + \frac{1}{2}\left(\widehat{V}_\tau^2 + \widehat{V}_n^2\right) = h_\infty + \frac{V_\infty^2}{2}.
\tag{1.10.37}
$$

---

[39] See equations (4.1.2) on page 235 in Part 1 of this book series.
[40] See equation (4.2.8) on page 241 and equation (4.5.14c) on page 276 in Part 1.

Taking into account that on the body surface $\widehat{V}_n = 0$, and using equation (1.10.30) for h_∞, we have

$$\hat{h} + \frac{1}{2}\widehat{V}_\tau^2 = \frac{a_\infty^2}{\gamma - 1} + \frac{V_\infty^2}{2}.$$

To express this equation in dimensionless form, we need to divide all the terms by V_∞^2, which leads to

$$h_e + \frac{1}{2}U_e^2 = \frac{1}{(\gamma - 1)M_\infty^2} + \frac{1}{2}. \tag{1.10.38}$$

Turning to the conditions on the body surface, we shall assume that the body is thermally isolated, and then we can write

$$U_0 = V_0 = 0, \quad \frac{\partial h_0}{\partial N} = 0 \quad \text{at} \quad N = 0. \tag{1.10.39}$$

Since equations (1.10.35a) and (1.10.35c) are parabolic, we also need to know the distributions of U_0 and h_0 across the boundary layer at the front stagnation point O. Guided by the Hiemenz solution for the incompressible flow near the front stagnation point (see Problem 1 in Exercises 3), we shall seek the corresponding solution of the compressible boundary-layer equations (1.10.35) in the form of the coordinate asymptotic expansions

$$\left.\begin{array}{ll} U_0 = sU_{01}(N) + \cdots, & V_0 = V_{00}(N) + \cdots, \\ P_0 = P_{00} + s^2 P_{02} + \cdots, & h_0 = h_{00}(N) + s^2 h_{02}(N) + \cdots, \\ \rho_0 = \rho_{00}(N) + s^2 \rho_{02}(N) + \cdots, & \mu_0 = \mu_{00}(N) + s^2 \mu_{02}(N) + \cdots \end{array}\right\} \tag{1.10.40}$$

valid in a small vicinity of point O, that is, in the limit

$$s \to 0, \quad N = O(1).$$

The substitution of (1.10.40) into the energy equation (1.10.35c) leads to the following equation for function $h_{00}(N)$:

$$\rho_{00} V_{00} \frac{dh_{00}}{dN} = \frac{1}{Pr} \frac{d}{dN} \left(\mu_{00} \frac{dh_{00}}{dN} \right). \tag{1.10.41a}$$

The boundary conditions for this equation are

$$h_{00} = \frac{1}{(\gamma - 1)M_\infty^2} + \frac{1}{2} \quad \text{at} \quad N = \infty, \tag{1.10.41b}$$

$$\frac{dh_{00}}{dN} = 0 \quad \text{at} \quad N = 0. \tag{1.10.41c}$$

The first condition (1.10.41b) specifies the enthalpy h_0 at the outer edge of the boundary layer. At the front stagnation point, it is given by the Bernoulli equation (1.10.38) with $U_e = 0$. The second condition (1.10.41c) is the statement of zero heat transfer on a thermally isolated wall.

Solution of the boundary-value problem (1.10.41) is given by

$$h_{00}(N) = \frac{1}{(\gamma - 1)M_\infty^2} + \frac{1}{2} \quad \text{for } N \in [0, \infty),$$

which signifies that, at the stagnation point, the gas temperature stays constant across the boundary layer. Hence, the sought initial conditions for the boundary-layer equations (1.10.35) are written as

$$U_0 = 0, \quad h_0 = \frac{1}{(\gamma - 1)M_\infty^2} + \frac{1}{2} \quad \text{at } s = 0. \tag{1.10.42}$$

Thus, to describe the flow in the boundary layer, one needs to solve equations (1.10.35) subject to the conditions (1.10.36), (1.10.39), and (1.10.42).

Theorem 1.2 *If the Prandtl number $Pr = 1$ and the body surface is thermally isolated, then the **total enthalpy** $H = h_0 + \frac{1}{2}U_0^2$ stays constant in the boundary layer, and is given by the equation*

$$H = \frac{1}{(\gamma - 1)M_\infty^2} + \frac{1}{2}, \tag{1.10.43}$$

*which is known as **Crocco's integral**.*

Proof Multiplying the momentum equation (1.10.35a) by U_0 and adding the result to the energy equation (1.10.35c), it is easily found that the total enthalpy H satisfies the equation

$$\rho_0 U_0 \frac{\partial H}{\partial s} + \rho_0 V_0 \frac{\partial H}{\partial N} = \frac{1}{Pr} \frac{\partial}{\partial N}\left(\mu_0 \frac{\partial H}{\partial N}\right) + \left(1 - \frac{1}{Pr}\right) \frac{\partial}{\partial N}\left[\mu_0 \frac{\partial}{\partial N}\left(\frac{U_0^2}{2}\right)\right]. \tag{1.10.44}$$

It further follows from (1.10.36), (1.10.38), (1.10.39), and (1.10.42) that the boundary conditions for equation (1.10.44) are

$$H = \frac{1}{(\gamma - 1)M_\infty^2} + \frac{1}{2} \quad \text{at } N = \infty, \tag{1.10.45a}$$

$$\frac{\partial H}{\partial N} = 0 \quad \text{at } N = 0, \tag{1.10.45b}$$

$$H = \frac{1}{(\gamma - 1)M_\infty^2} + \frac{1}{2} \quad \text{at } s = 0. \tag{1.10.45c}$$

Obviously, with $Pr = 1$, equation (1.10.44) is satisfied by (1.10.43), as are the boundary conditions (1.10.45). $\qquad\square$

Exercises 7

1. The proof of Theorem 1.2 was given for the flow past a body with rounded nose (Figure 1.39). Argue that the theorem is also applicable to the flow past a flat plate aligned with the oncoming flow.

   Hint: When dealing with the flow past a flat plate, one needs to substitute the initial condition (1.10.42) with (1.10.6).

2. Consider the flow past a flat plate again (see Figure 1.37), but this time assume linear dependence of the viscosity coefficient $\hat{\mu}$ on the temperature $\hat{T}$:

$$\hat{\mu} = \alpha \hat{T} = \frac{\alpha}{c_p} \hat{h}.$$

(a) Show that in this case $\mu\rho = 1$. Hence, conclude that the momentum equation (1.10.24a) reduces to the Blasius equation (1.1.47a):

$$-\frac{1}{2} f \frac{d^2 f}{d\xi^2} = \frac{d^3 f}{d\xi^3},$$

while the energy equation (1.10.24b) assumes the form

$$-\frac{1}{2} f \frac{dh}{d\xi} = \frac{1}{Pr} \frac{d^2 h}{d\xi^2} + \left(\frac{d^2 f}{d\xi^2} \right)^2. \tag{1.10.46}$$

(b) Now assume that the Prandtl number $Pr = 1$ and show that the general solution of (1.10.46) is given by

$$h = -\frac{1}{2} \left(\frac{df}{d\xi} \right)^2 + C_1 + C_2 \frac{df}{d\xi}, \tag{1.10.47}$$

where C_1 and C_2 are arbitrary constants.

Assume further that the (dimensional) temperature T_w of the plate surface is known, and show, using conditions (1.10.11) and (1.10.25a), that

$$C_1 = \frac{c_p T_w}{V_\infty^2}, \qquad C_2 = \frac{1}{(\gamma - 1)M_\infty^2} + \frac{1}{2} - \frac{c_p T_w}{V_\infty^2}. \tag{1.10.48}$$

(c) Finally, substitute (1.10.48) into (1.10.47) and show that the heat flux through the plate surface

$$\hat{q} = \hat{\kappa} \frac{\partial \hat{T}}{\partial \hat{y}} \bigg|_{\hat{y}=0} = \frac{\hat{\kappa}}{c_p} \frac{\partial \hat{h}}{\partial \hat{y}} \bigg|_{\hat{y}=0}$$

is given by

$$\frac{\hat{q}}{\rho_\infty V_\infty^3} = \frac{Re^{-1/2}}{\sqrt{x}} \frac{\lambda}{Pr} \left[\frac{1}{(\gamma - 1)M_\infty^2} + \frac{1}{2} - \frac{c_p T_w}{V_\infty^2} \right], \tag{1.10.49}$$

where λ is the skin-friction parameter in the incompressible Blasius solution; see equation (1.1.49).

Hint: Remember that[41]

$$\frac{\hat{\kappa}}{c_p} = \frac{\hat{\mu}}{Pr}.$$

---

[41] See equation (1.6.32) on page 59 in Part 1 of this book series.

3. Consider the flow past a body with a rounded nose (Figure 1.39). Assume that the Prandtl number $Pr = 1$ and that the body surface is thermally isolated. Show that in this case the temperature remains constant along the body surface, and may be calculated as

$$T_w = \frac{V_\infty^2}{c_p} \left[\frac{1}{(\gamma - 1)M_\infty^2} + \frac{1}{2} \right].$$

   Confirm this result with the help of equation (1.10.49).

4. Consider supersonic flow past a wedge with the front shock attached to the wedge tip (see Figure 1.40).

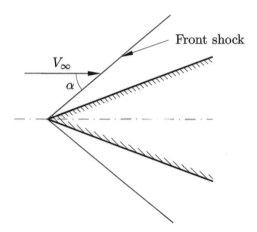

Fig. 1.40: Supersonic flow past a wedge with attached front shock.

What can you say about the boundary layer that forms on the wedge surface?

1.11 Hypersonic Boundary Layers

Hypersonic flows are the flows with large values of the free-stream Mach number M_∞. The inviscid theory of hypersonic flows was discussed in Chapter 5 in Part 2 of this book series. There, we saw that the behaviour of hypersonic flows is different for different body shapes. We shall first consider the flow past a body with a rounded nose; see Figure 1.41.

1.11.1 Hypersonic flow past a body with a rounded nose

According to Prandtl's hierarchical strategy, when dealing with a high-Reynolds-number flow, one needs to start the flow analysis with the inviscid region. We know that for a body with rounded nose the solution for the inviscid flow is represented as[42]

$$\widehat{\mathbf{V}} = V_\infty \Big[\mathbf{V}_0(x,y) + O(M_\infty^{-2}) \Big], \qquad \hat{p} = \rho_\infty V_\infty^2 \Big[p_0(x,y) + O(M_\infty^{-2}) \Big], \left.\begin{array}{c} \\ \\ \end{array}\right\}$$
$$\hat{\rho} = \rho_\infty \Big[\rho_0(x,y) + O(M_\infty^{-2}) \Big], \qquad \hat{h} = V_\infty^2 \Big[h_0(x,y) + O(M_\infty^{-2}) \Big], \qquad (1.11.1)$$

---

[42]See Section 5.1 in Part 2.

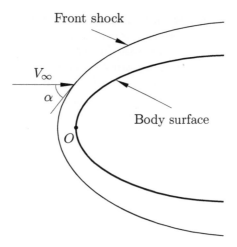

Fig. 1.41: Hypersonic flow past a body with a rounded nose.

with the independent variables made dimensionless in the usual way:

$$\hat{x} = Lx, \qquad \hat{y} = Ly.$$

Here, L is a suitably chosen length scale, say, the radius of curvature of the body contour at the front stagnation point O.

The leading-order terms in (1.11.1) satisfy the Euler equations, which are written in vector form as

$$\left.\begin{aligned}
\rho_0 (\mathbf{V}_0 \cdot \nabla) \mathbf{V}_0 &= -\nabla p_0, \\
\rho_0 \mathbf{V}_0 \cdot \nabla h_0 &= \mathbf{V}_0 \cdot \nabla p_0, \\
\operatorname{div}(\rho_0 \mathbf{V}_0) &= 0, \\
h_0 &= \frac{\gamma}{\gamma - 1} \frac{p_0}{\rho_0}.
\end{aligned}\right\} \tag{1.11.2}$$

These equations have to be solved with the impermeability condition on the body surface S

$$\left.(\mathbf{V}_0 \cdot \mathbf{n})\right|_S = 0, \tag{1.11.3}$$

and the following conditions on the front shock[43]

$$\left.\begin{aligned}
V_{0n} &= \frac{\gamma - 1}{\gamma + 1} \sin \alpha, \\
V_{0\tau} &= \cos \alpha, \\
\rho_0 &= \frac{\gamma + 1}{\gamma - 1}, \\
p_0 &= \frac{2}{\gamma + 1} \sin^2 \alpha.
\end{aligned}\right\} \tag{1.11.4}$$

[43]See equations (5.1.8)–(5.1.10) on page 249 in Part 2 of this book series.

Equations (1.11.4) specify the normal and tangential velocity components $(V_{0n}, V_{0\tau})$, the gas density ρ_0, and the pressure p_0 immediately behind the shock; α is the angle between the free-stream velocity vector and the tangent to the shock; see Figure 1.41.

The solution of the boundary-value problem (1.11.2)–(1.11.4) allows us to find distribution of fluid-dynamic functions in the flow field between the shock and the boundary layer. Of particular interest are the tangential velocity $U_e(s)$ and the enthalpy $h_e(s)$ on the body contour; we need to know these functions to calculate the flow in the boundary layer. Before turning to the boundary-layer analysis, the following comments about the inviscid flow behaviour are appropriate. When passing through the shock, the gas experiences a deceleration, and its density increases. However, according to (1.11.1) and (1.11.4), both the velocity and density remain comparable to their values in the unperturbed flow before the shock:

$$\widehat{V} \sim V_\infty, \qquad \widehat{\rho} \sim \rho_\infty.$$

At the same time, the gas temperature increases significantly. In order to estimate its value behind the shock, we can use the Bernoulli equation (1.10.37) that holds everywhere in the inviscid part of the flow. Taking into account that[44]

$$\widehat{h} = c_p \widehat{T}, \qquad h_\infty = \frac{a_\infty^2}{\gamma - 1}, \tag{1.11.5}$$

we can write the Bernoulli equation as

$$c_p \widehat{T} + \frac{1}{2} \widehat{V}^2 = \frac{a_\infty^2}{\gamma - 1} + \frac{1}{2} V_\infty^2, \tag{1.11.6}$$

where $\widehat{V}$ denotes the modulus of the velocity vector. We know that in a hypersonic flow $a_\infty \ll V_\infty$ which allows us to disregard the first term on the right-hand side of (1.11.6). We have

$$c_p \widehat{T} = \frac{1}{2} \left(V_\infty^2 - \widehat{V}^2 \right).$$

Now, taking into account that the shock reduces the gas velocity, but $\widehat{V}$ still remains comparable to V_∞, we see that the following order-of-magnitude equation holds:

$$c_p \widehat{T} \sim \frac{1}{2} V_\infty^2. \tag{1.11.7}$$

Dividing both sides of (1.11.7) by $h_\infty = c_p T_\infty$, and using the second equation in (1.11.5), we find that

$$\frac{\widehat{T}}{T_\infty} = O(M_\infty^2).$$

The increase of the gas temperature makes the speed of sound $\widehat{a}$ behind the shock much larger than it is before the shock. Indeed, using again equation (4.3.3) of Part 1, we can write

$$\widehat{a} = \sqrt{(\gamma - 1)\widehat{h}} = \sqrt{(\gamma - 1) c_p \widehat{T}}, \tag{1.11.8}$$

---

[44]See equation (1.3.28) on page 23 and equation (4.3.3) on page 250, both in Part 1 of this book series.

which, being combined with (1.11.7), allows us to conclude that in the flow behind the shock

$$\hat{a} \sim V_\infty.$$

Hence, the local Mach number $M = \widehat{V}/\hat{a}$ is an order one quantity. This suggests that the boundary layer, that forms on the body with rounded nose in a hypersonic flow, should behave similarly to the boundary layer in a finite Mach number flow; see Section 1.10.3. Of course, we must remember that, in the hypersonic flow, the gas temperature is very high. Therefore, the free-stream viscosity μ_∞ can no longer be used in (1.10.34) as a measure of the gas viscosity in the boundary layer. Instead, the viscosity μ_* at the stagnation point O is commonly used; see Figure 1.41. Remember that $\hat{\mu}$ is a function of temperature $\widehat{T}$. The value of the latter at the stagnation point is found by setting $\widehat{V} = 0$ in the Bernoulli equation (1.11.6). This leads to the following equation for T_*:

$$c_p T_* = \frac{a_\infty^2}{\gamma - 1} + \frac{1}{2} V_\infty^2. \tag{1.11.9}$$

Replacing μ_∞ in (1.10.34) with μ_*, and taking into account that in a hypersonic flow $p_\infty \ll \rho_\infty V_\infty^2$, we represent the fluid-dynamic functions in the boundary layer in the form

$$\left.\begin{aligned}
\widehat{V}_\tau &= V_\infty U_0(s, N) + \cdots, & \widehat{V}_n &= V_\infty Re_*^{-1/2} V_0(s, N) + \cdots, \\
\hat{\rho} &= \rho_\infty \rho_0(s, N) + \cdots, & \hat{p} &= \rho_\infty V_\infty^2 P_0(s, N) + \cdots, \\
\hat{h} &= V_\infty^2 h_0(s, N) + \cdots, & \hat{\mu} &= \mu_* \mu_0(s, N) + \cdots.
\end{aligned}\right\} \tag{1.11.10}$$

Here the body-fitted coordinates $(\hat{s}, \hat{n})$ are scaled as

$$\hat{s} = Ls, \qquad \hat{n} = LRe_*^{-1/2} N,$$

with the Reynolds number being

$$Re_* = \frac{\rho_\infty V_\infty L}{\mu_*}.$$

The flow in the boundary layer is described by the equations

$$\rho_0 U_0 \frac{\partial U_0}{\partial s} + \rho_0 V_0 \frac{\partial U_0}{\partial N} = -\frac{\partial P_0}{\partial s} + \frac{\partial}{\partial N}\left(\mu_0 \frac{\partial U_0}{\partial N}\right), \tag{1.11.11a}$$

$$\frac{\partial P_0}{\partial N} = 0, \tag{1.11.11b}$$

$$\rho_0 U_0 \frac{\partial h_0}{\partial s} + \rho_0 V_0 \frac{\partial h_0}{\partial N} = U_0 \frac{\partial P_0}{\partial s} + \frac{1}{Pr}\frac{\partial}{\partial N}\left(\mu_0 \frac{\partial h_0}{\partial N}\right) + \mu_0\left(\frac{\partial U_0}{\partial N}\right)^2, \tag{1.11.11c}$$

$$\frac{\partial(\rho_0 U_0)}{\partial s} + \frac{\partial(\rho_0 V_0)}{\partial N} = 0, \tag{1.11.11d}$$

$$h_0 = \frac{\gamma}{\gamma - 1}\frac{P_0}{\rho_0}. \tag{1.11.11e}$$

These have to be solved subject to the boundary conditions

$$U_0 = U_e(s), \qquad h_0 = h_e(s) \quad \text{at} \quad N = \infty, \tag{1.11.12a}$$

$$U_0 = V_0 = 0, \quad \frac{\partial h_0}{\partial N} = 0 \qquad \text{at} \quad N = 0, \tag{1.11.12b}$$

$$U_0 = 0, \qquad h_0 = \frac{1}{2} \qquad \text{at} \quad s = 0. \tag{1.11.12c}$$

The boundary-value problem (1.11.11), (1.11.12) may be obtained either by substituting (1.11.10) into the Navier–Stokes equations, or by simply setting $M_\infty = \infty$ in the state equation (1.10.35e) and the initial condition (1.10.42). As mentioned before, the behaviour of the hypersonic boundary layer on a body with a rounded nose is analogous to its finite Mach number counterpart. However, due to high temperature, the boundary-layer thickness, being proportional to $Re_*^{-1/2}$, is relatively large.

1.11.2   Hypersonic boundary layers on thin bodies

When analysing the compressible boundary-layer flow on a flat plate surface in Section 1.10, we found that the thickness of the boundary layer increases rapidly with the Mach number M_∞. As a consequence, the displacement effect of the boundary layer becomes stronger. This is further enhanced by specific properties of inviscid hypersonic flows past thin bodies. Remember that at large values of the free-stream Mach number, even a thin body can produce significant pressure perturbations in the flow.[45] In fact, in the inviscid theory of hypersonic flows past thin bodies, three flow regimes are distinguished depending on the hypersonic interaction parameter

$$\chi = M_\infty \varepsilon, \tag{1.11.13}$$

where ε is the dimensionless thickness of the body. If $\chi \ll 1$, then the pressure perturbations $\Delta \hat{p}$ produced by the body in the flow are small compared to p_∞, which is why this flow regime is referred to as *weak interaction*. If χ is an order one quantity, then $\Delta \hat{p} \sim p_\infty$, and we are dealing with *moderate interaction*. Finally, the regime of *strong interaction* corresponds to $\chi \gg 1$, in which case the pressure perturbations $\Delta \hat{p}$ are much larger than p_∞.

We shall see that this classification is also applicable when dealing with hypersonic boundary-layers.

Weak interaction

To describe the week interaction flow regime, consider hypersonic flow of a perfect gas past a flat plate that is aligned with the oncoming flow; see Figure 1.42. We shall use Cartesian coordinates $(\hat{x}, \hat{y})$ with $\hat{x}$ measured along the plate surface from its leading edge O, and $\hat{y}$ in the perpendicular direction. As before, we denote the free-stream velocity by V_∞, the pressure in the oncoming flow by p_∞, and the gas density by ρ_∞. The flow is described by the Navier–Stokes equations (1.10.1). When performing the asymptotic analysis of these equations, we need to take into account the following.

[45]See Section 5.3 in Part 2 of this book series.

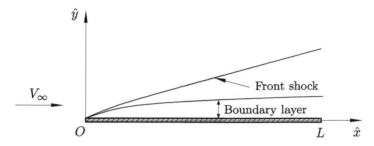

Fig. 1.42: Hypersonic boundary layer with weak interaction.

In the case of weak interaction, the flow outside the boundary layer is almost unperturbed, with the pressure, density, and enthalpy given by

$$\hat{p} = p_\infty, \qquad \hat{\rho} = \rho_\infty, \qquad \hat{h} = h_\infty = \frac{\gamma}{\gamma - 1} \frac{p_\infty}{\rho_\infty}. \tag{1.11.14}$$

Of course, the pressure penetrates the boundary layer and, in the leading-order approximation, it remains equal to p_∞. As far as the enthalpy $\hat{h}$ is concerned, it increases dramatically due to the conversion of the kinetic energy into heat as the fluid particles decelerate in the boundary layer. In this process the convective terms on the left-hand side of the energy equation (1.10.1c) maintain balance with the mechanical energy dissipation term on the right-hand side of this equation:[46]

$$\hat{\rho}\hat{u}\frac{\partial \hat{h}}{\partial \hat{x}} \sim \hat{\mu}\left(\frac{\partial \hat{u}}{\partial \hat{y}}\right)^2.$$

This leads to the following order-of-magnitude equation:

$$\hat{\rho}V_\infty \frac{\hat{h}}{L} \sim \hat{\mu}\frac{V_\infty^2}{\hat{\delta}^2}, \tag{1.11.15}$$

where $\hat{\delta}$ is the boundary-layer thickness and L is the plate length; see Figure 1.42.

We also know that in the boundary layer the convective terms on the left-hand side of the momentum equation (1.10.1a) have to be in balance with the main viscous term on the right-hand side of (1.10.1a):

$$\hat{\rho}\hat{u}\frac{\partial \hat{u}}{\partial \hat{x}} \sim \frac{\partial}{\partial \hat{y}}\left(\hat{\mu}\frac{\partial \hat{u}}{\partial \hat{y}}\right).$$

Hence, we can write

$$\hat{\rho}\frac{V_\infty^2}{L} \sim \hat{\mu}\frac{V_\infty}{\hat{\delta}^2}. \tag{1.11.16}$$

It immediately follows from (1.11.15) and (1.11.16) that in the boundary layer

$$\hat{h} \sim V_\infty^2, \tag{1.11.17}$$

which means that despite the flow at hand not having a stagnation point, the gas temperature in the boundary layer is comparable with the stagnation temperature T_*

----

[46]Notice that the continuity equation (1.10.1d) 'makes' the two terms on the left-hand side of (1.10.1c) same order quantities.

defined by equation (1.11.9). Denoting as before the viscosity coefficient at stagnation temperature by μ_*, we can express equation (1.11.16) as

$$\hat{\rho}\frac{V_\infty}{L} \sim \frac{\mu_*}{\hat{\delta}^2}. \tag{1.11.18}$$

It remains to find an estimate for the density $\hat{\rho}$. For this purpose we can use the state equation (1.10.1e). We have

$$\hat{\rho} \sim \frac{\hat{p}}{\hat{h}} \sim \frac{p_\infty}{V_\infty^2},$$

which may be also written as

$$\frac{\hat{\rho}}{\rho_\infty} \sim \frac{1}{M_\infty^2}. \tag{1.11.19}$$

Now we substitute (1.11.19) into (1.11.18) and solve the resulting equation for $\hat{\delta}$. We find that in the weak interaction regime the thickness of the boundary layer is estimated as

$$\hat{\delta} \sim L M_\infty Re_*^{-1/2}, \tag{1.11.20}$$

where

$$Re_* = \frac{\rho_\infty V_\infty L}{\mu_*}.$$

Using (1.11.20), the parameter ε in (1.11.13) is calculated as $\varepsilon = \hat{\delta}/L = M_\infty Re_*^{-1/2}$, and the hypersonic interaction parameter assumes the form

$$\chi = M_\infty^2 Re_*^{-1/2}.$$

It should be small for the weak interaction flow regime, considered here, to realize.

Guided by (1.11.17), (1.11.19), and (1.11.20), we represent the fluid-dynamic functions in the boundary layer in the form of the asymptotic expansions

$$\left.\begin{aligned}
\hat{u} &= V_\infty U_0(x,Y) + \cdots, & \hat{v} &= V_\infty M_\infty Re_*^{-1/2} V_0(x,Y) + \cdots, \\
\hat{\rho} &= \rho_\infty M_\infty^{-2} \rho_0(x,Y) + \cdots, & \hat{p} &= p_\infty + \cdots, \\
\hat{h} &= V_\infty^2 h_0(x,Y) + \cdots, & \hat{\mu} &= \mu_* \mu_0(x,Y) + \cdots,
\end{aligned}\right\} \tag{1.11.21a}$$

where the non-dimensional independent variables are defined by

$$\hat{x} = Lx, \qquad \hat{y} = L M_\infty Re_*^{-1/2} Y. \tag{1.11.21b}$$

The substitution of (1.11.21) into the Navier–Stokes equations (1.10.1) yields

$$\rho_0 U_0 \frac{\partial U_0}{\partial x} + \rho_0 V_0 \frac{\partial U_0}{\partial Y} = \frac{\partial}{\partial Y}\left(\mu_0 \frac{\partial U_0}{\partial Y}\right), \tag{1.11.22a}$$

$$\rho_0 U_0 \frac{\partial h_0}{\partial x} + \rho_0 V_0 \frac{\partial h_0}{\partial Y} = \frac{1}{Pr}\frac{\partial}{\partial Y}\left(\mu_0 \frac{\partial h_0}{\partial Y}\right) + \mu_0\left(\frac{\partial U_0}{\partial Y}\right)^2, \tag{1.11.22b}$$

$$\frac{\partial(\rho_0 U_0)}{\partial x} + \frac{\partial(\rho_0 V_0)}{\partial Y} = 0, \tag{1.11.22c}$$

$$h_0 = \frac{1}{(\gamma-1)\rho_0}. \tag{1.11.22d}$$

When formulating the boundary conditions for these equations, one needs to take into account that the free-stream enthalpy $h_\infty = a_\infty^2/(\gamma - 1)$ is much smaller than V_∞^2. Therefore, the leading-order term h_0 in the asymptotic expansion of the enthalpy $\hat{h}$ in (1.11.21) should become zero at the outer edge of the boundary layer. In these circumstances the boundary layer appears to have a well-defined outer edge that may be expressed in non-dimensional variables by the equation[47]

$$Y = \delta(x). \tag{1.11.23}$$

Here, function $\delta(x)$ is not known in advance, and should be found as part of the solution of the problem at hand. Using (1.11.23), the boundary conditions for (1.11.22) are written as

$$U_0 = 1, \qquad\qquad h_0 = 0 \quad \text{at} \quad Y = \delta(x), \tag{1.11.24a}$$

$$U_0 = V_0 = 0, \qquad \frac{\partial h_0}{\partial Y} = 0 \quad \text{at} \quad Y = 0. \tag{1.11.24b}$$

In (1.11.24b), we assume the plate surface to be thermally isolated. An alternative formulation corresponds to the situation where the wall temperature $\hat{T} = T_w$ is known. In this case, we can write

$$h_0 = \frac{c_p T_w}{V_\infty^2} \quad \text{at} \quad Y = 0.$$

There is an obvious similarity between the boundary-value problem (1.11.22), (1.11.24) and its finite Mach number counterpart (1.10.5), (1.10.6)–(1.10.8), (1.10.10a). Neither involve the pressure gradient in the momentum and energy equations. This suggests that the solution of (1.11.22), (1.11.24) may be sought in the self-similar form

$$\left.\begin{aligned} U_0(x, Y) &= U(\eta), & V_0(x, Y) &= \frac{1}{\sqrt{x}} V(\eta), & \rho_0(x, Y) &= \rho(\eta), \\ h_0(x, Y) &= h(\eta), & \mu_0(x, Y) &= \mu(\eta), & \delta(x) &= d\sqrt{x}, \end{aligned}\right\} \tag{1.11.25}$$

where

$$\eta = \frac{Y}{\sqrt{x}},$$

and d is a constant to be found. The substitution of (1.11.25) into the boundary-layer equations (1.11.22) reduces these to the following set of ordinary differential equations:

$$-\frac{1}{2}\eta\rho\, U\frac{dU}{d\eta} + \rho V\frac{dU}{d\eta} = \frac{d}{d\eta}\left(\mu\frac{dU}{d\eta}\right), \tag{1.11.26a}$$

$$-\frac{1}{2}\eta\rho\, U\frac{dh}{d\eta} + \rho V\frac{dh}{d\eta} = \frac{1}{Pr}\frac{d}{d\eta}\left(\mu\frac{dh}{d\eta}\right) + \mu\left(\frac{dU}{d\eta}\right)^2, \tag{1.11.26b}$$

$$-\frac{1}{2}\eta\frac{d}{d\eta}(\rho U) + \frac{d}{d\eta}(\rho V) = 0, \tag{1.11.26c}$$

$$h = \frac{1}{(\gamma - 1)\rho}. \tag{1.11.26d}$$

---

[47]See Problem 1 in Exercises 8.

They have to be solved with the boundary conditions

$$U = 1, \qquad h = 0 \quad \text{at} \quad \eta = d, \tag{1.11.27a}$$

$$U = V = 0, \qquad \frac{dh}{d\eta} = 0 \quad \text{at} \quad \eta = 0 \tag{1.11.27b}$$

that are obtained by substituting (1.11.25) into (1.11.24).

If we now introduce function $f(\eta)$ such that

$$\frac{df}{d\eta} = \rho U, \qquad f(0) = 0,$$

and perform the Dorodnitsyn–Howarth transformation

$$\xi = \int\limits_0^\eta \rho(\eta') \, d\eta', \tag{1.11.28}$$

then equations (1.11.26a)–(1.11.26c) turn into[48]

$$-\frac{1}{2} f \frac{dU}{d\xi} = \frac{d}{d\xi} \left(\mu\rho \frac{dU}{d\xi} \right), \tag{1.11.29a}$$

$$-\frac{1}{2} f \frac{dh}{d\xi} = \frac{1}{Pr} \frac{d}{d\xi} \left(\mu\rho \frac{dh}{d\xi} \right) + \mu\rho \left(\frac{dU}{d\xi} \right)^2, \tag{1.11.29b}$$

$$\frac{df}{d\xi} = U, \tag{1.11.29c}$$

and boundary conditions (1.11.27) assume the form

$$U = 1, \qquad h = 0 \quad \text{at} \quad \xi = \infty, \tag{1.11.30a}$$

$$U = f = 0, \qquad \frac{dh}{d\xi} = 0 \quad \text{at} \quad \xi = 0. \tag{1.11.30b}$$

When formulating the boundary conditions (1.11.30a) at the outer edge of the boundary layer, it was taken into account that, according to the state equation (1.11.26d), the gas density ρ becomes infinitely large as $\eta \to d$. Consequently, it was assumed, subject to subsequent confirmation (see Problem 1 in Exercises 8), that the Dorodnitsyn–Howarth transformation (1.11.28) maps a finite interval $\eta \in [0, d]$ onto a semi-infinite interval $\xi \in [0, \infty)$.

To solve equations (1.11.26), we need to specify the viscosity/temperature law. We shall assume, for simplicity, that it is given by the linear function

$$\hat{\mu} = C\widehat{T} = \frac{C}{c_p} \hat{h} = \frac{C}{c_p} V_\infty^2 h. \tag{1.11.31}$$

Choosing $\widehat{T}$ in (1.11.31) to coincide with the stagnation temperature T_*, we can write

$$\mu_* = CT_*.$$

---

[48] For details see equations (1.10.17)–(1.10.24) on page 102.

The stagnation temperature T_* is defined by equation (1.11.9). In a hypersonic flow, the first term on the right-hand side of (1.11.9) is M_∞^2 times smaller than the second term and may be disregarded, which gives

$$T_* = \frac{V_\infty^2}{2c_p}.$$

Consequently, the dimensionless viscosity coefficient is calculated as

$$\mu = \frac{\hat{\mu}}{\mu_*} = 2h. \tag{1.11.32}$$

It remains to combine (1.11.32) with the state equation (1.11.26d), and we can see that

$$\mu\rho = \frac{2}{\gamma - 1}. \tag{1.11.33}$$

With (1.11.33), the momentum equation (1.11.29a) does not involve the enthalpy h, and hence separates from the energy equation (1.11.29b). Substitution of (1.11.33) and (1.11.29c) into (1.11.29a) yields the following equation for function $f(\xi)$:

$$\frac{d^3 f}{d\xi^3} + \frac{\gamma - 1}{4} f \frac{d^2 f}{d\xi^2} = 0. \tag{1.11.34a}$$

It should be solved subject to the boundary conditions

$$\left.\begin{aligned}
\frac{df}{d\xi} &= f = 0 & \text{at} \quad \xi = 0, \\
\frac{df}{d\xi} &= 1 & \text{at} \quad \xi = \infty.
\end{aligned}\right\} \tag{1.11.34b}$$

When $f(\xi)$ is found, we can turn to the energy equation (1.11.29b):

$$\frac{1}{Pr}\frac{d^2 h}{d\xi^2} + \frac{\gamma - 1}{4} f \frac{dh}{d\xi} + \left(\frac{d^2 f}{d\xi^2}\right)^2 = 0. \tag{1.11.35a}$$

The boundary conditions for this equation are

$$\left.\begin{aligned}
\frac{dh}{d\xi} &= 0 & \text{at} \quad \xi = 0, \\
h &= 0 & \text{at} \quad \xi = \infty.
\end{aligned}\right\} \tag{1.11.35b}$$

After the boundary-value problem (1.11.35) is solved, the density distribution across the boundary layer can be found using the state equation (1.11.26d). Finally, we can return to the original similarity variable η, then the inverse Dorodnitsyn–Howarth transformation

$$\eta = \int_0^\xi \frac{d\xi'}{\rho(\xi')} \tag{1.11.36}$$

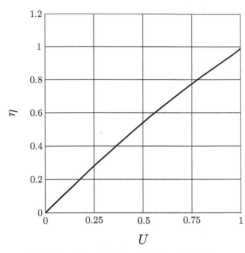

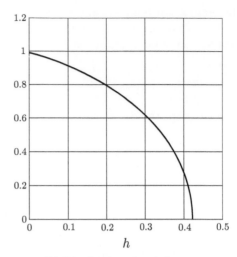

(a) Velocity profile in the boundary layer.   (b) Distribution of enthalpy.

Fig. 1.43: Results of numerical solution of equations (1.11.34)–(1.11.36) with $Pr = 0.713$ and $\gamma = 1.4$.

should be used. In particular, setting $\xi = \infty$ in (1.11.36), we have

$$d = \int\limits_{0}^{\infty} \frac{d\xi}{\rho(\xi)}, \qquad (1.11.37)$$

which allows us to write the equation for the outer edge of the boundary layer in the form

$$\delta(x) = d\sqrt{x} = \sqrt{x} \int\limits_{0}^{\infty} \frac{d\xi}{\rho(\xi)}.$$

To express this equation in dimensional variables, we need to remember that

$$\hat{\delta} = L M_\infty Re_*^{-1/2} \delta, \qquad \hat{x} = Lx.$$

Since equation (1.11.34a) does not have an analytic solution, the above tasks should be performed numerically. The results of the calculations are displayed in Figure 1.43. We found that the velocity U shows almost linear growth from $U = 0$ on the plate surface to $U = 1$ at the outer edge of the boundary layer $\eta = d = 0.9902$. Interestingly enough, the derivative $dU/d\eta$ does not become zero at $\eta = d$. Of course, everywhere outside the boundary layer, the flow is uniform in the leading-order approximation with $U \equiv 1$. This suggests that the solution presented above loses its validity in a small vicinity of the outer edge of the boundary layer. To smooth out the velocity profile, an additional layer should be introduced between the 'main part' of the boundary layer and the inviscid flow.

Moderate and strong interaction flow regimes

When the hypersonic interaction parameter χ becomes an order one quantity or larger, the pressure perturbations produced in the flow by the displacement effect of the boundary layer are no longer small compared with the free-stream pressure p_∞. This means that the arguments used on pages 115 and 116 to predict the form of the solution in the boundary layer (1.11.21) should be revised. Conveniently, for both moderate and strong interaction flow regimes, the pressure may be estimated as[49]

$$\hat{p} \sim \rho_\infty V_\infty^2 \varepsilon^2 \sim \rho_\infty V_\infty^2 \left(\hat{\delta}/L\right)^2. \tag{1.11.38}$$

Of course, in the boundary layer the first convective term on the left-hand side the $\hat{x}$-momentum equation (1.10.3a) should be in balance with leading-order viscous term, which allows us to use again equation (1.11.16):

$$\hat{\rho}\frac{V_\infty^2}{L} \sim \hat{\mu}\frac{V_\infty}{\hat{\delta}^2}. \tag{1.11.39}$$

Also, in the energy equation (1.10.3c), the first convective term should be in balance with the mechanical energy dissipation term. Hence, equation (1.11.15) also remains valid:

$$\hat{\rho}V_\infty\frac{\hat{h}}{L} \sim \hat{\mu}\frac{V_\infty^2}{\hat{\delta}^2}. \tag{1.11.40}$$

It follows from (1.11.39) and (1.11.40) that

$$\hat{h} \sim V_\infty^2, \tag{1.11.41}$$

and we can conclude that the gas temperature in the boundary layer is again comparable with the stagnation temperature T_* and

$$\hat{\mu} \sim \mu_*. \tag{1.11.42}$$

However, now the pressure is given by (1.11.38). Using (1.11.38) together with (1.11.41) in the state equation (1.10.3e), we find that the density can be estimated as

$$\hat{\rho} \sim \frac{\hat{p}}{\hat{h}} \sim \rho_\infty \left(\frac{\hat{\delta}}{L}\right)^2. \tag{1.11.43}$$

It remains to substitute (1.11.43) and (1.11.42) into (1.11.39) and solve the resulting equation for $\hat{\delta}$. We find that the thickness of the boundary layer is estimated as

$$\hat{\delta} \sim LRe_*^{-1/4}, \tag{1.11.44}$$

where

$$Re_* = \frac{\rho_\infty V_\infty L}{\mu_*}.$$

---

[49]See equation (5.3.28) on page 267 in Part 1 of this book series.

Thus, when dealing with the moderate or strong interaction flow regimes, we have to represent the fluid-dynamic functions in the boundary layer in the form

$$
\left.
\begin{aligned}
\hat{u} &= V_\infty U_0(x, Y) + \cdots\,, & \hat{v} &= V_\infty Re_*^{-1/4} V_0(x, Y) + \cdots\,, \\
\hat{\rho} &= \rho_\infty Re_*^{-1/2} \rho_0(x, Y) + \cdots\,, & \hat{p} &= \rho_\infty V_\infty^2 Re_*^{-1/2} P_0(x) + \cdots\,, \\
\hat{h} &= V_\infty^2 h_0(x, Y) + \cdots\,, & \hat{\mu} &= \mu_* \mu_0(x, Y) + \cdots\,,
\end{aligned}
\right\}
\qquad (1.11.45)
$$

where the non-dimensional independent variables are defined by

$$
\hat{x} = Lx, \qquad \hat{y} = L Re_*^{-1/4} Y.
$$

The substitution of (1.11.45) into the Navier–Stokes equations (1.10.1) yields

$$
\rho_0 U_0 \frac{\partial U_0}{\partial x} + \rho_0 V_0 \frac{\partial U_0}{\partial Y} = -\frac{dP_0}{dx} + \frac{\partial}{\partial Y}\left(\mu_0 \frac{\partial U_0}{\partial Y}\right), \qquad (1.11.46a)
$$

$$
\rho_0 U_0 \frac{\partial h_0}{\partial x} + \rho_0 V_0 \frac{\partial h_0}{\partial Y} = U_0 \frac{dP_0}{dx} + \frac{1}{Pr}\frac{\partial}{\partial Y}\left(\mu_0 \frac{\partial h_0}{\partial Y}\right) + \mu_0 \left(\frac{\partial U_0}{\partial Y}\right)^2, \qquad (1.11.46b)
$$

$$
\frac{\partial(\rho_0 U_0)}{\partial x} + \frac{\partial(\rho_0 V_0)}{\partial Y} = 0, \qquad (1.11.46c)
$$

$$
h_0 = \frac{\gamma}{\gamma - 1}\frac{P_0}{\rho_0}. \qquad (1.11.46d)
$$

If we assume that the plate is thermally isolated, then the boundary conditions for (1.11.46) are written as

$$
U_0 = 1, \qquad\qquad h_0 = 0 \quad \text{at} \quad Y = \delta(x), \qquad (1.11.47a)
$$

$$
U_0 = V_0 = 0, \qquad \frac{\partial h_0}{\partial Y} = 0 \quad \text{at} \quad Y = 0. \qquad (1.11.47b)
$$

It should be noted that in the boundary-value problem (1.11.46), (1.11.47), not only the function $\delta(x)$, representing the outer edge of the boundary layer, but also the pressure $P_0(x)$ are not known. This is a new situation for us. In all the examples considered so far, the flow analysis was conducted using Prandtl's hierarchical strategy. Remember that in this strategy the outer inviscid flow is studied first. This is done by disregarding the existence of the boundary layer and solving the Euler equations with the impermeability condition on the body surface. The solution in the inviscid flow gives a distribution of the pressure on the body surface, which is then used in the analysis of the boundary-layer flow. Here, we are dealing with the so-called *viscous-inviscid interaction*, where the inviscid flow is strongly influenced by the boundary layer.[50] Therefore, the boundary-layer equations (1.11.46) have to be solved simultaneously with the equations describing the inviscid part of the flow. The latter were deduced in Section 5.3.2 in Part 2 of this book series. By substituting ε in

---

[50]In subsequent chapters and in Part 4 of this book series, we will see that the viscous-inviscid interaction is common in fluid dynamics, and governs a variety of important phenomena.

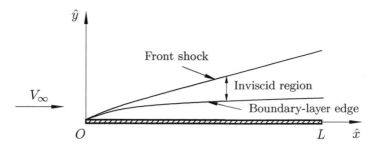

Fig. 1.44: Hypersonic boundary layer with strong interaction.

equations (5.3.30) in Part 2 with $Re_*^{-1/4}$, we express the asymptotic expansions of the fluid-dynamic functions in the inviscid flow region (see Figure 1.44) in the form

$$\left.\begin{aligned}
&\hat{u} = V_\infty + V_\infty Re_*^{-1/2} u_1(x,Y) + \cdots, \qquad \hat{v} = V_\infty Re_*^{-1/4} v_1(x,Y) + \cdots, \\
&\hat{p} = \rho_\infty V_\infty^2 Re_*^{-1/2} p_1(x,Y) + \cdots, \qquad \hat{\rho} = \rho_\infty \rho_1(x,Y) + \cdots, \\
&\hat{h} = V_\infty^2 Re_*^{-1/2} h_1(x,Y) + \cdots,
\end{aligned}\right\} \quad (1.11.48)$$

where

$$\hat{x} = Lx, \qquad \hat{y} = LRe_*^{-1/4}Y. \quad (1.11.49)$$

The substitution of (1.11.48), (1.11.49) into the Navier–Stokes equations (1.10.1) results in

$$\left.\begin{aligned}
&\rho_1\left(\frac{\partial u_1}{\partial x} + v_1 \frac{\partial u_1}{\partial Y}\right) = -\frac{\partial p_1}{\partial x}, \\
&\rho_1\left(\frac{\partial v_1}{\partial x} + v_1 \frac{\partial v_1}{\partial Y}\right) = -\frac{\partial p_1}{\partial Y}, \\
&\rho_1\left(\frac{\partial h_1}{\partial x} + v_1 \frac{\partial h_1}{\partial Y}\right) = \frac{\partial p_1}{\partial x} + v_1 \frac{\partial p_1}{\partial Y}, \\
&\frac{\partial \rho_1}{\partial x} + \frac{\partial(\rho_1 v_1)}{\partial Y} = 0, \\
&h_1 = \frac{\gamma}{\gamma-1}\frac{p_1}{\rho_1}.
\end{aligned}\right\} \quad (1.11.50)$$

To formulate the boundary conditions for equations (1.11.50), we notice that the gas density in the boundary layer is much smaller than it is in the inviscid part of the flow. This means that mass flux from the inviscid region through the outer edge of the boundary layer, $Y = \delta(x)$, is negligibly small, which allows us to use the impermeability condition[51]

$$v_1 = \frac{d\delta}{dx} \quad \text{at} \quad Y = \delta(x). \quad (1.11.51)$$

To complete the formulation of the inviscid flow problem, we also need to pose the conditions on the front shock:[52]

---

[51] For a derivation of this condition that is based on the flow analysis in the boundary layer, see Problem 2 in Exercises 8.

[52] See equations (5.3.42) on page 269 in Part 2 of this book series.

$$\left.\begin{aligned}
\rho_1\left[Y_s'(x) - v_1\right]^2 + p_1 &= \left[Y_s'(x)\right]^2 + \frac{1}{\gamma\chi^2}, \\
\frac{\gamma}{\gamma-1}\frac{p_1}{\rho_1} + \frac{1}{2}\left\{\left[Y_s'(x) - v_1\right]^2 - \left[Y_s'(x)\right]^2\right\} &= \frac{1}{(\gamma-1)\chi^2}, \\
\rho_1\left[Y_s'(x) - v_1\right] &= Y_s'(x)
\end{aligned}\right\} \quad \text{at } Y = Y_s(x). \quad (1.11.52)$$

In view of the fact that now $\varepsilon = Re_*^{-1/4}$, the hypersonic interaction parameter χ is calculated as

$$\chi = M_\infty\varepsilon = M_\infty Re_*^{-1/4}.$$

Solving the viscous-inviscid interaction problem, where the boundary-layer equations (1.11.46) are coupled with the inviscid region equations (1.11.50), is a rather difficult task. To simplify the analysis, we shall assume that $\chi \gg 1$.

Strong interaction; self-similar solution

A detailed discussion of the properties of inviscid hypersonic flows in the regime of strong interaction was given in Section 5.3.3 in Part 2 of this book series. It was shown that in the limit $\chi \to \infty$ equations (1.11.50) preserved their form and so did the impermeability boundary condition (1.11.51). As far as the shock conditions were concerned (1.11.52), they turned into

$$\left.\begin{aligned}
\rho_1\left[Y_s'(x) - v_1\right]^2 + p_1 &= \left[Y_s'(x)\right]^2, \\
\frac{\gamma}{\gamma-1}\frac{p_1}{\rho_1} + \frac{1}{2}\left\{\left[Y_s'(x) - v_1\right]^2 - \left[Y_s'(x)\right]^2\right\} &= 0, \\
\rho_1\left[Y_s'(x) - v_1\right] &= Y_s'(x)
\end{aligned}\right\} \quad \text{at } Y = Y_s(x). \quad (1.11.53)$$

The boundary-value problem (1.11.50), (1.11.51), (1.11.53) was shown to admit a family of self-similar solutions. These are applicable to situations when function $\delta(x)$, which represents the boundary-layer edge, may be expressed in the form

$$\delta(x) = bx^k. \quad (1.11.54)$$

In our analysis in Part 2, we assumed that factor b in (1.11.54) equals unity, and then we found that the pressure on contour (1.11.54) is given by

$$p_1\Big|_{Y=\delta(x)} = P_0(x) = P_b x^{2k-2}. \quad (1.11.55)$$

Here, P_b depends on k as shown in Figure 1.45; it reproduces Figure 5.14(b) of Part 2. Using invariant transformations (5.3.57) on page 272 in Part 2, this result is easily extended to arbitrary b in (1.11.54). We find that

$$P_0(x) = P_b b^2 x^{2k-2}. \quad (1.11.56)$$

Now we are prepared to turn to the boundary layer. Remember that the flow in the boundary layer is described by equations (1.11.46), which should be solved subject to

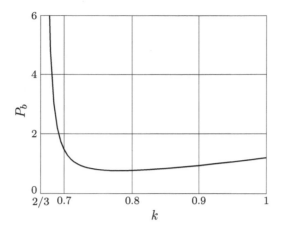

Fig. 1.45: Factor P_b in (1.11.55) as a function of parameter k; the calculations were performed for $\gamma = 1.4$.

boundary conditions (1.11.47). We shall seek their solution in the following self-similar form:

$$\left.\begin{aligned}U_0(x,Y) &= U(\eta), & V_0(x,Y) &= x^\alpha V(\eta),\\ \rho_0(x,Y) &= x^\beta \rho(\eta), & h_0(x,Y) &= h(\eta), & \mu_0(x,Y) &= \mu(\eta),\end{aligned}\right\} \tag{1.11.57}$$

where

$$\eta = \frac{Y}{x^k}.$$

The substitution of (1.11.57) and (1.11.56) into (1.11.46) results in

$$x^{\beta-1}\left(-k\eta\rho U\frac{dU}{d\eta}\right) + x^{\alpha+\beta-k}\left(\rho V\frac{dU}{d\eta}\right)$$
$$= -(2k-2)P_b b^2 x^{2k-3} + x^{-2k}\frac{d}{d\eta}\left(\mu\frac{dU}{d\eta}\right), \tag{1.11.58a}$$

$$x^{\beta-1}\left(-k\eta\rho U\frac{dh}{d\eta}\right) + x^{\alpha+\beta-k}\left(\rho V\frac{dh}{d\eta}\right)$$
$$= (2k-2)P_b b^2 x^{2k-3}U + x^{-2k}\frac{1}{Pr}\frac{d}{d\eta}\left(\mu\frac{dh}{d\eta}\right) + x^{-2k}\mu\left(\frac{dU}{d\eta}\right)^2, \tag{1.11.58b}$$

$$x^{\beta-1}\left[\beta\rho U - k\eta\frac{d}{d\eta}(\rho U)\right] + x^{\alpha+\beta-k}\frac{d}{d\eta}(\rho V) = 0, \tag{1.11.58c}$$

$$h = x^{2k-2-\beta}\frac{\gamma}{\gamma-1}\frac{P_b b^2}{\rho}. \tag{1.11.58d}$$

Since functions U, V, ρ, h, and μ depend on η only, equations (1.11.58) should not involve x. This is ensured by choosing

$$k = \frac{3}{4}, \qquad \alpha = -\frac{1}{4}, \qquad \beta = -\frac{1}{2}. \tag{1.11.59}$$

Using (1.11.59) in (1.11.57) and (1.11.58), we conclude that in the regime of strong interaction, the solution in the boundary layer has the form

$$
\left.\begin{array}{cc}
U_0(x, Y) = U(\eta), & V_0(x, Y) = x^{-1/4}V(\eta), \\
\rho_0(x, Y) = x^{-1/2}\rho(\eta), \quad h_0(x, Y) = h(\eta), & \mu_0(x, Y) = \mu(\eta), \\
\eta = \dfrac{Y}{x^{3/4}},
\end{array}\right\}
\tag{1.11.60}
$$

where functions $U(\eta)$, $V(\eta)$, $\rho(\eta)$, and $h(\eta)$ should be found by solving the following set of ordinary differential equations:

$$
-\frac{3}{4}\eta\rho U\frac{dU}{d\eta} + \rho V\frac{dU}{d\eta} = \frac{1}{2}P_b b^2 + \frac{d}{d\eta}\left(\mu\frac{dU}{d\eta}\right),
\tag{1.11.61a}
$$

$$
-\frac{3}{4}\eta\rho U\frac{dh}{d\eta} + \rho V\frac{dh}{d\eta} = -\frac{1}{2}P_b b^2 U + \frac{1}{Pr}\frac{d}{d\eta}\left(\mu\frac{dh}{d\eta}\right) + \mu\left(\frac{dU}{d\eta}\right)^2,
\tag{1.11.61b}
$$

$$
-\frac{3}{4}\eta\frac{d}{d\eta}(\rho U) - \frac{1}{2}\rho U + \frac{d}{d\eta}(\rho V) = 0,
\tag{1.11.61c}
$$

$$
h = \frac{\gamma}{\gamma - 1}\frac{P_b b^2}{\rho}.
\tag{1.11.61d}
$$

The boundary conditions for these equations

$$
U = 1, \qquad h = 0 \quad \text{at} \quad \eta = b,
\tag{1.11.62a}
$$

$$
U = V = 0, \qquad \frac{dh}{d\eta} = 0 \quad \text{at} \quad \eta = 0
\tag{1.11.62b}
$$

are deduced by substituting (1.11.60) into (1.11.47). Remember that factor P_b in (1.11.56) is determined by the inviscid flow analysis. In particular, for $k = 3/4$, it assumes the value[53]

$$
P_b = 0.7995.
$$

As far as the boundary-layer thickness parameter b is concerned, we expect it to be found as part of solution of the boundary-value problem (1.11.61), (1.11.62).

Let us introduce function $f(\eta)$ such that

$$
\frac{df}{d\eta} = \rho U, \qquad f(0) = 0,
$$

and perform the Dorodnitsyn–Howarth transformation

$$
\xi = \int_0^\eta \rho(\eta')\, d\eta'.
\tag{1.11.63}
$$

Then equations (1.11.61) turn into

[53]See equation (5.3.59) on page 273 in Part 2.

$$-\frac{1}{4}f\frac{dU}{d\xi} = \frac{P_b b^2}{2\rho} + \frac{d}{d\xi}\left(\mu\rho\frac{dU}{d\xi}\right), \qquad (1.11.64a)$$

$$-\frac{1}{4}f\frac{dh}{d\xi} = -\frac{P_b b^2}{2\rho}U + \frac{1}{Pr}\frac{d}{d\xi}\left(\mu\rho\frac{dh}{d\xi}\right) + \mu\rho\left(\frac{dU}{d\xi}\right)^2, \qquad (1.11.64b)$$

$$\frac{df}{d\xi} = U, \qquad (1.11.64c)$$

$$h = \frac{\gamma}{\gamma-1}\frac{P_b b^2}{\rho}, \qquad (1.11.64d)$$

and boundary conditions (1.11.62) assume the form

$$U = 1, \qquad h = 0 \quad \text{at} \quad \xi = \infty, \qquad (1.11.65a)$$

$$U = f = 0, \qquad \frac{dh}{d\xi} = 0 \quad \text{at} \quad \xi = 0. \qquad (1.11.65b)$$

In these variables, the boundary-layer thickness parameter b is calculated as

$$b = \int_0^\infty \frac{d\xi}{\rho(\xi)}. \qquad (1.11.66)$$

To conduct numerical solution of the boundary-value problem (1.11.64), (1.11.65) we assume that the viscosity coefficient is a linear function of temperature, which is expressed in dimensionless variables by equation (1.11.32):

$$\mu = 2h. \qquad (1.11.67)$$

It then follows from (1.11.67) and (1.11.64d) that $\mu\rho$ is a constant given by

$$\mu\rho = \frac{2\gamma}{\gamma-1}P_b b^2. \qquad (1.11.68)$$

Using (1.11.68) in the momentum (1.11.64a) and energy (1.11.64b) equations, and eliminating the density ρ with the help of the state equation (1.11.64d), we see that function U, h, and f can be found by solving the following set of ordinary differential equations:

$$-\frac{1}{4}f\frac{dU}{d\xi} = \frac{\gamma-1}{2\gamma}h + \frac{2\gamma}{\gamma-1}P_b b^2\frac{d^2U}{d\xi^2}, \qquad (1.11.69a)$$

$$-\frac{1}{4}f\frac{dh}{d\xi} = -\frac{\gamma-1}{2\gamma}hU + \frac{2\gamma}{\gamma-1}P_b b^2\frac{1}{Pr}\frac{d^2h}{d\xi^2} + \frac{2\gamma}{\gamma-1}P_b b^2\left(\frac{dU}{d\xi}\right)^2, \qquad (1.11.69b)$$

$$\frac{df}{d\xi} = U. \qquad (1.11.69c)$$

We performed the calculations using the following iterative procedure. On each iteration the momentum equation (1.11.69a) was solved for U with functions f and h, and constant b taken from the previous iteration. We used a uniform mesh:

$$\xi_j = \Delta\xi\,j,$$

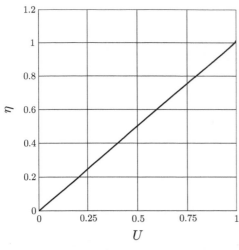

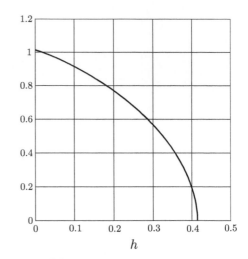

(a) Velocity profile in the boundary layer.   (b) Distribution of enthalpy.

Fig. 1.46: Results of numerical solution of the boundary-value problem (1.11.69), (1.11.65) with $Pr = 0.713$ and $\gamma = 1.4$.

and approximated the derivatives of U by the central differences:

$$\frac{dU}{d\xi} = \frac{U_{j+1} - U_{j-1}}{2\Delta\xi}, \qquad \frac{d^2U}{d\xi^2} = \frac{U_{j+1} - 2U_j + U_{j-1}}{(\Delta\xi)^2}.$$

The resulting set of the algebraic equations was solved using the Thomas method.[54] The energy equation (1.11.69b) was treated in the same way, and was used to update the distribution of the enthalpy h across the boundary layer. To find function f, we integrated equation (1.11.69c). Finally, to calculate the boundary-layer thickness parameter b, we used equation (1.11.66). To make the iterations converge, an 'under-relaxation' was employed when updating b:

$$b_{\text{new}} = r\bar{b} + (1 - r)\, b_{\text{old}},$$

The results of the calculations are shown in Figure 1.46. It is interesting that, similar to the hypersonic boundary-layer flow with weak interaction (see Figure 1.43a), the velocity U grows almost linearly with the distance η from the plate surface. The boundary-layer thickness parameter was found to be

$$b = 1.01346.$$

1.11.3   Upstream influence through hypersonic boundary layers

The self-similar solution for the hypersonic boundary layer in the regime of strong interaction, presented in the previous section, was first published by Stewartson (1955), but later, Neiland (1970b) demonstrated that this solution is not unique.[55] Neiland's

---

[54]See equations (1.9.5), (1.9.6) on pages 93, 94.

[55]See also Kozlova and Mikhailov (1970), and Ruban and Sychev (1973).

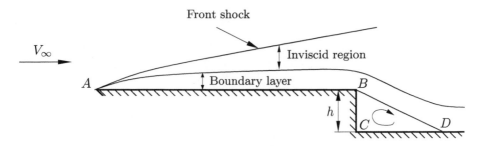

Fig. 1.47: Upstream influence in hypersonic boundary layer.

work was motivated by the development in 1969 of the theory of viscous-inviscid interaction, also known as the triple-deck theory. It was put forward simultaneously by Neiland (1969a) and Stewartson and Williams (1969) to describe the boundary-layer separation in supersonic flow. A detailed discussion of this theory is presented in Section 2.2. It describes the flow in a small $O(Re^{-3/8})$ vicinity of the separation point, where the boundary layer comes into interaction with the inviscid part of the flow, and can no longer be treated as parabolic. Instead it acquires some elements of ellipticity, which allows for perturbations to propagate against the flow (see Section 2.2.5).

To illustrate the situation with hypersonic flows, consider, for example, the hypersonic boundary layer on a flat surface AB with a backward facing step BCD; see Figure 1.47. Experimental observations show that the boundary layer separates at the corner point B, and then reattaches at point D to form a recirculation region at the base of the step. The pressure appears to be almost constant over the recirculation region, and is called the base pressure $\hat{p}_b$. Its value decreases as the step height h increases relative to the boundary-layer thickness at the separation point B. Here we shall show that the solution for the boundary layer on the flat surface AB depends on $\hat{p}_b$. Since in the regimes of moderate and strong interactions, the hypersonic boundary layer interacts with the inviscid flow on the entire surface AB, we expect the perturbations to propagate from point B through the boundary layer up to the leading edge A. To study this phenomenon, we use the following formulation of the viscous-inviscid interaction problem.

Problem formulation

We assume that the hypersonic interaction parameter $\chi = M_\infty Re_*^{-1/4}$ is large, and use for the boundary layer equations (1.11.46):

$$\rho_0 U_0 \frac{\partial U_0}{\partial x} + \rho_0 V_0 \frac{\partial U_0}{\partial Y} = -\frac{dP_0}{dx} + \frac{\partial}{\partial Y}\left(\mu_0 \frac{\partial U_0}{\partial Y}\right), \tag{1.11.70a}$$

$$\rho_0 U_0 \frac{\partial h_0}{\partial x} + \rho_0 V_0 \frac{\partial h_0}{\partial Y} = U_0 \frac{dP_0}{dx} + \frac{1}{Pr} \frac{\partial}{\partial Y}\left(\mu_0 \frac{\partial h_0}{\partial Y}\right) + \mu_0 \left(\frac{\partial U_0}{\partial Y}\right)^2, \tag{1.11.70b}$$

$$\frac{\partial(\rho_0 U_0)}{\partial x} + \frac{\partial(\rho_0 V_0)}{\partial Y} = 0, \tag{1.11.70c}$$

$$h_0 = \frac{\gamma}{\gamma - 1} \frac{P_0}{\rho_0}. \tag{1.11.70d}$$

Here, the pressure $P_0(x)$ is not know in advance, and must be found by solving equations (1.11.50) for the inviscid flow region; see Figure 1.44 on page 123. To simplify the analysis, we approximate the solution of (1.11.50) by the *tangent-wedge formula*:

$$P_0 = \varkappa \left(\frac{d\delta}{dx} \right)^2, \qquad (1.11.70e)$$

where we choose the factor $\varkappa$ to be $\varkappa = 1.421$ (see Problem 4b in Exercises 8).

We assume that surface AB is thermally isolated, which allows us to use boundary conditions (1.11.47):

$$U_0 = 1, \qquad h_0 = 0 \quad \text{at} \quad Y = \delta(x), \qquad (1.11.70f)$$

$$U_0 = V_0 = 0, \quad \frac{\partial h_0}{\partial Y} = 0 \quad \text{at} \quad Y = 0. \qquad (1.11.70g)$$

Dorodnitsyn–Howarth transformation

Unlike in Section 1.10.2, we do not expect the solution to be self-similar. Therefore, we apply the Dorodnitsyn–Howarth transformation directly to the viscous-inviscid problem (1.11.70). Instead of x and Y, we introduce new independent variables $s = s(x, Y)$ and $\xi = \xi(x, Y)$ defined through the equations

$$s(x, Y) = x, \qquad \xi(x, Y) = \int\limits_0^Y \rho_0(x, Y')\, dY'. \qquad (1.11.71)$$

Function $\tilde{U}(s, \xi)$, which represents the longitudinal velocity component in the new variables, is defined as

$$U_0(x, Y) = \tilde{U}\big[s(x, Y), \xi(x, Y)\big]. \qquad (1.11.72)$$

If we differentiate (1.11.72) using the chain rule, then we will have

$$\frac{\partial U_0}{\partial x} = \frac{\partial \tilde{U}}{\partial s}\frac{\partial s}{\partial x} + \frac{\partial \tilde{U}}{\partial \xi}\frac{\partial \xi}{\partial x} = \frac{\partial \tilde{U}}{\partial s} + \frac{\partial \xi}{\partial x}\frac{\partial \tilde{U}}{\partial \xi}, \qquad (1.11.73)$$

$$\frac{\partial U_0}{\partial Y} = \frac{\partial \tilde{U}}{\partial s}\frac{\partial s}{\partial Y} + \frac{\partial \tilde{U}}{\partial \xi}\frac{\partial \xi}{\partial Y} = \rho_0 \frac{\partial \tilde{U}}{\partial \xi}. \qquad (1.11.74)$$

Let us start with the continuity equation (1.11.70c). It may be expressed in the form

$$\rho_0 \frac{\partial U_0}{\partial x} + U_0 \frac{\partial \rho_0}{\partial x} + \frac{\partial (\rho_0 V_0)}{\partial Y} = 0.$$

Taking into account that $\rho_0 = \partial \xi / \partial Y$, we can write

$$\rho_0 \frac{\partial U_0}{\partial x} + U_0 \frac{\partial^2 \xi}{\partial x \partial Y} + \frac{\partial (\rho_0 V_0)}{\partial Y} = 0,$$

or, equivalently,

$$\rho_0 \frac{\partial U_0}{\partial x} - \frac{\partial \xi}{\partial x}\frac{\partial U_0}{\partial Y} + \frac{\partial}{\partial Y}\left(\rho_0 V_0 + U_0 \frac{\partial \xi}{\partial x} \right) = 0. \qquad (1.11.75)$$

Now, if we substitute $\partial U_0/\partial x$ and $\partial U_0/\partial Y$ in (1.11.75) by their expressions given by (1.11.73) and (1.11.74), then we will have

$$\rho_0 \frac{\partial \widetilde{U}}{\partial s} + \frac{\partial}{\partial Y}\left(\rho_0 V_0 + U_0 \frac{\partial \xi}{\partial x}\right) = 0. \tag{1.11.76}$$

This suggests that the lateral velocity component $\widetilde{V}$ should be defined as

$$\rho_0 V_0 + U_0 \frac{\partial \xi}{\partial x} = \widetilde{V}\big[s(x,Y), \xi(x,Y)\big]. \tag{1.11.77}$$

The differentiation of (1.11.77) with respect to Y results in

$$\frac{\partial}{\partial Y}\left(\rho_0 V_0 + U_0 \frac{\partial \xi}{\partial x}\right) = \rho_0 \frac{\partial \widetilde{V}}{\partial \xi}. \tag{1.11.78}$$

It remains to substitute (1.11.78) into (1.11.76) and we arrive at a conclusion that, in the new variables, the continuity equation takes the 'incompressible form':

$$\frac{\partial \widetilde{U}}{\partial s} + \frac{\partial \widetilde{V}}{\partial \xi} = 0. \tag{1.11.79}$$

Let us now turn to the momentum equation (1.11.70a). In addition to the velocity components that are defined in the new variables by equations (1.11.72) and (1.11.77), we shall write

$$\left.\begin{array}{ll} \rho_0(x,Y) = \tilde{\rho}\big[s(x,Y), \xi(x,Y)\big], & P_0(x) = \widetilde{P}\big[s(x,Y), \xi(x,Y)\big], \\[2mm] h_0(x,Y) = \tilde{h}\big[s(x,Y), \xi(x,Y)\big], & \mu_0(x,Y) = \tilde{\mu}\big[s(x,Y), \xi(x,Y)\big]. \end{array}\right\} \tag{1.11.80}$$

We start by differentiating P_0 in (1.11.80). We have

$$\frac{\partial P_0}{\partial x} = \frac{\partial \widetilde{P}}{\partial s} + \frac{\partial \widetilde{P}}{\partial \xi}\frac{\partial \xi}{\partial x}, \qquad \frac{\partial P_0}{\partial Y} = \rho_0 \frac{\partial \widetilde{P}}{\partial \xi}. \tag{1.11.81}$$

Since P_0 is independent of Y, it follows from the second equation in (1.11.81) that $\partial \widetilde{P}/\partial \xi$ is zero, and the first equation reduces to

$$\frac{\partial P_0}{\partial x} = \frac{\partial \widetilde{P}}{\partial s}. \tag{1.11.82}$$

Substitution of (1.11.72)–(1.11.74), (1.11.77), (1.11.80), and (1.11.82) into (1.11.70a) results in

$$\widetilde{U}\frac{\partial \widetilde{U}}{\partial s} + \widetilde{V}\frac{\partial \widetilde{U}}{\partial \xi} = -\frac{1}{\tilde{\rho}}\frac{d\widetilde{P}}{ds} + \frac{\partial}{\partial \xi}\left(\tilde{\mu}\tilde{\rho}\frac{\partial \widetilde{U}}{\partial \xi}\right). \tag{1.11.83}$$

The energy equation (1.11.70b) is transformed in the same way leading to

$$\widetilde{U}\frac{\partial \tilde{h}}{\partial s} + \widetilde{V}\frac{\partial \tilde{h}}{\partial \xi} = \frac{\widetilde{U}}{\tilde{\rho}}\frac{d\widetilde{P}}{ds} + \frac{1}{Pr}\frac{\partial}{\partial \xi}\left(\tilde{\mu}\tilde{\rho}\frac{\partial \tilde{h}}{\partial \xi}\right) + \tilde{\mu}\tilde{\rho}\left(\frac{\partial \widetilde{U}}{\partial \xi}\right)^2. \tag{1.11.84}$$

Obviously, the state equation (1.11.70d) remains unchanged

$$\tilde{h} = \frac{\gamma}{\gamma - 1} \frac{\tilde{P}}{\tilde{\rho}},$$

(1.11.85)

while the interaction law (1.11.70e) assumes the form

$$\tilde{P} = \varkappa \left(\frac{d\delta}{ds}\right)^2.$$

(1.11.86)

Finally, we need to consider boundary conditions (1.11.70f) and (1.11.70g). It follows from the second equation in (1.11.71) that $\xi = 0$ at $Y = 0$, which allows us to write (1.11.70g) as

$$\tilde{U} = \tilde{V} = 0, \quad \frac{\partial \tilde{h}}{\partial \xi} = 0 \quad \text{at} \quad \xi = 0.$$

(1.11.87)

Also, since ρ_0 tends to infinity when the outer edge of the boundary layer is approached, we expect that $\xi \to \infty$ as $Y \to \delta$, which means that the boundary conditions (1.11.70f) should be written as

$$\tilde{U} = 1, \quad \tilde{h} = 0 \quad \text{at} \quad \xi = \infty.$$

(1.11.88)

It remains to deduce an equation that would allow us to calculate the boundary-layer thickness $\delta(s)$. For this purpose, we consider a pair of functions

$$x = x(s, \xi), \qquad Y = Y(s, \xi),$$

(1.11.89)

that are inverse to functions (1.11.71). We write the latter as

$$s = s(x, Y), \qquad \xi = \xi(x, Y).$$

(1.11.90)

By definition of inverse functions, the substitution of (1.11.90) into (1.11.89) should lead to the identities

$$x \equiv x\big[s(x, Y), \xi(x, Y)\big], \qquad Y \equiv Y\big[s(x, Y), \xi(x, Y)\big].$$

(1.11.91)

The differentiation of the second equation in (1.11.91) with respect to Y results in

$$1 = \frac{\partial Y}{\partial s} \frac{\partial s}{\partial Y} + \frac{\partial Y}{\partial \xi} \frac{\partial \xi}{\partial Y}.$$

(1.11.92)

It follows from (1.11.71) that $\partial s/\partial Y = 0$ and $\partial \xi/\partial Y = \rho_0$, which reduces (1.11.92) to

$$1 = \tilde{\rho} \frac{\partial Y}{\partial \xi}.$$

We write this equation as

$$\frac{\partial Y}{\partial \xi} = \frac{1}{\tilde{\rho}},$$

and integrate it with respect to ξ, taking into account that $Y = 0$ at $\xi = 0$. We have

$$Y(s,\xi) = \int_0^\xi \frac{d\xi'}{\tilde{\rho}(s,\xi')}. \tag{1.11.93}$$

Setting $\xi \to \infty$ in (1.11.93) gives the boundary-layer thickness

$$\delta(s) = \int_0^\infty \frac{d\xi}{\tilde{\rho}(s,\xi)}. \tag{1.11.94}$$

Thus, in Dorodnitsyn–Howarth variables, the viscous-inviscid interaction problem is written as follows. The boundary-layer equations have the form:

$$\tilde{U}\frac{\partial \tilde{U}}{\partial s} + \tilde{V}\frac{\partial \tilde{U}}{\partial \xi} = -\frac{1}{\tilde{\rho}}\frac{d\tilde{P}}{ds} + \frac{\partial}{\partial \xi}\left(\tilde{\mu}\tilde{\rho}\frac{\partial \tilde{U}}{\partial \xi}\right), \tag{1.11.95a}$$

$$\tilde{U}\frac{\partial \tilde{h}}{\partial s} + \tilde{V}\frac{\partial \tilde{h}}{\partial \xi} = \frac{\tilde{U}}{\tilde{\rho}}\frac{d\tilde{P}}{ds} + \frac{1}{Pr}\frac{\partial}{\partial \xi}\left(\tilde{\mu}\tilde{\rho}\frac{\partial \tilde{h}}{\partial \xi}\right) + \tilde{\mu}\tilde{\rho}\left(\frac{\partial \tilde{U}}{\partial \xi}\right)^2, \tag{1.11.95b}$$

$$\frac{\partial \tilde{U}}{\partial s} + \frac{\partial \tilde{V}}{\partial \xi} = 0, \tag{1.11.95c}$$

$$\tilde{h} = \frac{\gamma}{\gamma-1}\frac{\tilde{P}}{\tilde{\rho}}. \tag{1.11.95d}$$

The pressure $\tilde{P}$ is found using the tangent-wedge formula:

$$\tilde{P} = \varkappa\left(\frac{d\delta}{ds}\right)^2, \tag{1.11.95e}$$

where the boundary-layer thickness $\delta(s)$ is calculated as

$$\delta(s) = \int_0^\infty \frac{d\xi}{\tilde{\rho}(s,\xi)}. \tag{1.11.95f}$$

The boundary conditions

$$\tilde{U} = 1, \qquad \tilde{h} = 0 \quad \text{at} \quad \xi = \infty, \tag{1.11.95g}$$

$$\tilde{U} = \tilde{V} = 0, \quad \frac{\partial \tilde{h}}{\partial \xi} = 0 \quad \text{at} \quad \xi = 0 \tag{1.11.95h}$$

close the formulation of the interaction problem.

Non-uniqueness of the solution

To simplify the analysis of the interaction problem (1.11.95), we assume that the Prandtl number $Pr = 1$, in which case the energy equation (1.11.95b) may be substituted by Crocco's integral (1.10.43). The hypersonic version of (1.10.43) is written as

$$\tilde{h} = \frac{1}{2} - \frac{\tilde{U}^2}{2}. \tag{1.11.96}$$

We further assume that the viscosity coefficient is a linear function of temperature, that is, equation (1.11.32) is valid. It is written in the variables used here as

$$\tilde{\mu} = 2\tilde{h}. \tag{1.11.97}$$

By combining (1.11.96) and (1.11.97) with the state equation, we find that

$$\frac{1}{\tilde{\rho}} = \frac{\gamma - 1}{2\gamma \tilde{P}}(1 - \tilde{U}^2), \qquad \tilde{\mu}\tilde{\rho} = \frac{2\gamma}{\gamma - 1}\tilde{P}. \tag{1.11.98}$$

The continuity equation (1.11.95c) allows us to introduce the stream function $\tilde{\Psi}(s, \xi)$ such that

$$\tilde{U} = \frac{\partial \tilde{\Psi}}{\partial \xi}, \qquad \tilde{V} = -\frac{\partial \tilde{\Psi}}{\partial s}. \tag{1.11.99}$$

The substitution of (1.11.98) and (1.11.99) into the momentum equation (1.11.95a) turns it into

$$\frac{\partial \tilde{\Psi}}{\partial \xi}\frac{\partial^2 \tilde{\Psi}}{\partial s \partial \xi} - \frac{\partial \tilde{\Psi}}{\partial s}\frac{\partial^2 \tilde{\Psi}}{\partial \xi^2} = -\frac{\gamma - 1}{2\gamma \tilde{P}}\left[1 - \left(\frac{\partial \tilde{\Psi}}{\partial \xi}\right)^2\right]\frac{d\tilde{P}}{ds} + \frac{2\gamma}{\gamma - 1}\tilde{P}\frac{\partial^3 \tilde{\Psi}}{\partial \xi^3}. \tag{1.11.100a}$$

Equation (1.11.100a) involves two unknown functions $\tilde{\Psi}$ and $\tilde{P}$, and should be solved together with the interaction law (1.11.95e):

$$\tilde{P} = \varkappa \left(\frac{d\delta}{ds}\right)^2. \tag{1.11.100b}$$

The substitution of the first equation in (1.11.98) into (1.11.95f) allows us to express the boundary-layer thickness $\delta(s)$ in terms of $\tilde{\Psi}$ and $\tilde{P}$:

$$\delta = \frac{\gamma - 1}{2\gamma \tilde{P}}\int_0^\infty \left[1 - \left(\frac{\partial \tilde{\Psi}}{\partial \xi}\right)^2\right]d\xi. \tag{1.11.100c}$$

To close the boundary-value problem (1.11.100a)–(1.11.100c), we reformulate the boundary conditions for $\tilde{U}$ and $\tilde{V}$ (1.11.95g), (1.11.95h) in terms of the stream function:

$$\frac{\partial \tilde{\Psi}}{\partial \xi} = 1 \qquad \text{at} \quad \xi = \infty, \tag{1.11.100d}$$

$$\tilde{\Psi} = \frac{\partial \tilde{\Psi}}{\partial \xi} = 0 \quad \text{at} \quad \xi = 0. \tag{1.11.100e}$$

The non-uniqueness of the solution of the boundary-value problem (1.11.100) may be demonstrated by analysing the flow in the vicinity of the leading edge of the plate. We seek functions $\widetilde{\Psi}$, $\widetilde{P}$, and $\widetilde{\delta}$ in this vicinity in the form of asymptotic expansions

$$
\left.
\begin{aligned}
\widetilde{\Psi}(s,\xi) &= s^{1/4} f_0(\zeta) + s^{1/4+\lambda} f_1(\zeta) + \cdots, \\
\widetilde{P}(s) &= s^{-1/2} P_0 + s^{-1/2+\lambda} P_1 + \cdots, \\
\widetilde{\delta}(s) &= s^{3/4} \delta_0 + s^{3/4+\lambda} \delta_1 + \cdots
\end{aligned}
\right\}
\quad \text{as} \quad s \to 0,
\qquad (1.11.101)
$$

with

$$
\zeta = \xi/s^{1/4} = O(1).
$$

The leading-order terms in (1.11.101) represent Stewartson's (1955) self-similar solution (1.11.60) that has been studied in the previous section.[56] Here, we concentrate on the perturbation terms in (1.11.101). Our task is to prove that there exists a positive value of λ for which $f_1(\zeta)$, P_1, and δ_1 are non-zero.

By substituting (1.11.101) into (1.11.100) and working with the leading-order terms, we find that

$$
\left.
\begin{aligned}
&\frac{2\gamma}{\gamma-1} P_0 f_0''' + \frac{1}{4} f_0 f_0'' + \frac{\gamma-1}{4\gamma}\left(1 - f_0'^2\right) = 0, \\
&P_0 = \frac{9}{16}\varkappa\delta_0^2, \qquad \delta_0 = \frac{\gamma-1}{2\gamma}\frac{1}{P_0}\int_0^\infty \left(1 - f_0'^2\right) d\zeta, \\
&f_0(0) = f_0'(0) = 0, \qquad f_0'(\infty) = 1.
\end{aligned}
\right\}
\qquad (1.11.102)
$$

The results of the numerical solution of (1.11.102) are displayed in Figure 1.48, where the longitudinal velocity component f_0' is plotted as a function of the similarity

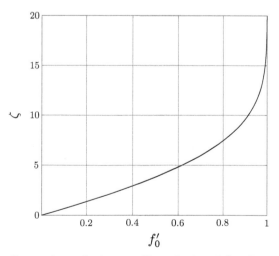

Fig. 1.48: Leading-order velocity profile calculated for $Pr = 1$ and $\gamma = 1.4$.

---

[56] We have to recalculate this solution taking into account that now the Prandtl number $Pr = 1$.

variable ζ. The calculations also yield

$$\delta_0 = 1.052, \qquad P_0 = 0.8847.$$

Now, we turn our attention to the perturbation terms. Substituting (1.11.101) into the momentum equation (1.11.100a) and working with the $O(s^{-1-\lambda})$ terms, we find that

$$\frac{2\gamma}{\gamma - 1} P_0 f_1''' + \frac{1}{4} f_0 f_1'' - \left(\lambda + \frac{\gamma - 1}{2\gamma}\right) f_0' f_1' + \left(\lambda + \frac{1}{4}\right) f_0'' f_1$$
$$= \left(\frac{\gamma - 1}{2\gamma} \lambda \frac{1 - f_0'^2}{P_0} - \frac{2\gamma}{\gamma - 1} f_0'''\right) P_1. \tag{1.11.103}$$

The interaction law (1.11.100b) gives

$$P_1 = \frac{3}{2}\left(\lambda + \frac{3}{4}\right) \varkappa \delta_0 \delta_1, \tag{1.11.104}$$

and it follows from (1.11.100c) that

$$\delta_1 = -\frac{\gamma - 1}{2\gamma} \frac{1}{P_0} \left[2 \int_0^\infty f_0' f_1' \, d\zeta + \frac{P_1}{P_0} \int_0^\infty (1 - f_0'^2) \, d\zeta\right]. \tag{1.11.105}$$

The boundary conditions

$$f_1(0) = f_1'(0) = f_1'(\infty) = 0 \tag{1.11.106}$$

are obtained by substituting the asymptotic expansion for $\widetilde{\Psi}$ in (1.11.101) into (1.11.100d) and (1.11.100e).

We can see that the boundary-value problem (1.11.103)–(1.11.106) admits a trivial solution

$$f_1 = 0, \quad P_1 = 0, \quad \delta_1 = 0.$$

Hence, it may be treated as an eigenvalue problem. When solving this problem, it is convenient to eliminate δ_1 from (1.11.104) and (1.11.105). Substituting (1.11.105) into (1.11.104) and solving the resulting equation for P_1, we have

$$P_1 = \Lambda \int_0^\infty f_0' f_1' \, d\zeta, \tag{1.11.107}$$

where

$$\Lambda = -\frac{\gamma - 1}{\gamma P_0} \left[\frac{8}{3(3 + 4\lambda)\varkappa\delta_0} + \frac{\gamma - 1}{2\gamma P_0^2} \int_0^\infty (1 - f_0'^2) \, d\zeta\right]^{-1}.$$

Now we write

$$f_1(\zeta) = P_1 \bar{f}_1(\zeta),$$

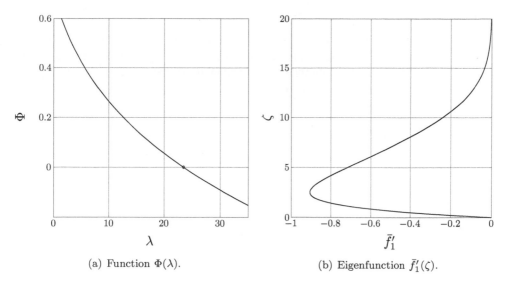

(a) Function $\Phi(\lambda)$.

(b) Eigenfunction $\bar{f}_1'(\zeta)$.

Fig. 1.49: Function $\Phi(\lambda)$ and the solution of the boundary-value problem (1.11.108) for $\lambda = \lambda_*$ and $\gamma = 1.4$.

which turns equation (1.11.103) and boundary conditions (1.11.106) into

$$
\left.
\begin{aligned}
\frac{2\gamma}{\gamma-1}P_0\bar{f}_1''' + \frac{1}{4}f_0\bar{f}_1'' - \left(\lambda+\frac{\gamma-1}{2\gamma}\right)f_0'\bar{f}_1' + \left(\lambda+\frac{1}{4}\right)f_0''\bar{f}_1 \\
= \left(\frac{\gamma-1}{2\gamma}\lambda\frac{1-f_0'^2}{P_0} - \frac{2\gamma}{\gamma-1}f_0'''\right), \\
\bar{f}_1(0) = \bar{f}_1'(0) = \bar{f}_1'(\infty) = 0,
\end{aligned}
\right\}
\tag{1.11.108}
$$

while equation (1.11.107) assumes the form

$$
\Phi(\lambda) = \Lambda \int\limits_0^\infty f_0'\bar{f}_1\, d\zeta - 1 = 0.
\tag{1.11.109}
$$

Given λ, function $\bar{f}_1(\zeta)$ is found by numerical solution of (1.11.108), and then function $\Phi(\lambda)$ can be calculated. We repeated this procedure for a set of values of λ, and the results of the calculations are displayed in Figure 1.49(a). Equation (1.11.109) is satisfied at the point shown by a circle, where

$$
\lambda = \lambda_* = 23.508.
$$

Figure 1.49(b) also shows the solution of the boundary-value problem (1.11.108) for $\lambda = \lambda_*$ which is an eigensolution of problem (1.11.103)–(1.11.106).

To summarize, remember that parameter P_1 remains arbitrary from the viewpoint of the 'local' analysis of the flow near the leading edge A of the plate; see Figure 1.47. Obviously, P_1 depends on an additional boundary condition that should be imposed on the solution of (1.11.95) at point B.

Exercises 8

1. Consider a hypersonic boundary layer on a flat plate surface in the regime of weak interaction. Your task is to study the flow behaviour near the outer edge of the boundary layer. Perform this task in the following steps:

   (a) Show that the affine transformations

   $$f = A\bar{f}, \qquad \xi = B\bar{\xi},$$

   with suitably chosen factors A and B, allow us to reduce the weak-interaction hypersonic boundary-layer problem (1.11.34a), (1.11.34b):

   $$\frac{d^3 f}{d\xi^3} + \frac{\gamma - 1}{4} f \frac{d^2 f}{d\xi^2} = 0,$$

   $$\frac{df}{d\xi} = f = 0 \quad \text{at} \quad \xi = 0,$$

   $$\frac{df}{d\xi} = 1 \quad \text{at} \quad \xi = \infty$$

   to the classical Blasius problem (1.1.47a), (1.1.47b) for the incompressible boundary layer on a flat plate surface:

   $$\frac{d^3 \bar{f}}{d\bar{\xi}^3} + \frac{1}{2} \bar{f} \frac{d^2 \bar{f}}{d\bar{\xi}^2} = 0,$$

   $$\frac{d\bar{f}}{d\bar{\xi}} = \bar{f} = 0 \quad \text{at} \quad \bar{\xi} = 0,$$

   $$\frac{d\bar{f}}{d\bar{\xi}} = 1 \quad \text{at} \quad \bar{\xi} = \infty.$$

   (b) Show that near the outer edge of the boundary layer

   $$f(\xi) = \xi - \tilde{A} + \frac{\tilde{C}}{(\xi - \tilde{A})^2} e^{-(\gamma-1)(\xi - \tilde{A})^2/8} + \cdots \quad \text{as} \quad \xi \to \infty, \quad (1.11.110)$$

   where

   $$\tilde{C} = C \left(\frac{2}{\gamma - 1} \right)^{2/3}, \qquad \tilde{A} = A_- \sqrt{\frac{2}{\gamma - 1}}, \qquad A_- = 1.7208.$$

   Suggestion: You may use without proof equation (1.1.53) that is written in the variables used here as

   $$\bar{f} = \bar{\xi} - A_- + \frac{C}{(\bar{\xi} - A_-)^2} e^{-(\bar{\xi} - A_-)^2/4} + \cdots \quad \text{as} \quad \bar{\xi} \to \infty.$$

(c) Assume that the Prandtl number $Pr = 1$ and the plate surface is thermally isolated. By direct substitution into the energy equation (1.11.35a)

$$\frac{1}{Pr}\frac{d^2h}{d\xi^2} + \frac{\gamma - 1}{4}f\frac{dh}{d\xi} + \left(\frac{d^2f}{d\xi^2}\right)^2 = 0,$$

and boundary conditions (1.11.35b)

$$\frac{dh}{d\xi} = 0 \quad \text{at} \quad \xi = 0,$$
$$h = 0 \quad \text{at} \quad \xi = \infty,$$

or otherwise, show that everywhere inside the boundary layer

$$h = \frac{1}{2} - \frac{1}{2}\left(\frac{df}{d\xi}\right)^2. \tag{1.11.111}$$

(d) Substitute (1.11.110) into (1.11.111) and make use of the state equation (1.11.26d)

$$h = \frac{1}{(\gamma - 1)\rho}, \tag{1.11.112}$$

to show that near the outer edge of the boundary layer

$$\rho = \frac{4}{\widetilde{C}(\gamma - 1)^2}\left(\xi - \widetilde{A}\right)e^{(\gamma - 1)(\xi - \widetilde{A})^2/8} + \cdots \quad \text{as} \quad \xi \to \infty. \tag{1.11.113}$$

Hence, conclude that the integral (1.11.37) does converge to give a finite value of the boundary-layer thickness parameter

$$d = \int\limits_{0}^{\infty} \frac{d\xi}{\rho(\xi)}.$$

Also substitute (1.11.113) into the inverse Dorodnitsyn–Howarth transformation (1.11.36)

$$\eta = \int\limits_{0}^{\xi} \frac{d\xi'}{\rho(\xi')},$$

and using the integration-by-parts technique,[57] show that

$$\eta = d - \frac{\widetilde{C}(\gamma - 1)}{\left(\xi - \widetilde{A}\right)^2}e^{-(\gamma - 1)(\xi - \widetilde{A})^2/8} + \cdots \quad \text{as} \quad \xi \to \infty. \tag{1.11.114}$$

Substitute (1.11.113) into (1.11.112), and using (1.11.114), show that near the outer edge of the boundary layer

[57]See Section 1.1.2 in Part 2 of this book series

$$h = \frac{1}{\sqrt{2(\gamma-1)}}(d-\eta)\sqrt{|\ln(d-\eta)|} + \cdots \quad \text{as} \quad \eta \to d. \qquad (1.11.115)$$

Show further that the longitudinal velocity U behaves as

$$U = \frac{df}{d\xi} = 1 - \frac{1}{\sqrt{2(\gamma-1)}}(d-\eta)\sqrt{|\ln(d-\eta)|} + \cdots \quad \text{as} \quad \eta \to d.$$

(e) Remember that in the boundary layer the dimensional enthalpy is given by[58]

$$\hat{h} = V_\infty^2 h. \qquad (1.11.116)$$

Cast equation (1.11.116) in the form

$$\hat{h} = h_\infty(\gamma-1)M_\infty^2 h$$

to show that in the boundary layer the enthalpy is M_∞^2 times larger than outside the boundary layer. Use (1.11.115) to show that near the outer edge of the boundary layer there is a thinner layer where $\hat{h} \sim h_\infty$. Estimate the thickness of this layer.

2. Consider a hypersonic boundary layer on a flat plate surface in the regime of weak interaction. You can use without proof the fact that in this flow the continuity equation is expressed in the similarity variables (1.11.25) in the form of equation (1.11.26c), that is

$$-\frac{1}{2}\eta\frac{d}{d\eta}(\rho U) + \frac{d}{d\eta}(\rho V) = 0.$$

You need to perform the following tasks:

(a) Introduce function $f(\eta)$ such that

$$\frac{df}{d\eta} = \rho U, \qquad f(0) = 0,$$

and show that the velocity component normal to the plate surface can be expressed in the form

$$V = \frac{1}{2}\eta U - \frac{f}{2\rho}. \qquad (1.11.117)$$

(b) By substituting (1.11.110), (1.11.113), and (1.11.114) into (1.11.117), prove that the outer edge of the boundary layer, given by the equation

$$\delta(x) = d\sqrt{x},$$

represents a streamline.

---

[58]See equations (1.11.21) and (1.11.25).

3. Consider a hypersonic boundary layer on a flat plate surface in the regime of weak interaction again, but this time assume that the plate temperature is much smaller than the stagnation temperature T_* defined by equation (1.11.9). Assume further that the viscosity coefficient is a power function of the enthalpy, that is

$$\hat{\mu} = C\hat{h}^n.$$

Show that in this case the solution of the boundary-layer equations (1.11.22) displays the following behaviour at the bottom of the boundary layer:

$$\left.\begin{array}{l} U_0 = F(x)Y^{1/(n+1)} + \cdots, \\ h_0 = G(x)Y^{1/(n+1)} + \cdots \end{array}\right\} \quad \text{as} \quad Y \to 0.$$

Hint: Notice that for small values of Y the momentum (1.11.22a) and energy (1.11.22b) equations reduce to

$$\frac{\partial}{\partial Y}\left(\mu_0 \frac{\partial U_0}{\partial Y}\right) = 0, \qquad \frac{\partial}{\partial Y}\left(\mu_0 \frac{\partial h_0}{\partial Y}\right) = 0.$$

4. When dealing with inviscid hypersonic flow past a thin body in the regime of strong interaction, the *tangent-wedge formula*

$$\hat{p} = \rho_\infty V_\infty^2 \varkappa \left(\frac{d\hat{y}_b}{d\hat{x}}\right)^2 \qquad (1.11.118)$$

is often used to calculate the pressure $\hat{p}$ on the body surface. The equation of the body contour is written in dimensional variables as $\hat{y} = \hat{y}_b(\hat{x})$. Your task is to calculate factor $\varkappa$ in (1.11.118).

(a) First assume that the body placed in the flow is a wedge with small angle θ at the apex; see Figure 1.50. Applying the mass conservation law to the control volume $ABCDE$, show that the shock angle

$$\alpha = \frac{\gamma+1}{2}\theta.$$

Then show that behind the shock

$$\hat{p} = \rho_\infty V_\infty^2 \frac{\gamma+1}{2}\theta^2.$$

Hence, conclude that $\varkappa = (\gamma+1)/2$.

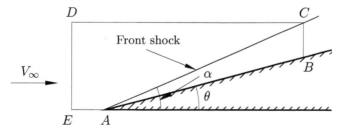

Fig. 1.50: Hypersonic flow past a wedge.

Hint: You may use without proof equations (4.5.32) and (4.5.33) from Part 1 of this book series. These give the density $\hat{\rho}$ and the pressure $\hat{p}$ behind an oblique shock:

$$\frac{\hat{\rho}}{\rho_\infty} = \frac{(\gamma+1)M_\infty^2 \sin^2\alpha}{2+(\gamma-1)M_\infty^2 \sin^2\alpha}, \qquad \frac{\hat{p}}{p_\infty} = \frac{2\gamma M_\infty^2 \sin^2\alpha - (\gamma-1)}{\gamma+1}.$$

Remember that in the regime of strong interaction $M_\infty \sin\alpha \gg 1$.

(b) When dealing a hypersonic boundary-layer flow with strong interaction, such as Stewartson's flow (see Section 1.11.2), it makes sense to calculate factor $\varkappa$ in (1.11.118) using the self-similar solution of the hypersonic inviscid flow theory. To perform this task, return to equations (1.11.48), (1.11.49) that are valid in the inviscid flow outside the boundary layer. Express the outer edge of the boundary layer by the equation

$$\hat{y} = L\varepsilon\delta(x), \qquad x = \hat{x}/L, \tag{1.11.119}$$

where $\varepsilon = Re_*^{-1/4}$, and write the asymptotic expansion for the pressure in the form

$$\hat{p} = \rho_\infty V_\infty^2 \varepsilon^2 p_1(x,Y) + \cdots . \tag{1.11.120}$$

You may further use without derivation the following result of the inviscid flow analysis.[59] If $\delta(x)$ is a power function, that is

$$\delta(x) = bx^k, \tag{1.11.121}$$

then at the outer edge of the boundary layer, the dimensionless pressure p_1 is given by equation (1.11.56):

$$p_1\big|_{Y=\delta(x)} = P_0(x) = P_b b^2 x^{2k-2}. \tag{1.11.122}$$

Substitute (1.11.121) into (1.11.119) and then into the tangent-wedge formula (1.11.118). Compare the resulting equation for $\hat{p}$ with the equation that is obtained by substituting (1.11.122) into (1.11.120), and deduce that

$$\varkappa = \frac{P_b}{k^2}.$$

Finally, calculate the numerical value for $\varkappa$, taking into account that in the Stewartson's (1955) solution, $k = 3/4$ and $P_b = 0.7995$.

---

[59]For a detailed discussion of this result, the reader is referred to Section 5.3.3 in Part 2 of this book series.

2
Boundary-Layer Separation

2.1 Experimental Evidence

Separation is a fluid-dynamic phenomenon that is observed in a wide variety of fluid flows. In fact, to avoid separation, rather strict restrictions have to be imposed of the shape of the body placed in flow. The difference between an attached flow and its separated counterpart is demonstrated in Figures 2.1 and 2.2 using as an example the flow of an incompressible fluid past a circular cylinder. A well-known solution of the Euler equations for this flow is written in terms of the complex potential $w(z)$ as[1]

$$w(z) = V_\infty \left(z + \frac{a^2}{z} \right). \tag{2.1.1}$$

Here, V_∞ is the free-stream velocity, a is the cylinder radius, and $z = x + iy$, with x and y being Cartesian coordinates. Figure 2.1 shows the streamline pattern plotted using solution (2.1.1). It is contrasted in Figure 2.2 with an experimental visualization of the flow. Clearly, there is a significant difference between the theory and the real flow. In the theoretically predicted form of the flow, the fluid particles follow closely the cylinder surface from the front stagnation point all the way to the rear stagnation point. In the experimentally observed flow, the fluid breaks away from the cylinder surface to form recirculating eddies in the wake behind the cylinder.

In addition to distorting of the streamline pattern, the separation also leads to a significant alteration of the fluid-dynamic forces acting on the body. In particular, in the flow past an aerofoil, the separation leads to a decrease of the lift force and a significant increase of the drag.

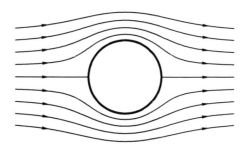

Fig. 2.1: Theoretical predictions based on the potential flow solution (2.1.1).

[1]See equation (3.4.28) on page 166 in Part 1 of this book series.

Fluid Dynamics: Part 3: Boundary Layers. © Anatoly I. Ruban, 2018. Published 2018 by Oxford University Press. 10.1093/oso/9780199681754.001.0001

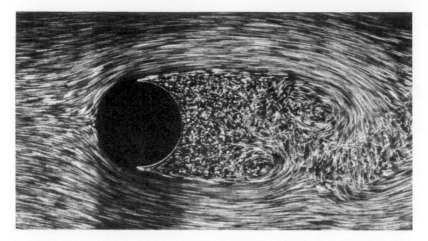

Fig. 2.2: Visualization of the flow past a circular cylinder by Werlé and Gallon (1972); $Re = 2000$.

Of course, solution (2.1.1) was obtained based on the inviscid flow assumption, and one might think that the attached form of the flow (Figure 2.1), would be recovered in the limit as the fluid viscosity tends to zero. However, this does not happen. Instead, the separation region extends further downstream as the Reynolds number increases. This behaviour is illustrated in Figure 2.3, where the results of a numerical solution of the Navier–Stokes equations are displayed.

The described 'contradiction' is resolved by the following important property of the Euler equations. In addition to attached flow solutions, the Euler equations also allow for solutions with separation. The first separated solution was constructed by Kirchhoff

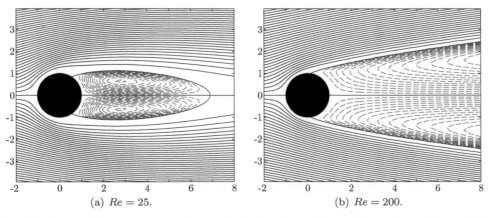

(a) $Re = 25$.   (b) $Re = 200$.

Fig. 2.3: Steady flow past a circular cylinder for Reynolds numbers $Re = 25$ and 200. The calculations were performed by J. S. B. Gajjar. The results for other values the Reynolds number are shown in Figures 2.19 and 2.20 on pages 125 and 126 in Part 1 of this book series.

(1869) who studied the flow past a flat plate perpendicular to the free-stream velocity with the separation taking place at the plate edges.[2] Later, the Kirchhoff model was applied to other body shapes. In particular, Levi-Civita (1907) used it to study the separated flow past a circular cylinder. An important conclusion of this work was that the Euler equations admit a family of solutions with the position of the separation point on the cylinder surface playing the role of a free parameter.

To find the location of the separation point, one needs to take into consideration the boundary layer that forms on the cylinder surface. According to Prandtl (1904), this is due to a specific behaviour of the flow in the boundary layer that separation takes place. Prandtl described the separation process as follows. Since the flow in the boundary layer has to satisfy the no-slip condition on the body surface, the fluid velocity decreases from the value dictated by the inviscid theory at the outer edge of the boundary layer to zero on the body surface. The slow-moving fluid near the body surface is very sensitive to pressure variations along the boundary layer. On the front part of the body, the pressure decreases in the downstream direction, and the pressure gradient is negative. This is referred to as the *favourable pressure gradient* because it acts to accelerate the flow keeping the boundary layer attached to the body surface. However, further downstream the pressure starts to rise, and the boundary layer finds itself under the action of a positive (*adverse*) pressure gradient. In these conditions the boundary layer tends to separate from the body surface. The reason for separation may be explained as follows. Since the velocity in the boundary layer decreases towards the wall, the kinetic energy of fluid particles inside the boundary layer appears to be less than that at the outer edge of the boundary layer. In fact, the closer a fluid particle is to the wall, the smaller its kinetic energy appears to be. This means that while the pressure rise in the outer flow may be quite significant, the fluid particles inside the boundary layer may not be able to get over it. Even a rather small increase of pressure may cause the fluid particles near the wall to stop and then turn back to form a reverse flow region characteristic of separated flows, as shown in Figure 2.4.

The separation point S may be identified as a point on the body contour where the skin friction turns zero:

$$\tau_w = \left.\frac{\partial U_0}{\partial N}\right|_{N=0} = 0. \tag{2.1.2}$$

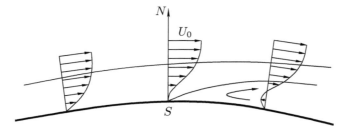

Fig. 2.4: Boundary-layer separation.

[2]See Section 3.8 in Part 1 of this book series.

Indeed, with τ_w being positive upstream of this point, the longitudinal velocity $U_0 > 0$, which means that the fluid particles in the boundary layer move downstream along the wall (see Figure 2.4) and the flow appears to be attached to the body surface. However, as soon as the skin friction τ_w turns negative, a layer of reversed flow ($U_0 < 0$) forms near the wall, giving rise to a region of recirculation which, obviously, originates from the point S where condition (2.1.2) holds.

We will see that Prandtl's description 'captured' the physical processes leading to separation rather well. Still, some important aspects of the separation phenomenon remained unexplained. In particular, it was not clear why the recirculation region could not be confined to the boundary layer whose thickness decreases with the Reynolds number as $O(Re^{-1/2})$. Instead, experiments invariably showed that the separation eddies were comparable in size with the body; see Figure 2.2. Howarth (1938) and Hartree (1939) were the first to address this question in their study of the boundary layer with an adverse pressure gradient. Specifically, they assumed the velocity $U_e(s)$ at the outer edge of the boundary layer to be a linearly decreasing function of the coordinate s measured along the body contour. They found that the solution developed a singularity at the point $s = s_0$ of zero skin friction (2.1.2). The form of this singularity was uncovered by Landau and Lifshitz (1944), who demonstrated that upstream of the separation, the skin friction decreases as the square root of the distance $s_0 - s$ from the separation point, that is[3]

$$\tau_w \sim \sqrt{s_0 - s} \quad \text{as} \quad s \to s_0 - 0,$$

and the velocity component normal to the body surface shows an unbounded growth:

$$V_0 \sim \frac{1}{\sqrt{s_0 - s}} \quad \text{as} \quad s \to s_0 - 0. \tag{2.1.3}$$

Equation (2.1.3) appeared to explain the observed eruption of eddies from the boundary layer. However, later Goldstein (1948) revisited the problem with the purpose of investigating the flow behind the zero skin friction point. He discovered that the singularity in the boundary layer precludes the solution to be continued into the reversed flow region, and it became clear that the mathematical description of the separation phenomenon had to be reexamined. In view of the importance of this result, the singularity at the point of zero skin friction is referred to as the *Goldstein singularity*.

At the time of Goldstein's discovery another important development was taking place. During 1940s and 1950s, a significant volume of experimental work was performed to study boundary-layer separation in supersonic flows. A review of these efforts may be found, for example, in Chapman *et al.* (1958). Most disputable was the phenomenon of upstream influence through the boundary layer prior to separation. It was observed in a number of physical situations. As an example, consider the boundary layer on the surface of a flat plate in supersonic flow; see Figure 2.5. If the plate is aligned with the oncoming flow, then the self-similar solution of Section 1.10.2 can be used to describe the flow.

Let us assume that the flow is perturbed with an oblique shock wave impinging on the boundary layer at point A. Applying Prandtl's hierarchical strategy, we first

---

[3]See equations (1.9.13).

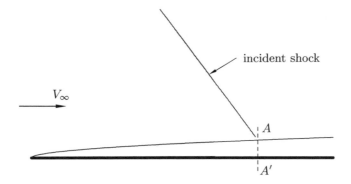

Fig. 2.5: Shock wave impinging upon the boundary layer.

need to consider the inviscid flow outside the boundary layer. Remember that at this stage of the flow analysis, the existence of the boundary layer can be ignored. We know that perturbations do not propagate upstream in supersonic flows. This is due to hyperbolic nature of the governing Euler equations.[4] Hence, we should expect the inviscid part of the flow to remain unperturbed everywhere in front of the incident shock. Turning to the second step in Prandtl's hierarchical strategy, we need to consider the boundary layer on the plate surface. With given (constant) pressure, the boundary-layer equations (1.10.5) are parabolic, and, therefore, the boundary layer also does not allow perturbations to propagate upstream of the cross section AA'.

Fig. 2.6: Oblique shock wave interacting boundary layer; flow from left to right. Visualization by Liepmann *et al.* (1952).

---

[4]See Sections 4.1 and 4.3 in Part 1 of this book series.

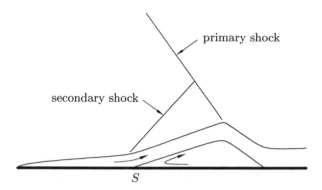

Fig. 2.7: Schematic representation of the flow in Figure 2.6. Here we also show the λ-structure that is hardly visible in Figure 2.6.

These theoretical arguments fail completely in predicting the real behaviour of the flow. The experiments invariably showed that, unless the incident shock was very weak, the flow was observed to separate from the plate surface some distance upstream of the incident shock; see Figure 2.6. It was also established that the boundary layer was perturbed even upstream of the separation point S. In fact, the distance through which the pressure perturbations were able to travel upstream from point S in the boundary layer was found to be significantly larger than the boundary-layer thickness. An increase of the pressure in the boundary layer in the vicinity of the separation point causes a thickening of the boundary layer. This gives rise to a secondary shock in the inviscid flow above the boundary layer (see Figure 2.7). Together with the primary shock they form a characteristic shock structure called the *λ-structure*.

To find an explanation to this unexpected behaviour, the experimental data were carefully analysed. Of particular interest was an observation that the boundary-layer separation process had a universal character, namely, it was found that the flow in the vicinity of the separation point S remained unchanged when instead of the impinging shock, another source of perturbations, for example, the forward-facing step (see Figure 2.8) was used to provoke the separation. Various authors (see, for example, Oswatitsch and Wieghardt, 1948) argued that this universality and the observed upstream influence through the boundary layer may be explained by an interaction between the boundary layer and external inviscid part of the flow. The impinging shock (Figure 2.7) or forward-facing step (Figure 2.8) serve to trigger the interaction, but once started, the process proceeds very much independently obeying its own laws. For this reason Chapman *et al.* (1958) termed the interaction of the boundary layer with the inviscid supersonic flow the *free interaction*.

The phenomenon of free interaction may be described vaguely in the following way. Let us suppose that, for some reason, the pressure in the boundary layer starts to rise. This would lead to a deceleration of the fluid particles inside the boundary layer and, as a consequence, to a displacement of the streamlines from the wall. The response of the external supersonic flow to the displacement effect of the boundary layer is such that it further increases the pressure, and the chain of events repeats itself. We will see that

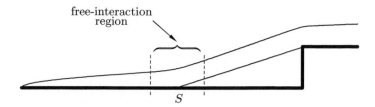

Fig. 2.8: Separation upstream of a forward-facing step.

this process, once initiated, leads to a formation of the free-interaction region where monotonic growth of the pressure ultimately results in the separation of the boundary layer. The asymptotic theory of this phenomenon, called the *self-induced separation*, was developed independently by Neiland (1969*a*) and Stewartson and Williams (1969). We shall now turn to the mathematical description of the separation process.

2.2   Self-Induced Separation of Supersonic Boundary Layer

2.2.1   Formulation of the problem

The theory presented here is applicable to a wide variety of physical situations. However, to describe the theory, it is convenient to consider a particular flow layout. We shall assume that the boundary-layer separation takes place at point S on the surface of a flat plate (see Figure 2.9) that is aligned with oncoming flow, which is supersonic with the Mach number $M_\infty > 1$. In what follows, we shall use Cartesian coordinates $(\hat{x}, \hat{y})$ with $\hat{x}$ measured along the flat plate surface from its leading edge, and $\hat{y}$ in the perpendicular direction. Assuming that the flow is steady and two-dimensional, we can use the Navier–Stokes equations in the form given by (1.10.1); see page 97.

We denote the distance between the leading edge of the plate and the separation point S by L. The gas velocity, density, dynamic viscosity coefficient, and the pressure in the unperturbed flow upstream of the plate are denoted as V_∞, ρ_∞, μ_∞, and p_∞, respectively. Using these quantities, the non-dimensional variables are introduced as follows:

$$\left.\begin{aligned}
\hat{x} = Lx, \qquad \hat{y} = Ly, \qquad\qquad &\hat{u} = V_\infty u, \qquad \hat{v} = V_\infty v, \\
\hat{\rho} = \rho_\infty \rho, \qquad \hat{p} = p_\infty + \rho_\infty V_\infty^2 p, \qquad &\hat{h} = V_\infty^2 h, \qquad \hat{\mu} = \mu_\infty \mu.
\end{aligned}\right\} \qquad (2.2.1)$$

Fig. 2.9: The flow layout.

These turn the Navier–Stokes equations (1.10.1) into the non-dimensional form:

$$\rho\left(u\frac{\partial u}{\partial x} + v\frac{\partial u}{\partial y}\right) = -\frac{\partial p}{\partial x} + \frac{1}{Re}\left\{\frac{\partial}{\partial x}\left[\mu\left(\frac{4}{3}\frac{\partial u}{\partial x} - \frac{2}{3}\frac{\partial v}{\partial y}\right)\right]\right.$$
$$\left. + \frac{\partial}{\partial y}\left[\mu\left(\frac{\partial u}{\partial y} + \frac{\partial v}{\partial x}\right)\right]\right\}, \quad (2.2.2a)$$

$$\rho\left(u\frac{\partial v}{\partial x} + v\frac{\partial v}{\partial y}\right) = -\frac{\partial p}{\partial y} + \frac{1}{Re}\left\{\frac{\partial}{\partial y}\left[\mu\left(\frac{4}{3}\frac{\partial v}{\partial y} - \frac{2}{3}\frac{\partial u}{\partial x}\right)\right]\right.$$
$$\left. + \frac{\partial}{\partial x}\left[\mu\left(\frac{\partial u}{\partial y} + \frac{\partial v}{\partial x}\right)\right]\right\}, \quad (2.2.2b)$$

$$\rho\left(u\frac{\partial h}{\partial x} + v\frac{\partial h}{\partial y}\right) = u\frac{\partial p}{\partial x} + v\frac{\partial p}{\partial y} + \frac{1}{Re}\left\{\frac{1}{Pr}\left[\frac{\partial}{\partial x}\left(\mu\frac{\partial h}{\partial x}\right) + \frac{\partial}{\partial y}\left(\mu\frac{\partial h}{\partial y}\right)\right]\right.$$
$$\left. + \mu\left(\frac{4}{3}\frac{\partial u}{\partial x} - \frac{2}{3}\frac{\partial v}{\partial y}\right)\frac{\partial u}{\partial x} + \mu\left(\frac{4}{3}\frac{\partial v}{\partial y} - \frac{2}{3}\frac{\partial u}{\partial x}\right)\frac{\partial v}{\partial y} + \mu\left(\frac{\partial u}{\partial y} + \frac{\partial v}{\partial x}\right)^2\right\}, \quad (2.2.2c)$$

$$\frac{\partial \rho u}{\partial x} + \frac{\partial \rho v}{\partial y} = 0, \quad (2.2.2d)$$

$$h = \frac{1}{(\gamma - 1)M_\infty^2}\frac{1}{\rho} + \frac{\gamma}{\gamma - 1}\frac{p}{\rho}. \quad (2.2.2e)$$

The asymptotic analysis of the Navier–Stokes equations (2.2.2) will be conducted assuming that the free-stream Mach number

$$M_\infty = \frac{V_\infty}{\sqrt{\gamma p_\infty/\rho_\infty}}$$

is a finite quantity larger than one, while the Reynolds number

$$Re = \frac{\rho_\infty V_\infty L}{\mu_\infty}$$

tends to infinity.

2.2.2 The flow upstream of the interaction region

Before analysing the free-interaction region that occupies a small vicinity of the separation point S, we need to know the state of the flow upstream of the interaction region. This flow can be described in the framework of Prandtl's hierarchical approach. As usual, we have to subdivide the flow field into two parts: the inviscid part of the flow and the boundary layer. We shall call the former region 1 and the latter region 2. In region 1 the flow is unperturbed in the leading-order approximation. The solution in the boundary layer (region 2) was discussed in detail in Section 1.10. Remember that the asymptotic analysis of the Navier–Stokes equations (2.2.2) in the boundary layer is conducted based on the limit

$$x = O(1), \quad Y = Re^{1/2}y = O(1), \quad Re \to \infty.$$

The fluid-dynamic functions are expressed in the form of asymptotic expansions (1.10.4):

$$
\left.
\begin{aligned}
u(x, y; Re) &= U_0(x, Y) + \cdots, & v(x, y; Re) &= Re^{-1/2}V_0(x, Y) + \cdots, \\
\rho(x, y; Re) &= \rho_0(x, Y) + \cdots, & p(x, y; Re) &= Re^{-1/2}P_1(x, Y) + \cdots, \\
h(x, y; Re) &= h_0(x, Y) + \cdots, & \mu(x, y; Re) &= \mu_0(x, Y) + \cdots.
\end{aligned}
\right\} \quad (2.2.3)
$$

Here functions U_0, V_0, ρ_0, and h_0 satisfy the classical boundary-layer equations (1.10.5). For the flow past a flat plate, these equations admit a self-similar solution (1.10.12) for a thermally isolated wall as well as in the case when the wall temperature is constant. However, we do not need to restrict ourselves to these particular flow conditions. To proceed further, we only need to know that the solution remains smooth when the interaction region is approached, that is, the functions U_0, h_0, ρ_0, and μ_0 may be represented in the form of Taylor expansions:

$$
\left.
\begin{aligned}
U_0(x, Y) &= U_{00}(Y) + (x - 1)U_{01}(Y) + \cdots, \\
h_0(x, Y) &= h_{00}(Y) + (x - 1)h_{01}(Y) + \cdots, \\
\rho_0(x, Y) &= \rho_{00}(Y) + (x - 1)\rho_{01}(Y) + \cdots, \\
\mu_0(x, Y) &= \mu_{00}(Y) + (x - 1)\mu_{01}(Y) + \cdots
\end{aligned}
\right\} \quad \text{as} \quad x \to 1. \quad (2.2.4)
$$

The leading-order terms in (2.2.4) exhibit the following behaviour near the plate surface:

$$
\left.
\begin{aligned}
U_{00}(Y) &= \lambda Y + \cdots, \\
h_{00}(Y) &= h_w + \cdots, \\
\rho_{00}(Y) &= \rho_w + \cdots, \\
\mu_{00}(Y) &= \mu_w + \cdots
\end{aligned}
\right\} \quad \text{as} \quad Y \to 0, \quad (2.2.5)
$$

where λ, h_w, ρ_w, and μ_w are positive constants representing the dimensionless skin friction, enthalpy, density, and the viscosity coefficient on the plate surface. For future reference, note that h_w and ρ_w are related to one another by the equation

$$
h_w = \frac{1}{(\gamma - 1)M_\infty^2} \frac{1}{\rho_w}, \quad (2.2.6)
$$

which can be obtained by setting $h_0 = h_w$ and $\rho_0 = \rho_w$ in the state equation (1.10.5d).

2.2.3 Inspection analysis of the interaction process

It is useful to precede the strict mathematical analysis of the separation process with the following 'physical' arguments, referred to as the *inspection analysis*. Assume that for some reason the pressure starts to rise at the outer edge of the boundary layer. Let us further assume that the pressure increment Δp is small, and is distributed over a short distance Δx (see Figure 2.10). The pressure rise will cause a deceleration of fluid particles inside the boundary layer. To estimate the corresponding velocity

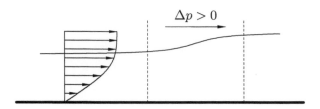

Fig. 2.10: Viscous-inviscid interaction region.

decrease Δu, we need to compare the convective terms on the left-hand side of the longitudinal momentum equation (2.2.2a) with the pressure gradient:[5]

$$\rho u \frac{\partial u}{\partial x} \sim \frac{\partial p}{\partial x}. \tag{2.2.7}$$

Symbol $\sim$ implies that the terms on the left- and right-hand sides of (2.2.7) are same order quantities (see Definition 1.6 on page 17 in Part 2 of this book series).

Since the perturbations are small, we can substitute the gas density ρ and the velocity u in (2.2.7) by the leading-order terms in their initial distributions (2.2.4). This gives

$$\rho_{00} U_{00} \frac{\partial u}{\partial x} \sim \frac{\partial p}{\partial x}.$$

Approximating the derivatives in the above equation by finite differences, we have

$$\rho_{00} U_{00} \frac{\Delta u}{\Delta x} \sim \frac{\Delta p}{\Delta x},$$

and it follows that

$$\Delta u \sim \frac{\Delta p}{\rho_{00} U_{00}}. \tag{2.2.8}$$

Since everywhere in the boundary layer, except near the wall, both ρ_{00} and U_{00} are order one quantities, we can conclude that the velocity decreases by a value

$$\Delta u \sim \Delta p. \tag{2.2.9}$$

Similarly, it can be found from the energy equation (2.2.2c) that

$$\Delta h \sim \Delta p,$$

---

[5]It should be noted that the two terms on the left-hand side of (2.2.2a) are same order quantities. Considered together they give the rate of change of u along a streamline. This may be more conveniently expressed in terms of the von Mises variables. Indeed, if instead of the Cartesian coordinates (x, y) we introduced the von Mises variables $(\tilde{x}, \psi)$, where $\tilde{x}(x, y) = x$ and $\psi(x, y)$ is the stream function such that

$$\frac{\partial \psi}{\partial x} = -\rho v, \qquad \frac{\partial \psi}{\partial y} = \rho u,$$

then we will have

$$u \frac{\partial u}{\partial x} + v \frac{\partial u}{\partial y} = u \frac{\partial u}{\partial \tilde{x}}.$$

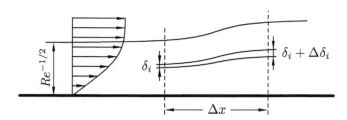

Fig. 2.11: Thickening of a stream filament in the main part of the boundary layer.

and then it follows from the state equation (2.2.2e) that the density variation is estimated as[6]

$$\Delta\rho \sim \Delta p. \qquad (2.2.10)$$

Let us now consider a thin fluid filament in the boundary layer confined between
two neighbouring streamlines; see Figure 2.11. Denoting the initial distance between
these streamlines by δ_i, and assuming that due to the fluid deceleration it increases
by $\Delta\delta_i$, we can write the mass conservation law as

$$\rho_{00}U_{00}\delta_i = (\rho_{00} + \Delta\rho)(U_{00} + \Delta u)(\delta_i + \Delta\delta_i). \qquad (2.2.11)$$

When performing multiplications on the right-hand side of (2.2.11), we neglect squares
of perturbations. We find that

$$\frac{\Delta\delta_i}{\delta_i} \sim \frac{\Delta\rho}{\rho_{00}} + \frac{\Delta u}{U_{00}}.$$

Since both ρ_{00} and U_{00} are order one quantities, and Δu and $\Delta\rho$ are given by (2.2.9)
and (2.2.10), we can conclude that the thickness of the fluid filament increases by an
amount

$$\Delta\delta_i \sim \delta_i \Delta p.$$

The integral effect of the thickening of all fluid filaments in the boundary layer is
calculated as

$$\Delta\delta = \sum_i \Delta\delta_i \sim \sum_i \delta_i \Delta p \sim \Delta p \sum_i \delta_i \sim Re^{-1/2}\Delta p. \qquad (2.2.12)$$

Here, it is taken into account that the total thickness of the boundary layer is an
$O(Re^{-1/2})$ quantity.

The above analysis relies on the assumption that the longitudinal velocity is perturbed by an amount Δu that is small compared to the initial velocity U_{00}. This
assumption holds everywhere in the boundary layer, but not near the body surface.
Indeed, we know from (2.2.5) that U_{00} tends to zero as $Y \to 0$, and therefore equation
(2.2.8) predicts an unbounded growth of Δu. Of course, equation (2.2.8) loses its validity when Δu becomes comparable with U_{00} and nonlinear effects start to influence
the flow behaviour. It should be noticed that the existence of the 'nonlinear region' is

---

[6]See Problem 1 in Exercises 9.

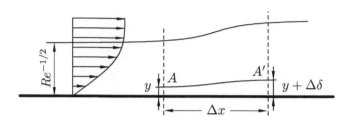

Fig. 2.12: Thickening of the near-wall sublayer.

essential for describing a separated flow where the velocity u changes its sign. Clearly, in a thin nonlinear layer near the plate surface, where $\Delta u \sim U_{00}$, equation (2.2.8) turns into the following order-of-magnitude equation:

$$\Delta u \sim U_{00} \sim \sqrt{\Delta p}. \tag{2.2.13}$$

Combining (2.2.13) with the asymptotic expansion for U_{00} in (2.2.5), we can deduce that

$$Y \sim \sqrt{\Delta p},$$

which means that the thickness of the nonlinear near-wall region is estimated as

$$y = Re^{-1/2} Y \sim Re^{-1/2} \sqrt{\Delta p}. \tag{2.2.14}$$

Since this region is much thinner than the original boundary layer, we shall call it the *near-wall sublayer*.

To estimate the displacement effect of the near-wall sublayer, we shall again use the mass conservation law. Let AA' be a streamline that lies inside the sublayer (see Figure 2.12). The initial distance between this streamline and the body surface is denoted by y. As a result of flow deceleration it increases to $y + \Delta \delta$. We shall apply the mass conservation law to the 'channel' formed by streamline AA' and the body surface. Of course, the fluid flux through the channel at the initial position A should coincide with that at A'. Since in the region considered $\Delta u \sim u$, this is only possible if $\Delta \delta \sim y$. Thus, using equation (2.2.14), we can conclude that

$$\Delta \delta \sim Re^{-1/2} \sqrt{\Delta p}. \tag{2.2.15}$$

The total displacement effect of the boundary layer is obtained by adding (2.2.15) and (2.2.12) together. However, it is apparent that for any $\Delta p \ll 1$ the contribution of the near-wall sublayer is significantly larger than that of the main part of the boundary layer. This is despite the sublayer occupies a relatively small portion of the boundary layer.[7] Hence, the streamline slope angle θ at the outer edge of the boundary layer can be calculated using (2.2.15) as follows:

$$\theta \sim \frac{\Delta \delta}{\Delta x} \sim \frac{Re^{-1/2} \sqrt{\Delta p}}{\Delta x}. \tag{2.2.16}$$

---

[7]It should be noted that this important result was known before the formal theory of the self-induced separation was developed; see, for example, Oswatitsch and Wieghardt (1948).

Now we turn to the inviscid flow outside the boundary layer. If $\theta \ll 1$, then the pressure perturbations produced in the inviscid flow by the displacement effect of the boundary layer can be calculated with the help of the Ackeret formula:[8]

$$\Delta p = \frac{\hat{p} - p_\infty}{\rho_\infty V_\infty^2} = \frac{\theta}{\sqrt{M_\infty^2 - 1}}. \tag{2.2.17}$$

Taking into account that in a supersonic flow $\sqrt{M_\infty^2 - 1} = O(1)$ and using equation (2.2.16) for θ, we can see that

$$\Delta p \sim \frac{Re^{-1/2}\sqrt{\Delta p}}{\Delta x}. \tag{2.2.18}$$

Finally, we need to discuss the role of viscosity in the boundary-layer separation process. We know that in the main part of the boundary layer the longitudinal velocity component u experiences only small perturbations, which means that the main viscous term in the x-momentum equation (2.2.2a)

$$\frac{1}{Re}\frac{\partial}{\partial y}\left(\mu\frac{\partial u}{\partial y}\right)$$

remains an order one quantity. Therefore, if we assume (subject to subsequent confirmation) that the pressure gradient $\partial p/\partial x$ is large in the interaction region, then the viscous terms in (2.2.2a) can be disregarded in the leading-order approximation, and the main part of the boundary layer can be treated as inviscid. However, the viscosity has to restore its action in the near-wall sublayer. Indeed, if the flow in the sublayer were inviscid, then the Bernoulli equation

$$\frac{u^2 + v^2}{2} + \frac{p}{\rho} = H(\psi) \tag{2.2.19}$$

would hold. Here we use the 'incompressible form' of the Bernoulli equation.[9] This is because the local Mach number in the near-wall sublayer is small, and therefore the flow in this region can be treated as incompressible.[10]

The Bernoulli function $H(\psi)$ on the right-hand side of (2.2.19) can be found using the fact that all streamlines originate from the boundary layer upstream of the interaction region, where the fluid-dynamic functions are represented by asymptotic expansions (2.2.3). We see that v^2 and p can be disregarded on the left-hand side of (2.2.19), while for u we can use the first of equations (2.2.5):

$$u = \lambda Y. \tag{2.2.20}$$

If we introduce the non-dimensional stream function ψ such that

$$\frac{\partial \psi}{\partial Y} = u, \qquad \psi\Big|_{Y=0} = 0,$$

---

[8]See equation (3.2.35) on page 174 in Part 2 of this book series.
[9]See equation (3.1.6) on page 130 in Part 1.
[10]See Problem 4 in Exercises 4 in Part 1.

then we will have $\psi = \frac{1}{2}\lambda Y^2$, and therefore equation (2.2.20) can be written as

$$u = \sqrt{2\lambda\psi}. \tag{2.2.21}$$

By substituting (2.2.21) into (2.2.19), we find that $H(\psi) = \lambda\psi$, which allows us to express the Bernoulli equation (2.2.19) in the form

$$\frac{u^2 + v^2}{2} + \frac{p}{\rho} = \lambda\psi. \tag{2.2.22}$$

It is now easy to see that if the flow was inviscid, then any rise of the pressure $(p > 0)$ in the interaction region would lead to a formation of a layer near the body surface where $u^2 + v^2 < 0$. Since this is impossible, we have to conclude that the flow in the sublayer should be viscous, that is, the convective terms in the x-momentum equation (2.2.2a) should be in balance with the viscous term:

$$\rho u \frac{\partial u}{\partial x} \sim \frac{1}{Re} \frac{\partial}{\partial y} \left(\mu \frac{\partial u}{\partial y} \right). \tag{2.2.23}$$

Taking into account that the density ρ and viscosity μ are order one quantities everywhere in the boundary layer, and approximating the derivatives in equation (2.2.23) by finite differences, we have

$$u \frac{\Delta u}{\Delta x} \sim \frac{1}{Re} \frac{\Delta u}{(\Delta y)^2}.$$

Since in the sublayer $\Delta y \sim y$, this equation may be also written as

$$\frac{u}{\Delta x} \sim \frac{1}{Re\, y^2}. \tag{2.2.24}$$

Now, let us summarize the results of the inspection analysis. We found that near the separation point the boundary layer splits into two regions: the main part of the boundary layer, and the near-wall nonlinear sublayer. The characteristic thickness of the sublayer is given by equation (2.2.14):

$$y \sim Re^{-1/2}\sqrt{\Delta p}. \tag{2.2.25}$$

We further found that the velocity perturbations in the sublayer are given by equation (2.2.13). Keeping in mind that in the region considered $\Delta u \sim u$, we can write

$$u \sim \sqrt{\Delta p}. \tag{2.2.26}$$

Also, when analysing the displacement effect of the boundary layer we deduced equation (2.2.18) for the pressure perturbations in the interaction region. It may be written as

$$\sqrt{\Delta p} \sim \frac{Re^{-1/2}}{\Delta x}. \tag{2.2.27}$$

Finally, analysis of the viscosity effects in the near-wall sublayer lead us to equation (2.2.24). Considered together, equations (2.2.24)–(2.2.27) form a closed set of order-of-magnitude equations for four unknown quantities: the velocity u in the near-wall sublayer, the characteristic thickness of the sublayer y, the induced pressure Δp, and the longitudinal extent Δx of the interaction region. These equations may be treated as algebraic equations. By solving these equations, we find that

$$u \sim Re^{-1/8}, \qquad y \sim Re^{-5/8}, \qquad \Delta p \sim Re^{-1/4}, \qquad \Delta x \sim Re^{-3/8}. \tag{2.2.28}$$

We are now in a position to confirm that the pressure gradient in the interaction region is indeed large:

$$\frac{\partial p}{\partial x} \sim \frac{\Delta p}{\Delta x} \sim Re^{1/8}. \tag{2.2.29}$$

2.2.4 Triple-deck model

The inspection analysis presented in Section 2.2.3 serves two purposes. First, it allows us to obtain estimates for fluid dynamic quantities in different parts of the flow. These are useful in predicting the form of the asymptotic expansions of the sought functions. Second, it reveals the physical nature of the processes involved. In particular, it shows that, under certain conditions, the boundary layer comes into interaction with external inviscid flow, termed the *viscous-inviscid interaction*. The interaction region occupies an $O(Re^{-3/8})$ vicinity of the separation point, and has a three-tiered structure (shown in Figure 2.13). It is composed of the near-wall nonlinear viscous sublayer (region 3), the main part of the boundary layer (region 4), and the inviscid potential flow (region 5) situated outside the boundary layer.

The characteristic thickness of the viscous sublayer (region 3) is estimated as an $O(Re^{-5/8})$ quantity, so that it occupies an $O(Re^{-1/8})$ portion of the boundary layer. The fluid velocity in this region is $O(Re^{-1/8})$ relative to the free-stream velocity. Due to the slow motion of the fluid in this layer, the flow exhibits high sensitivity to pressure variations. Even a small pressure rise causes significant deceleration of the fluid. As a result, the stream filaments increase their thickness, and the streamlines are displaced from the plate surface. Remember that this phenomenon is referred to as the *displacement effect of the boundary layer*.

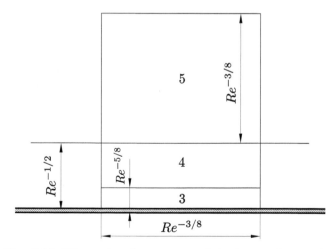

Fig. 2.13: Three-tiered structure of the interaction region.

The main part of the boundary layer (region 4) represents a continuation of the conventional boundary layer on the plate surface into the interaction region. Its thickness is estimated as $y \sim Re^{-1/2}$, and the non-dimensional velocity u is an order one quantity. The flow in middle tier is significantly less sensitive to the pressure variations. It does not produce any noticeable contribution to the displacement effect of the boundary layer. We will see that all the streamlines in the middle tier are parallel to each other and carry the deformation produced by the displacement effect of the viscous sublayer.

Finally, the upper tier (region 5) is situated in the inviscid potential flow outside the boundary layer. It serves to 'convert' the perturbations in the shape of the streamlines into perturbations of the pressure. These are then transmitted through the main part of the boundary layer back to the sublayer, enhancing the process of fluid deceleration. This process is self-sustained, and it drives the boundary layer towards the separation.

Viscous sublayer (region 3)

Estimates (2.2.28) suggest that asymptotic analysis of the Navier–Stokes equations in the viscous sublayer (region 3) should be conducted based on the limit procedure, where

$$x_* = \frac{x-1}{Re^{-3/8}} = O(1), \quad Y_* = \frac{y}{Re^{-5/8}} = O(1), \quad Re \to \infty. \qquad (2.2.30)$$

The asymptotic expansions of the fluid-dynamic functions in this region are written as

$$\left.\begin{array}{ll} u = Re^{-1/8}U^*(x_*, Y_*) + \cdots, & h = h^*(x_*, Y_*) + \cdots, \\ v = Re^{-3/8}V^*(x_*, Y_*) + \cdots, & \rho = \rho^*(x_*, Y_*) + \cdots, \\ p = Re^{-1/4}P^*(x_*, Y_*) + \cdots, & \mu = \mu^*(x_*, Y_*) + \cdots. \end{array}\right\} \qquad (2.2.31)$$

The form of the asymptotic expansions for u and p directly follows from estimates (2.2.28). To obtain an estimate for v, the continuity equation (2.2.2d) is used. The two terms in this equations should be in balance with one another:

$$\frac{\partial \rho u}{\partial x} \sim \frac{\partial \rho v}{\partial y}.$$

Expressing the above equation in the finite-difference from

$$\frac{\rho u}{\Delta x} \sim \frac{\rho v}{y},$$

we find that

$$v \sim y \frac{u}{\Delta x} \sim Re^{-3/8}.$$

As far as the enthalpy h, density ρ, and the viscosity coefficient μ are concerned, in normal circumstances (when there is no extreme heating or cooling of the plate surface), they remain order one functions throughout the boundary layer.

The substitution of (2.2.31) into the Navier–Stokes equations (2.2.2) yields

$$\rho^* U^* \frac{\partial U^*}{\partial x_*} + \rho^* V^* \frac{\partial U^*}{\partial Y_*} = -\frac{\partial P^*}{\partial x_*} + \frac{\partial}{\partial Y_*}\left(\mu^* \frac{\partial U^*}{\partial Y_*}\right), \tag{2.2.32a}$$

$$\frac{\partial P^*}{\partial Y_*} = 0, \tag{2.2.32b}$$

$$\rho^* U^* \frac{\partial h^*}{\partial x_*} + \rho^* V^* \frac{\partial h^*}{\partial Y_*} = -\frac{1}{Pr}\frac{\partial}{\partial Y_*}\left(\mu^* \frac{\partial h^*}{\partial Y_*}\right), \tag{2.2.32c}$$

$$\frac{\partial \rho^* U^*}{\partial x_*} + \frac{\partial \rho^* V^*}{\partial Y_*} = 0, \tag{2.2.32d}$$

$$h^* = \frac{1}{(\gamma - 1)M_\infty^2}\frac{1}{\rho^*}. \tag{2.2.32e}$$

Remember that when performing the inspection analysis, we speculated that, since the flow in the near-wall sublayer is slow, it should behave as incompressible. We shall now confirm this proposition mathematically. For this purpose, we will use the energy equation (2.2.32c). This is a parabolic equation, and, therefore, it requires an initial condition at $x_* = -\infty$, a condition on the plate surface ($Y_* = 0$), and a condition at the outer edge of the sublayer ($Y_* \to \infty$). We start with the initial condition. It may be formulated by matching with the solution in the boundary layer (region 2) upstream of the interaction region. According to (2.2.3), the asymptotic expansion for the enthalpy in region 2 has the form

$$h(x, y; Re) = h_0(x, Y) + \cdots. \tag{2.2.33}$$

To perform the matching, we have to re-expand (2.2.33) in variables (2.2.30) of region 3. We start with the longitudinal coordinate

$$x = 1 + Re^{-3/8}x_*.$$

Since $x - 1$ is small, we can use the Taylor expansion (2.2.4) for the enthalpy:

$$h_0(x, Y) = h_{00}(Y) + (x - 1)h_{01}(Y) + \cdots.$$

Being substituted into (2.2.33), it gives

$$h(x, y; Re) = h_{00}(Y) + O(Re^{-3/8}). \tag{2.2.34}$$

We further know that region 3 is much thinner than the original boundary layer (region 2). Indeed, in region 2

$$y = Re^{-1/2}Y,$$

while in region 3

$$y = Re^{-5/8}Y_*.$$

Comparing these equations, we see that

$$Y = Re^{-1/8}Y_*,$$

which means that Y is small, and we can use the asymptotic expansion for $h_{00}(Y)$ given by (2.2.5):

$$h_{00}(Y) = h_w + \cdots \quad \text{as} \quad Y \to 0. \tag{2.2.35}$$

The substitution of (2.2.35) into (2.2.34) yields

$$h(x, y; Re) = h_w + \cdots . \tag{2.2.36}$$

It remains to compare (2.2.36) with the asymptotic expansion (2.2.31) of the enthalpy in region 3, and we can conclude that the sought matching condition is

$$h^* \to h_w \quad \text{as} \quad x_* \to -\infty. \tag{2.2.37}$$

On the plate surface, we can either prescribe the wall temperature

$$h^* = h_w \quad \text{at} \quad Y_* = 0, \tag{2.2.38a}$$

or assume that the wall is thermally isolated

$$\frac{\partial h^*}{\partial Y_*} = 0 \quad \text{at} \quad Y_* = 0. \tag{2.2.38b}$$

Finally, we need to analyse the behaviour of the enthalpy near the outer edge of region 3. In the overlap region between region 3 and the inviscid main part of the boundary layer (region 4), the viscous term on the right-hand side of the energy equation (2.2.32c) is expected to become negligible, which leads to

$$U^* \frac{\partial h^*}{\partial x_*} + V^* \frac{\partial h^*}{\partial Y_*} = 0. \tag{2.2.39}$$

The above equation shows that the enthalpy h^* does not change along streamlines. Here, our interest is in the streamlines that lie above the separation region. All of them originate from an upstream location, where condition (2.2.37) holds. Integrating (2.2.39) with (2.2.37), we arrive at a conclusion that the sought boundary condition at the outer edge of region 3 may be written as

$$h^* \to h_w \quad \text{as} \quad Y_* \to \infty. \tag{2.2.40}$$

The solution of the energy equation (2.2.32c), satisfying conditions (2.2.37), (2.2.38), and (2.2.40), is given by

$$h^* = h_w. \tag{2.2.41}$$

We have demonstrated that the enthalpy h^* is constant everywhere in the viscous sublayer (region 3). Now, substituting (2.2.41) into the state equation (2.2.32e) and comparing the result with (2.2.6), we can see that the density ρ^* is also constant and equals ρ_w. This confirms that the flow in region 3 may indeed be treated as incompressible. Finally, we know that the viscosity coefficient μ^* is a function of the enthalpy only. According to (2.2.5), it assumes the value $\mu^* = \mu_w$ when $h^* = h_w$.

Keeping this in mind, we can write the momentum (2.2.32a) and continuity (2.2.32d) equations as

$$\rho_w U^* \frac{\partial U^*}{\partial x_*} + \rho_w V^* \frac{\partial U^*}{\partial Y_*} = -\frac{dP^*}{dx_*} + \mu_w \frac{\partial^2 U^*}{\partial Y_*^2}, \qquad (2.2.42a)$$

$$\frac{\partial U^*}{\partial x_*} + \frac{\partial V^*}{\partial Y_*} = 0. \qquad (2.2.42b)$$

Notice that according to (2.2.32b) the pressure P^* is a function of x_* only.

Equations (2.2.42) should be solved with the no-slip condition on the plate surface

$$U^* = V^* = 0 \quad \text{at} \quad Y_* = 0. \qquad (2.2.43)$$

We also need to formulate an initial condition for U^* at $x_* = -\infty$. We do this by matching with the solution in the boundary layer upstream of the interaction region (region 2). According to (2.2.3), the longitudinal velocity component u is represented in region 2 by the asymptotic expansion

$$u(x, y; Re) = U_0(x, Y) + \cdots . \qquad (2.2.44)$$

To perform the matching, we need to re-expand (2.2.44) in terms of the variables (2.2.30) of region 3. We write

$$u(x, y; Re) = U_0(1 + Re^{-3/8} x_*, Re^{-1/8} Y_*) + \cdots .$$

Taking into account that the first argument of function U_0 is close to one, and the second is small, we can use asymptotic expansions (2.2.4) and (2.2.5), which yields

$$u(x, y; Re) = Re^{-1/8} \lambda Y_* + \cdots . \qquad (2.2.45)$$

Comparing (2.2.45) with the asymptotic expansion (2.2.31) for u in region 3, we see that the sought initial condition is written as

$$U^* = \lambda Y_* \quad \text{at} \quad x_* = -\infty. \qquad (2.2.46)$$

Now we shall study the asymptotic behaviour of the solution of equations (2.2.42) near the outer edge of the viscous sublayer, that is in the limit

$$x_* = O(1), \qquad Y_* \to \infty. \qquad (2.2.47)$$

For this purpose it is convenient to introduce the stream function $\Psi^*(x_*, Y_*)$. Its existence follows from the continuity equation (2.2.42b), and we can write

$$U^* = \frac{\partial \Psi^*}{\partial Y_*}, \qquad V^* = -\frac{\partial \Psi^*}{\partial x_*}. \qquad (2.2.48)$$

We shall seek the asymptotic expansion of stream function with respect to the limit (2.2.47) in the form

$$\Psi^*(x_*, Y_*) = A_0(x_*) Y_*^\alpha + \cdots \quad \text{as} \quad Y_* \to \infty. \qquad (2.2.49)$$

Here, parameter α and function $A_0(x_*)$ are expected to be found by using the momentum equation (2.2.42a).

The substitution of (2.2.49) into (2.2.48) yields

$$U^* = \alpha A_0(x_*)Y_*^{\alpha-1} + \cdots, \qquad V^* = -\frac{dA_0}{dx_*}Y_*^{\alpha} + \cdots. \qquad (2.2.50)$$

Therefore, the convective terms on the left-hand side of equation (2.2.42a) and the viscous term on its right-hand side are calculated as

$$\rho_w U^* \frac{\partial U^*}{\partial x_*} = \rho_w \alpha^2 A_0 \frac{dA_0}{dx_*} Y_*^{2\alpha-2} + \cdots,$$

$$\rho_w V^* \frac{\partial U^*}{\partial Y_*} = -\rho_w \alpha(\alpha-1)A_0 \frac{dA_0}{dx_*} Y_*^{2\alpha-2} + \cdots,$$

$$\mu_w \frac{\partial^2 U^*}{\partial Y_*^2} = \alpha(\alpha-1)(\alpha-2)\mu_w A_0 Y_*^{\alpha-3} + \cdots.$$

We see that if we assume, subject to subsequent confirmation, that $\alpha > 1$, then the convective terms appear to be much larger than the viscous term as well as the pressure gradient; the latter remains finite as $Y_* \to \infty$. Consequently, in the leading-order approximation, the momentum equation (2.2.42a) reduces to

$$A_0 \frac{dA_0}{dx_*} = 0.$$

The initial condition for this equation may be obtained by substituting the first of equations (2.2.50) into (2.2.46). We see that

$$A_0(-\infty) = \begin{cases} \lambda/\alpha & \text{if} \quad \alpha = 2, \\ 0 & \text{if} \quad \alpha \neq 2. \end{cases}$$

Hence, a non-trivial solution exists only if $\alpha = 2$, in which case $A_0 = \frac{1}{2}\lambda$ and (2.2.49) turns into

$$\Psi^*(x_*, Y_*) = \frac{1}{2}\lambda Y_*^2 + \cdots \quad \text{as} \quad Y_* \to \infty.$$

Let us now find the second term in this expansion. We write

$$\Psi^*(x_*, Y_*) = \frac{1}{2}\lambda Y_*^2 + A_1(x_*)Y_*^{\beta} + \cdots \quad \text{as} \quad Y_* \to \infty. \qquad (2.2.51)$$

To ensure that the second term in (2.2.51) is smaller than the first one, we have to assume that $\beta < 2$. The substitution of (2.2.51) into (2.2.48) yields

$$U^* = \lambda Y_* + \beta A_1(x_*)Y_*^{\beta-1} + \cdots, \qquad V^* = -\frac{dA_1}{dx_*}Y_*^{\beta} + \cdots.$$

Correspondingly, the convective and viscous terms in (2.2.42a) are calculated as

$$\rho_w U^* \frac{\partial U^*}{\partial x_*} = \rho_w \lambda \beta \frac{dA_1}{dx_*} Y_*^{\beta} + \cdots,$$

$$\rho_w V^* \frac{\partial U^*}{\partial Y_*} = -\rho_w \lambda \frac{dA_1}{dx_*} Y_*^{\beta} + \cdots,$$

$$\mu_w \frac{\partial^2 U^*}{\partial Y_*^2} = \beta(\beta-1)(\beta-2)\mu_w A_1 Y_*^{\beta-3} + \cdots.$$

We see again that the convective terms are much larger than the viscous term. If we assume that $\beta > 0$, then the pressure gradient will be small compared to the convective terms. This means that in the limit $Y_* \to \infty$, the momentum equation (2.2.42a) reduces to

$$(\beta - 1)\frac{dA_1}{dx_*} = 0.$$

Since the initial profile (2.2.46) does not contain any terms except for the one that matches with the leading-order term in (2.2.51), we have to conclude that

$$A_1(-\infty) = 0.$$

We see that a non-trivial solution for A_1 is only possible if $\beta = 1$. The function $A_1(x_*)$ remains arbitrary from the viewpoint of the asymptotic analysis of equations (2.2.42) as $Y_* \to \infty$. However, we expect to find this function as a part of the solution of the problem as a whole.

Redenoting $A_1(x_*)$ as $A(x_*)$ renders (2.2.51) in the form

$$\Psi^* = \frac{1}{2}\lambda Y_*^2 + A(x_*)Y_* + \cdots \quad \text{as} \quad Y_* \to \infty. \tag{2.2.52}$$

The physical meaning of function $A(x_*)$ may be clarified by calculating the streamline slope angle

$$\vartheta = \arctan \frac{v}{u}. \tag{2.2.53}$$

In the viscous sublayer the velocity components u and v are given by the asymptotic expansions (2.2.31):

$$u = Re^{-1/8}U^*(x_*, Y_*) + \cdots, \qquad v = Re^{-3/8}V^*(x_*, Y_*) + \cdots. \tag{2.2.54}$$

The substitution of (2.2.52) into (2.2.48) shows that at the outer edge of the viscous sublayer

$$U^*(x_*, Y_*) = \lambda Y_* + A(x_*) + \cdots, \qquad V^*(x_*, Y_*) = -\frac{dA}{dx_*}Y_* + \cdots. \tag{2.2.55}$$

Now, we substitute (2.2.55) into (2.2.54). We find that

$$u = Re^{-1/8}\lambda Y_* + Re^{-1/8}A(x_*) + \cdots, \qquad v = Re^{-3/8}\left(-\frac{dA}{dx_*}Y_*\right) + \cdots. \tag{2.2.56}$$

It remains to substitute (2.2.56) into (2.2.53), and we will have

$$\vartheta = \arctan \frac{v}{u} = Re^{-1/4}\frac{V^*}{U^*}\bigg|_{Y_* \to \infty} + \cdots = Re^{-1/4}\left(-\frac{1}{\lambda}\frac{dA}{dx_*}\right) + \cdots. \tag{2.2.57}$$

In view of this, function $A(x_*)$ is referred to as the *displacement function*.

Main part of the boundary layer (region 4)

Region 4, the middle tier of the triple-deck structure (see Figure 2.13), represents a continuation of the conventional boundary layer (region 2) into the interaction region. The thickness of region 4 is estimated as $y \sim Re^{-1/2}$. Of course, the longitudinal extent of region 4 coincides with that of the interaction region as a whole, being estimated as $|x - 1| = O(Re^{-3/8})$. Consequently, the asymptotic analysis of the Navier–Stokes equations (2.2.2) in region 4 has to be performed based on the limit

$$x_* = \frac{x-1}{Re^{-3/8}} = O(1), \quad Y_* = \frac{y}{Re^{-1/2}} = O(1), \quad Re \to \infty. \tag{2.2.58}$$

The form of the asymptotic expansions of the fluid-dynamic functions in region 4 may be predicted by analysing the solution in the overlap region that lies between regions 3 and 4. At the outer edge of region 3, the velocity components are given by the asymptotic expansions (2.2.56). If we express (2.2.56) in term of variables (2.2.58) of region 4, using the fact that $Y = Re^{-1/8}Y_*$, then we will find that at the 'bottom' of region 4

$$u = \lambda Y + Re^{-1/8}A(x_*) + \cdots, \qquad v = Re^{-1/4}\left(-\frac{dA}{dx_*}Y\right) + \cdots. \tag{2.2.59}$$

This suggests that the solution in region 4 should be sought in the form

$$\left. \begin{aligned} u &= U_{00}(Y) + Re^{-1/8}\tilde{U}_1(x_*,Y) + \cdots, \\ v &= Re^{-1/4}\tilde{V}_1(x_*,Y) + \cdots. \end{aligned} \right\} \tag{2.2.60}$$

The leading-order term $U_{00}(Y)$ in the expansion for u coincides with the velocity profile (2.2.4) in the boundary layer immediately before the interaction region. According to (2.2.5)

$$U_{00} = \lambda Y + \cdots \quad \text{as} \quad Y \to 0. \tag{2.2.61}$$

It further follows from (2.2.59) that the perturbation terms $\tilde{U}_1(x_*,Y)$ and $\tilde{V}_1(x_*,Y)$ in (2.2.60) should exhibit the following behaviour at the 'bottom' of region 4:

$$\left. \begin{aligned} \tilde{U}_1 &= A(x_*) + \cdots, \\ \tilde{V}_1 &= -\frac{dA}{dx_*}Y + \cdots \end{aligned} \right\} \quad \text{as} \quad Y \to 0. \tag{2.2.62}$$

By analogy with the longitudinal velocity component u in (2.2.60), we shall seek the enthalpy h, the density ρ, and the viscosity μ in region 4 in the form of the asymptotic expansions

$$\left. \begin{aligned} h &= h_{00}(Y) + Re^{-1/8}\tilde{h}_1(x_*,Y) + \cdots, \\ \rho &= \rho_{00}(Y) + Re^{-1/8}\tilde{\rho}_1(x_*,Y) + \cdots, \\ \mu &= \mu_{00}(Y) + Re^{-1/8}\tilde{\mu}_1(x_*,Y) + \cdots. \end{aligned} \right\} \tag{2.2.63}$$

Finally, we expect the pressure p to remain unchanged across the boundary layer. Consequently, the asymptotic representation of p in region 4 should have the same form

$$p = Re^{-1/4}\widetilde{P}_1(x_*, Y) + \cdots \qquad (2.2.64)$$

as that in region 3; see equation (2.2.31).

The substitution of (2.2.60), (2.2.63), and (2.2.64) into the Navier–Stokes equations (2.2.2) results in

$$U_{00}(Y)\frac{\partial \widetilde{U}_1}{\partial x_*} + \widetilde{V}_1\frac{dU_{00}}{dY} = 0, \qquad (2.2.65a)$$

$$\frac{\partial \widetilde{P}_1}{\partial Y} = 0, \qquad (2.2.65b)$$

$$U_{00}(Y)\frac{\partial \widetilde{h}_1}{\partial x_*} + \widetilde{V}_1\frac{dh_{00}}{dY} = 0, \qquad (2.2.65c)$$

$$\rho_{00}(Y)\frac{\partial \widetilde{U}_1}{\partial x_*} + U_{00}(Y)\frac{\partial \widetilde{\rho}_1}{\partial x_*} + \rho_{00}(Y)\frac{\partial \widetilde{V}_1}{\partial Y} + \widetilde{V}_1\frac{d\rho_{00}}{dY} = 0, \qquad (2.2.65d)$$

$$h_{00} = \frac{1}{(\gamma-1)M_\infty^2}\frac{1}{\rho_{00}}, \qquad \widetilde{h}_1 = -\frac{1}{(\gamma-1)M_\infty^2}\frac{\widetilde{\rho}_1}{\rho_{00}^2}. \qquad (2.2.65e)$$

It is interesting to note that the above equations do not involve viscous terms, which means that the flow in region 4 may be treated as inviscid in the leading-order approximation.

Equations (2.2.65) are easily solved using the following elimination procedure. We first substitute (2.2.65e) into the energy equation (2.2.65c). This leads to

$$U_{00}(Y)\frac{\partial \widetilde{\rho}_1}{\partial x_*} + \widetilde{V}_1\frac{d\rho_{00}}{dY} = 0,$$

which shows that the continuity equation (2.2.65d) may be written as

$$\frac{\partial \widetilde{U}_1}{\partial x_*} + \frac{\partial \widetilde{V}_1}{\partial Y} = 0. \qquad (2.2.66)$$

Now, using (2.2.66), we can eliminate $\partial \widetilde{U}_1/\partial x_*$ from the longitudinal momentum equation (2.2.65a). This results in

$$U_{00}(Y)\frac{\partial \widetilde{V}_1}{\partial Y} - \widetilde{V}_1\frac{dU_{00}}{dY} = 0. \qquad (2.2.67)$$

Dividing both terms in (2.2.67) by U_{00}^2, we have

$$\frac{1}{U_{00}(Y)}\frac{\partial \widetilde{V}_1}{\partial Y} - \frac{\widetilde{V}_1}{U_{00}^2}\frac{dU_{00}}{dY} = 0,$$

or, equivalently,

$$\frac{\partial}{\partial Y}\left(\frac{\widetilde{V}_1}{U_{00}}\right) = 0.$$

We see that the ratio $\tilde{V}_1/U_{00}$ is a function of x_* only, say $G(x_*)$:

$$\frac{\tilde{V}_1}{U_{00}} = G(x_*). \tag{2.2.68}$$

This function may be found by making use of equations (2.2.61), (2.2.62) describing the behaviour of U_{00} and $\tilde{V}_1$ at the bottom of region 4. We see that

$$G(x_*) = -\frac{1}{\lambda}\frac{dA}{dx_*}. \tag{2.2.69}$$

The substitution of (2.2.69) back into (2.2.68) yields the equation

$$\frac{\tilde{V}_1}{U_{00}} = -\frac{1}{\lambda}\frac{dA}{dx_*}, \tag{2.2.70}$$

valid everywhere in region 4.

Now let us return to the asymptotic expansions (2.2.60) of the velocity components, and calculate the streamline slope angle in region 4:

$$\vartheta = \arctan\frac{v}{u} = Re^{-1/4}\frac{\tilde{V}_1}{U_{00}} + \cdots = Re^{-1/4}\left(-\frac{1}{\lambda}\frac{dA}{dx_*}\right) + \cdots. \tag{2.2.71}$$

We see that ϑ does not depend on Y, which means that it stays unchanged across the main part of the boundary layer. This confirms an important result of the inspection analysis, i.e., that the displacement effect of the main part of the boundary layer is relatively small, and may be disregarded in the leading-order analysis. The streamline slope angle (2.2.57), produced by the viscous sublayer, is simply 'transported' by the main part of the boundary layer towards the bottom of the upper tier of the triple-deck structure, shown as region 5 in Figure 2.13.

Upper tier (region 5)

The upper tier lies in the inviscid supersonic flow outside the boundary layer. The flow in this region can be described in the framework of the thin aerofoil theory.[11] We start by noticing that the longitudinal extent of region 5 should be the same as that of regions 1 and 2:

$$\Delta x \sim Re^{-3/8}.$$

To avoid degeneration of the governing equations, we choose the lateral size of region 5 to be[12]

$$y \sim \Delta x \sim Re^{-3/8}.$$

Correspondingly, the asymptotic analysis of the Navier–Stokes equations in region 5 will be performed based on the limit

$$x_* = \frac{x-1}{Re^{-3/8}} = O(1), \quad y_* = \frac{y}{Re^{-3/8}} = O(1), \quad Re \to \infty. \tag{2.2.72}$$

---

[11]A detailed discussion of this theory may be found in Chapter 3 of Part 2 of this book series.

[12]We shall see that the flow in region 5 is governed by the wave equation (2.2.78). With the freestream Mach number $M_\infty = O(1)$, it can only have a meaningful solution if $x_* \sim y_*$, that is, if the longitudinal and lateral sizes of region 3 are same order quantities.

To predict the form of the asymptotic expansions of the fluid-dynamic functions in region 5, we note that if the flow in this region was not affected by the displacement of the boundary layer (2.2.71), then we would have

$$u = 1, \quad v = 0, \quad \rho = 1, \quad p = 0, \quad h = \frac{1}{(\gamma - 1)M_\infty^2}, \tag{2.2.73}$$

that is, the non-dimensional velocity components u, v, the density ρ, the pressure p, and the enthalpy h would assume their free-stream values. The displacement effect of the boundary layer leads to $O(Re^{-1/4})$ perturbations in (2.2.73), which suggests that the solution in region 5 should be sought in the form

$$\left.\begin{aligned}
u(x, y; Re) &= 1 + Re^{-1/4}u_*(x_*, y_*) + \cdots, \\
v(x, y; Re) &= Re^{-1/4}v_*(x_*, y_*) + \cdots, \\
\rho(x, y; Re) &= 1 + Re^{-1/4}\rho_*(x_*, y_*) + \cdots, \\
p(x, y; Re) &= Re^{-1/4}p_*(x_*, y_*) + \cdots, \\
h(x, y; Re) &= \frac{1}{(\gamma - 1)M_\infty^2} + Re^{-1/4}h_*(x_*, y_*) + \cdots.
\end{aligned}\right\} \tag{2.2.74}$$

By substituting (2.2.74) into the Navier–Stokes equations (2.2.2) and working with the $O(Re^{-1/4})$ terms, we find that

$$\frac{\partial u_*}{\partial x_*} = -\frac{\partial p_*}{\partial x_*} \qquad \text{(x-momentum equation)}, \tag{2.2.75a}$$

$$\frac{\partial v_*}{\partial x_*} = -\frac{\partial p_*}{\partial y_*} \qquad \text{(y-momentum equation)}, \tag{2.2.75b}$$

$$\frac{\partial h_*}{\partial x_*} = \frac{\partial p_*}{\partial x_*} \qquad \text{(energy equation)}, \tag{2.2.75c}$$

$$\frac{\partial u_*}{\partial x_*} + \frac{\partial \rho_*}{\partial x_*} + \frac{\partial v_*}{\partial y_*} = 0 \qquad \text{(continuity equation)}, \tag{2.2.75d}$$

$$h_* = \frac{\gamma}{\gamma - 1}p_* - \frac{1}{(\gamma - 1)M_\infty^2}\rho_* \qquad \text{(state equation)}. \tag{2.2.75e}$$

This set of equations can be reduced to a single equation for the pressure p_* using the following elimination procedure. To eliminate the enthalpy h_*, we substitute the state equation (2.2.75e) into the energy equation (2.2.75e), which results in

$$\frac{\partial \rho_*}{\partial x_*} = M_\infty^2 \frac{\partial p_*}{\partial x_*}. \tag{2.2.76}$$

Now the density ρ_* and the longitudinal velocity u_* can be eliminated by substituting (2.2.76) and (2.2.75a) into the continuity equation (2.2.75d). This gives

$$\left(M_\infty^2 - 1\right)\frac{\partial p_*}{\partial x_*} - \frac{\partial v_*}{\partial y_*} = 0. \tag{2.2.77}$$

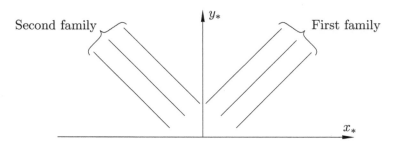

Fig. 2.14: Characteristics of equation (2.2.78).

It remains to eliminate the lateral velocity v_* from equations (2.2.77) and (2.2.75b), which is done by cross-differentiation of these equations. The resulting equation for the pressure p_* has the form

$$\left(M_\infty^2 - 1\right)\frac{\partial^2 p_*}{\partial x_*^2} - \frac{\partial^2 p_*}{\partial y_*^2} = 0. \tag{2.2.78}$$

Equation (2.2.78) is easily solved by introducing the *characteristic variables*

$$\xi = x_* - \sqrt{M_\infty^2 - 1}\, y_*, \qquad \eta = x_* + \sqrt{M_\infty^2 - 1}\, y_*. \tag{2.2.79}$$

If we use ξ and η as new independent variables, then the equation (2.2.78) assumes the form

$$\frac{\partial^2 p_*}{\partial \xi \partial \eta} = 0.$$

Its general solution is written as

$$p_* = f(\xi) + g(\eta), \tag{2.2.80}$$

where $f(\xi)$ and $g(\eta)$ are arbitrary functions of their respective arguments.

Assuming ξ or η constant yields the equations for the characteristics of the first and second families, respectively (see Figure 2.14):

$$y_* = \frac{x_* - \xi}{\sqrt{M_\infty^2 - 1}}, \qquad y_* = -\frac{x_* - \eta}{\sqrt{M_\infty^2 - 1}}.$$

Correspondingly, functions $f(\xi)$ and $g(\eta)$ in (2.2.80) represent the pressure perturbations propagating along the characteristics. Remember that in supersonic flows, perturbations can only propagate in the downstream direction and are confined to the Mach cone, as shown in Figure 4.2 on page 234 in Part 1. Hence, $g(\eta)$ could only be non-zero if there were perturbations emanating from a source situated upstream of region 5.[13] In the case of self-induced separation considered here, the perturbations in region 5 are produced solely by the displacement effect of the boundary layer only. Consequently, we have to set $g(\eta) = 0$.

---

[13]An example of such situation is discussed in Problem 4 in Exercises 12 where we analyse the boundary-layer separation caused by an impinging shock.

To find function $f(\xi)$, we need a boundary condition that accounts for the displacement of the boundary layer. We can deduce it by matching the streamline slope angle ϑ in regions 4 and 5. It follows from (2.2.74) that in region 5,

$$\vartheta = \arctan \frac{v}{u} = Re^{-1/4}v_*(x_*, y_*) + \cdots .$$ (2.2.81)

Comparing (2.2.81) with (2.2.71), we see that the sought boundary condition is

$$v_*\Big|_{y_*=0} = -\frac{1}{\lambda}\frac{dA}{dx_*}.$$ (2.2.82)

This condition is easily converted into a condition for the pressure p_*. Setting $y_* = 0$ in the y-momentum equation (2.2.75b) and using (2.2.82) on the left-hand side of (2.2.75b), we see that

$$\frac{\partial p_*}{\partial y_*}\Big|_{y_*=0} = \frac{1}{\lambda}\frac{d^2 A}{dx_*^2}.$$ (2.2.83)

The substitution of (2.2.80) and (2.2.79) into (2.2.83) results in the equation

$$f'(x_*) = -\frac{1}{\lambda\sqrt{M_\infty^2 - 1}}\frac{d^2 A}{dx_*^2},$$

which is easily integrated to yield

$$f(\xi) = -\frac{1}{\lambda\sqrt{M_\infty^2 - 1}}\frac{dA}{dx_*}(\xi) + C.$$ (2.2.84)

Clearly, the constant of integration C in (2.2.84) has to be set to zero since no pressure perturbations can be produced in region 5 when streamline slope angle (2.2.71) is zero. It remains to substitute (2.2.84) into (2.2.80), and we can conclude that in region 5

$$p_* = -\frac{1}{\lambda\sqrt{M_\infty^2 - 1}}\frac{dA}{dx_*}(\xi), \qquad \xi = x_* - \sqrt{M_\infty^2 - 1}\, y_*.$$ (2.2.85)

Remember that the pressure does not change across the main part of the boundary layer (region 4), as well as across the viscous sublayer (region 3). Consequently, the pressure P^* in region 3 is found by simply setting $y_* = 0$ in (2.2.85):

$$P^* = -\frac{1}{\lambda\sqrt{M_\infty^2 - 1}}\frac{dA}{dx_*}.$$ (2.2.86)

Equation (2.2.86) establishes a relationship between the displacement effect of the boundary layer and the induced pressure in the interaction region. It therefore is referred to as the *interaction law*.

Formulation of the interaction problem

Let us now return to the viscous sublayer (region 3) where the flow is described by equations (2.2.42):

$$\rho_w U^* \frac{\partial U^*}{\partial x_*} + \rho_w V^* \frac{\partial U^*}{\partial Y_*} = -\frac{dP^*}{dx_*} + \mu_w \frac{\partial^2 U^*}{\partial Y_*^2}, \qquad (2.2.87)$$

$$\frac{\partial U^*}{\partial x_*} + \frac{\partial V^*}{\partial Y_*} = 0. \qquad (2.2.88)$$

These have to be solved subject to the no-slip conditions (2.2.43) on the plate surface

$$U^* = V^* = 0 \quad \text{at} \quad Y_* = 0, \qquad (2.2.89)$$

and the initial condition (2.2.46):

$$U^* = \lambda Y_* \quad \text{at} \quad x_* = -\infty. \qquad (2.2.90)$$

It should be noted that while equations (2.2.87), (2.2.88) are the same as in the classical boundary-layer theory, they should now be treated differently. Unlike in Prandtl's formulation, the pressure gradient in (2.2.87) is not known in advance. Instead, the interaction law (2.2.86) should be used:

$$P^* = -\frac{1}{\lambda \sqrt{M_\infty^2 - 1}} \frac{dA}{dx_*}, \qquad (2.2.91)$$

with the displacement function $A(x_*)$ being defined by the first equation in (2.2.55):

$$U^*(x_*, Y_*) = \lambda Y_* + A(x_*) + \cdots \quad \text{as} \quad Y_* \to \infty. \qquad (2.2.92)$$

Considered together, equations (2.2.87)–(2.2.92) constitute the *viscous-inviscid interaction problem*.

2.2.5   Upstream influence

The phenomenon of upstream propagation of perturbations through the boundary layer in supersonic flows was discovered experimentally, and became a main motivation for the development of the theory of self-induced separation.[14] The first theoretical description of the phenomenon was given by Lighthill (1953), whose approach will be adopted here.

We can see that the boundary-value problem (2.2.87)–(2.2.92) admits a trivial solution

$$U^*(x_*, Y_*) = \lambda Y_*, \quad V^*(x_*, Y_*) = 0, \quad P^*(x_*) = 0, \quad A(x_*) = 0,$$

that corresponds to an unperturbed flow in the interaction region. Perturbations may be introduced in the flow by various means, for example, by a shock wave impinging on

---

[14]See Section 2.1.

the boundary layer. If the shock is strong enough, then the boundary-layer separation will take place as shown in Figures 2.6 and 2.7. According to experimental observations, not only the separation takes place before the impinging shock, but also the boundary layer upstream of the separation point S appears to be perturbed. To see how the perturbations decay with the distance $|x_*|$ from the separation point, we shall seek the solution of the interaction problem (2.2.87)–(2.2.92) in the form

$$\left.\begin{aligned}
U^*(x_*, Y_*) &= \lambda Y_* + e^{\kappa x_*} f(Y_*) + \cdots, \\
V^*(x_*, Y_*) &= e^{\kappa x_*} g(Y_*) + \cdots, \\
P^*(x_*) &= e^{\kappa x_*} P_0 + \cdots, \\
A(x_*) &= e^{\kappa x_*} A_0 + \cdots
\end{aligned}\right\} \quad \text{as} \quad x_* \to -\infty. \tag{2.2.93}$$

Here κ, P_0, and A_0 are constants. Of course, the perturbations can only decay if κ is positive.

Substituting (2.2.93) into the momentum (2.2.87) and continuity (2.2.88) equations, and disregarding squares of perturbations, we find that

$$\rho_w \lambda \kappa Y_* f + \rho_w \lambda g = -\kappa P_0 + \mu_w \frac{d^2 f}{dY_*^2}, \tag{2.2.94a}$$

$$\kappa f + \frac{dg}{dY_*} = 0. \tag{2.2.94b}$$

The substitution of (2.2.93) into the boundary conditions (2.2.89) and (2.2.92) gives

$$f = g = 0 \quad \text{at} \quad Y_* = 0, \tag{2.2.94c}$$
$$f = A_0 \quad \text{at} \quad Y_* = \infty. \tag{2.2.94d}$$

Finally, it follows from the interaction law (2.2.91) that

$$P_0 = -\frac{\kappa}{\lambda \sqrt{M_\infty^2 - 1}} A_0. \tag{2.2.94e}$$

Since the boundary-value problem (2.2.94) allows for a trivial solution $f = g = P_0 = A_0 = 0$, it should be treated as an eigenvalue problem for κ. When solving this problem it is convenient to start with elimination of $g(Y_*)$ from equations (2.2.94a) and (2.2.94b). We differentiate (2.2.94a) with respect to Y_*

$$\rho_w \lambda k f + \rho_w \lambda k Y_* \frac{df}{dY_*} + \rho_w \lambda \frac{dg}{dY_*} = \mu_w \frac{d^3 f}{dY_*^3}, \tag{2.2.95}$$

and use (2.2.94b) to eliminate dg/dY_* from (2.2.95). This results in the following equation

$$\mu_w \frac{d^2 h}{dY_*^2} - \rho_w \lambda k Y_* h = 0 \tag{2.2.96}$$

for function

$$h = \frac{df}{dY_*}. \tag{2.2.97}$$

The first boundary condition for equation (2.2.96)

$$\left. \frac{dh}{dY_*} \right|_{Y_*=0} = \frac{\kappa}{\mu_w} P_0$$

is obtained by setting $Y_* = 0$ in (2.2.94a) and using the no-slip boundary conditions (2.2.94c). With the help of the interaction law (2.2.94e) this condition may be expressed in the form

$$\left. \frac{dh}{dY_*} \right|_{Y_*=0} = -\frac{\kappa^2}{\lambda \mu_w \sqrt{M_\infty^2 - 1}} A_0. \tag{2.2.98}$$

To formulate the second condition, we integrate (2.2.97) from $Y_* = 0$, where function f satisfies the no-slip condition (2.2.94c). We have

$$f(Y_*) = \int_0^{Y_*} h(Y_*') \, dY_*'. \tag{2.2.99}$$

Now, to use the boundary condition (2.2.94d), we set $Y_* = \infty$ in (2.2.99). This yields

$$\int_0^\infty h(Y_*) \, dY_* = A_0. \tag{2.2.100}$$

Equation (2.2.96) can be turned into the Airy equation

$$\frac{d^2 h}{dz^2} - zh = 0. \tag{2.2.101a}$$

We do this by introducing a new independent variable $z = (\rho_w \lambda \kappa / \mu_w)^{1/3} Y_*$. Boundary conditions (2.2.98) and (2.2.100) are written in terms of z as

$$\left. \frac{dh}{dz} \right|_{z=0} = -\frac{\kappa^{5/3}}{\lambda^{4/3} \rho_w^{1/3} \mu_w^{2/3} \sqrt{M_\infty^2 - 1}} A_0, \tag{2.2.101b}$$

$$\int_0^\infty h(z) \, dz = \left(\frac{\rho_w \lambda \kappa}{\mu_w} \right)^{1/3} A_0. \tag{2.2.101c}$$

The general solution of equation (2.2.101a) may be written in the form

$$h = C_1 Ai(z) + C_2 Bi(z), \tag{2.2.102}$$

where $Ai(z)$ and $Bi(z)$ are two complementary solutions of the Airy equation, the properties of which are well known (see, for example, Abramowitz and Stegun, 1965).

In particular, it is known that at large values of z the Airy function $Ai(z)$ decays exponentially, while $Bi(z)$ grows exponentially, that is,

$$Ai(z) = \frac{z^{-1/4}}{2\sqrt{\pi}} e^{-\zeta} + \cdots, \qquad Bi(z) = \frac{z^{-1/4}}{\sqrt{\pi}} e^{\zeta} + \cdots,$$

where $\zeta = \frac{2}{3} z^{3/2}$.

Now we need to consider boundary conditions (2.2.101b) and (2.2.101c). Clearly, the integral on the left-hand side of (2.2.101c) should be convergent, which is only possible if the exponentially growing function $Bi(z)$ is excluded from (2.2.102). Hence, we set $C_2 = 0$, which turns (2.2.102) into

$$h = C_1 Ai(z). \tag{2.2.103}$$

The substitution of (2.2.103) into (2.2.101b) and (2.2.101c) results in the following set of linear algebraic equations for C_1 and A_0:[15]

$$\left.\begin{aligned} \frac{1}{3^{1/3}\Gamma(1/3)} C_1 &= \frac{\kappa^{5/3}}{\lambda^{4/3} \rho_w^{1/3} \mu_w^{2/3} \sqrt{M_\infty^2 - 1}} A_0, \\ \frac{1}{3} C_1 &= \left(\frac{\rho_w \lambda \kappa}{\mu_w}\right)^{1/3} A_0. \end{aligned}\right\} \tag{2.2.104}$$

A nontrivial solution of (2.2.104) exists if and only if

$$\begin{vmatrix} \dfrac{1}{3^{1/3}\Gamma(1/3)} & -\dfrac{\kappa^{5/3}}{\lambda^{4/3} \rho_w^{1/3} \mu_w^{2/3} \sqrt{M_\infty^2 - 1}} \\[2ex] \dfrac{1}{3} & -\left(\dfrac{\rho_w \lambda \kappa}{\mu_w}\right)^{1/3} \end{vmatrix}$$

$$= -\frac{(\rho_w \lambda \kappa)^{1/3}}{(3\mu_w)^{1/3}\Gamma(1/3)} + \frac{\kappa^{5/3}}{3\lambda^{4/3} \rho_w^{1/3} \mu_w^{2/3} \sqrt{M_\infty^2 - 1}} = 0. \tag{2.2.105}$$

Equation (2.2.105) has a positive solution

$$\kappa = \lambda^{5/4} \rho_w^{1/2} \mu_w^{1/4} (M_\infty^2 - 1)^{3/8} \frac{\sqrt{3}}{[\Gamma(1/3)]^{3/4}}.$$

This proves that the interaction problem (2.2.87)–(2.2.92) does admit the solution in form of asymptotic expansions (2.2.93) describing the flow upstream of the separation.

---

[15] Here it is taken into account that

$$Ai'(0) = -\frac{1}{3^{1/3}\Gamma(1/3)}, \qquad \int\limits_0^\infty Ai(z)\, dz = \frac{1}{3},$$

with Γ denoting the gamma function.

We were able to find parameter κ in (2.2.93), but the amplitude of the perturbations remained arbitrary. Thus, the asymptotic solution (2.2.93) appears to be non-unique. To clarify the nature of this non-uniqueness, let us consider, say, the pressure:

$$P^*(x_*) = e^{\kappa x_*} P_0 + \cdots . \tag{2.2.106}$$

If we disregard the trivial case of unperturbed flow ($P_0 = 0$), then we can write

$$P_0 = \begin{cases} e^{-\kappa x'_*} & \text{if } P_0 > 0, \\ -e^{-\kappa x'_*} & \text{if } P_0 < 0, \end{cases} \tag{2.2.107}$$

where $x'_* = -\ln|P_0|/\kappa$. The substitution of (2.2.107) into (2.2.106) results in

$$P^*(x_*) = \begin{cases} e^{-\kappa(x_* - x'_*)} & \text{if } P_0 > 0, \\ -e^{-\kappa(x_* - x'_*)} & \text{if } P_0 < 0. \end{cases}$$

We see that there are two solution branches: a 'compression branch' ($P_0 > 0$), and an 'expansion branch' ($P_0 < 0$). Our interest here is in the 'compression branch' that describes the process of deceleration of the flow in the interaction region leading to the flow separation.[16] It should be noted that, apart from an arbitrary shift x'_* along the plate surface, the flow deceleration in the interaction region follows a universal law. Remember that this property of the viscous-inviscid interaction was first established experimentally.[17]

To determine the shift parameter x'_*, an additional boundary condition is required at a downstream location. From a 'physical viewpoint' this can be interpreted as an *upstream influence* through the boundary layer. The analysis presented in this section shows that the upstream propagation of the perturbations through the boundary layer is observed even if the boundary layer remains attached to the plate surface. Behind the separation point, one more upstream-influence mechanism comes into play.

2.2.6   Flow behind the interaction region

The separation leads to a formation of a reverse flow region where $U^* < 0$. Hence, if we use again the analogy with the heat-transfer process, then equation (1.1.27) should be written as

$$\frac{\partial T}{\partial t} = -a' \frac{\partial^2 T}{\partial x^2}, \tag{2.2.108}$$

with the constant a' being positive. It is easy to see that the original form of the heat-transfer equation (1.1.27) can be restored by the 'time inversion' transformation $t' = -t$. This means that equation (2.2.108) requires an 'initial condition in the future'. Correspondingly, when solving the boundary-layer equations (2.2.87), (2.2.88) we need to know the behaviour of the longitudinal velocity U^* in the separation flow region

---

[16]The 'expansion branch' corresponds to accelerating boundary layers that are observed, for example, near corners where the inviscid supersonic flow undergoes Prandtl–Meyer expansion. These were studied by Neiland (1969b) and Neiland and Sychev (1966).

[17]See the discussion on pages 147–149.

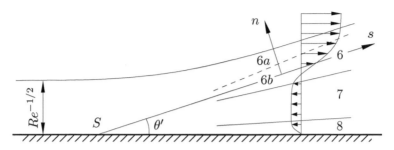

Fig. 2.15: Flow behind the separation point S.

behind the region of viscous-inviscid interaction. To describe this flow, we shall follow Neiland's original 1971 paper.

As the boundary layer separates from the plate surface, it turns into a shear layer. The latter is shown as region 6 in Figure 2.15. In Section 1.4 we studied the properties of shear-layer flows using Chapman's model, where the flow analysis was simplified by an assumption that the boundary layer upstream of the separation point did not exist. In the flow considered here, the shear layer 'originates' from the middle tier (region 4) of the interaction region (see Figure 2.13 on page 157). We shall start the analysis of the flow behind the interaction region by determining the shape of the *dividing streamline*. The latter is the streamline that originates from the separation point S on the plate surface.

It is known from experimental observations that in the viscous-inviscid interaction region the pressure rises monotonically, but then exhibits a characteristic *plateau* some distance downstream of the separation. This is because the fluid motion in the recirculation region is very slow. Let us denote the plateau value of the scaled pressure P^* by P^*_{pl}. The corresponding value of the dimensional pressure is calculated using (2.2.31) and (2.2.1), and may be written as

$$\hat{p} = p_\infty + \rho_\infty V_\infty^2 Re^{-1/4} P^*_{\text{pl}}. \tag{2.2.109}$$

At the outer edge of the shear layer, the Ackeret formula[18]

$$\Delta p = \frac{\hat{p} - p_\infty}{\rho_\infty V_\infty^2} = \frac{\theta}{\sqrt{M_\infty^2 - 1}} \tag{2.2.110}$$

holds. Here $\theta = \theta' + \tilde{\theta}$, with θ' being the slope angle of the dividing streamline and $\tilde{\theta}$ the contribution of the displacement effect of the shear layer. The latter is an order $Re^{-1/2}$ quantity, and can be disregarded in the leading-order approximation. Combining (2.2.110) with (2.2.109) and solving the resulting equation for θ', we can conclude that the dividing streamline is a straight line that makes an angle

$$\theta' = Re^{-1/4}\sqrt{M_\infty^2 - 1}\, P^*_{\text{pl}} \tag{2.2.111}$$

with the plate surface (see Figure 2.15).

[18]See equation (3.2.35) on page 174 in Part 2 of this book series.

Shear layer (region 6)

The asymptotic analysis of the Navier–Stokes equations (2.2.2) in the shear layer (region 6) is based on the limit

$$s = O(1), \quad N = Re^{1/2}n = O(1), \quad Re \to \infty.$$

Here (s, n) are Cartesian coordinates with s measured along the dividing streamline from the separation point S and n in the normal direction. We denote the velocity components in these coordinates as (V_τ, V_n) and represent the fluid-dynamic functions in region 6 in the form of asymptotic expansions

$$
\left.
\begin{aligned}
V_\tau(s, n; Re) &= U_0'(s, N) + \cdots, & V_n(s, n; Re) &= Re^{-1/2}V_0'(s, N) + \cdots, \\
\rho(s, n; Re) &= \rho_0(s, N) + \cdots, & p(s, n; Re) &= Re^{-1/4}P_{\mathrm{pl}}^* + \cdots, \\
h(s, n; Re) &= h_0(s, N) + \cdots, & \mu(s, n; Re) &= \mu_0(s, N) + \cdots.
\end{aligned}
\right\} \quad (2.2.112)
$$

We use the 'prime' in U_0', V_0' to distinguish these from the correspondent velocity components in the (x, y) coordinates. Of course, the Navier–Stokes equations in the (s, n) coordinates have the same form (2.2.2) as in the (x, y) coordinates. One simply needs to replace (x, y) and (u, v) in (2.2.2) by (s, n) and (V_τ, V_n), respectively. The substitution of (2.2.112) into the Navier–Stokes equations results in

$$\rho_0 U_0' \frac{\partial U_0'}{\partial s} + \rho_0 V_0' \frac{\partial U_0'}{\partial N} = \frac{\partial}{\partial N}\left(\mu_0 \frac{\partial U_0'}{\partial N}\right), \tag{2.2.113a}$$

$$\rho_0 U_0' \frac{\partial h_0}{\partial s} + \rho_0 V_0' \frac{\partial h_0}{\partial N} = \frac{1}{Pr}\frac{\partial}{\partial N}\left(\mu_0 \frac{\partial h_0}{\partial N}\right) + \mu_0\left(\frac{\partial U_0'}{\partial N}\right)^2, \tag{2.2.113b}$$

$$\frac{\partial(\rho_0 U_0')}{\partial s} + \frac{\partial(\rho_0 V_0')}{\partial N} = 0, \tag{2.2.113c}$$

$$h_0 = \frac{1}{(\gamma - 1)M_\infty^2}\frac{1}{\rho_0}. \tag{2.2.113d}$$

The initial conditions for these equations

$$
\left.
\begin{aligned}
U_0' &= U_{00}(N), \\
h_0 &= h_{00}(N)
\end{aligned}
\right\} \quad \text{at} \quad s = 0, \tag{2.2.114}
$$

are obtained by matching the asymptotic expansions for the longitudinal velocity V_τ and enthalpy h in (2.2.112) with the corresponding expansions (2.2.60), (2.2.63) in the middle tier of the interaction region.

When formulating the boundary conditions at the outer edge of the shear layer, we need to keep in mind that the angle θ' of deflection of the shear layer from the plate surface is an order $Re^{-1/4}$ quantity. It can only produce $O(Re^{-1/4})$ perturbations in the inviscid flow above the shear layer. Hence, in the leading-order approximation, the longitudinal velocity and enthalpy should tend to their free-stream values:

$$
\left.
\begin{aligned}
U_0' &= 1, \\
h_0 &= \frac{1}{(\gamma - 1)M_\infty^2}
\end{aligned}
\right\} \quad \text{as} \quad N \to \infty. \tag{2.2.115}
$$

At the lower edge of the shear layer, we shall require[19]

$$U_0' = 0, \left.\vphantom{\begin{matrix}a\\b\end{matrix}}\right\} \quad \text{at} \quad N = -\infty. \qquad (2.2.116)$$
$$h_0 = h_w$$

To close the boundary-value problem (2.2.113)–(2.2.116), we also need a boundary condition for V_0'. Since the s-axis lies along a streamline, we can write

$$V_0' = 0 \quad \text{at} \quad N = 0. \qquad (2.2.117)$$

Remember that our task here is to formulate the 'downstream' boundary condition for equations (2.2.87), (2.2.88) that describe the flow in the lower tier of the interaction region. To perform this task we need to know the behaviour of the solution to the boundary-value problem (2.2.113)–(2.2.117) for small values of s. We start with the main part of the shear layer, shown as region $6a$ in Figure 2.15. It is defined by the limit

$$N = O(1), \quad s \to 0.$$

In this region the longitudinal velocity component U_0' is represented by the initial velocity profile $U_{00}(N)$:

$$U_0' = U_{00}(N) + \cdots. \qquad (2.2.118)$$

Notice that, according to (2.2.5),

$$U_{00}(N) = \lambda N + \cdots \quad \text{as} \quad N \to 0, \qquad (2.2.119)$$

which precludes a smooth transition to the reverse flow below the shear layer. Consequently, in addition to region $6a$, we need to consider a viscous region $6b$, where N decreases with s, and the convective terms in the momentum equation (2.2.113a) are in balance with the viscous term:

$$\rho_0 U_0' \frac{\partial U_0'}{\partial s} \sim \frac{\partial}{\partial N}\left(\mu_0 \frac{\partial U_0'}{\partial N}\right). \qquad (2.2.120)$$

Representing the derivatives in (2.2.120) by finite differences and taking into account that the density ρ_0 and the viscosity μ_0 are order one quantities, we have

$$U_0' \frac{\Delta U_0'}{s} \sim \frac{\Delta U_0'}{N^2}. \qquad (2.2.121)$$

It follows from (2.2.118) and (2.2.119) that $U_0' \sim N$, which being substituted into (2.2.121) yields

$$N \frac{\Delta U_0'}{s} \sim \frac{\Delta U_0'}{N^2}.$$

We see that the thickness of region $6b$ tends to zero as

$$N \sim s^{1/3}. \qquad (2.2.122)$$

---

[19]The validity of these conditions will be confirmed by analysing the flow in the recirculation region 7.

An estimate for the longitudinal velocity in region 6b,

$$U_0' \sim s^{1/3}, \tag{2.2.123}$$

can now be obtained by substituting (2.2.122) into (2.2.119) and then into (2.2.118).

To progress further it is convenient to introduce the stream function ψ. Remember that its existence is guaranteed by continuity equation.[20] In the flow considered here the continuity equation (2.2.2d) is written as

$$\frac{\partial(\rho V_\tau)}{\partial s} + \frac{\partial(\rho V_n)}{\partial n} = 0.$$

We see that the stream function $\psi(s, n; Re)$ may be defined by the equations

$$V_\tau = \frac{1}{\rho}\frac{\partial\psi}{\partial n}, \qquad V_n = -\frac{1}{\rho}\frac{\partial\psi}{\partial s}. \tag{2.2.124}$$

Corresponding to (2.2.112), the asymptotic expansion of the stream function in region 6 should be sought in the form

$$\psi(s, n; Re) = Re^{-1/2}\Psi(s, N) + \cdots \quad \text{as} \quad Re \to \infty. \tag{2.2.125}$$

The substitution of (2.2.125) together with the asymptotic expansions for V_τ and V_n in (2.2.112) into equations (2.2.124) yields

$$U_0' = \frac{1}{\rho_0}\frac{\partial\Psi}{\partial N}, \qquad V_0' = -\frac{1}{\rho_0}\frac{\partial\Psi}{\partial s}. \tag{2.2.126}$$

The form of the solution for $\Psi(s, N)$ in region 6b is easily predicted by approximating the first of equations (2.2.126) as

$$U_0' \sim \frac{\Psi}{\rho_0 N},$$

and using estimates (2.2.122) and (2.2.123) for U_0' and N, respectively. Keeping in mind that the dimensionless density ρ_0 is an order one quantity everywhere in the flow field, we can see that $\Psi \sim s^{2/3}$. This suggests that in region 6b the stream function should be sought in the form

$$\Psi(s, N) = s^{2/3}\varphi(\eta) + \cdots \quad \text{as} \quad s \to 0, \tag{2.2.127a}$$

where

$$\eta = \frac{N}{s^{1/3}}. \tag{2.2.127b}$$

Since the flow in region 6b is slow, we can expect it to behave as incompressible. To prove this conjecture we write the asymptotic expansions of the enthalpy, density,

---

[20]See Section 3.3.1 in Part 1 of this book series, where the incompressible form of the continuity equation (3.3.7) was used for this purpose.

and viscosity coefficient as

$$\left.\begin{array}{l} h_0(s, N) = \check{h}(\eta) + \cdots, \\ \rho_0(s, N) = \check{\rho}(\eta) + \cdots, \\ \mu_0(s, N) = \check{\mu}(\eta) + \cdots \end{array}\right\} \quad \text{as} \quad s \to 0. \qquad (2.2.128)$$

The substitution of (2.2.127) together with the asymptotic expansion of ρ_0 in (2.2.128) into equations (2.2.126) yields

$$U_0' = s^{1/3}\frac{1}{\check{\rho}}\frac{d\varphi}{d\eta} + \cdots, \qquad V_0' = -s^{-1/3}\frac{1}{\check{\rho}}\left(\frac{2}{3}\varphi - \frac{1}{3}\eta\frac{d\varphi}{d\eta}\right) + \cdots. \qquad (2.2.129)$$

To evaluate the compressibility of the flow, we need to analyse the energy equation (2.2.113b). The substitution of (2.2.128) and (2.2.129) into (2.2.113b) results in

$$\frac{1}{Pr}\frac{d}{d\eta}\left(\check{\mu}\frac{d\check{h}}{d\eta}\right) + \frac{2}{3}\varphi\frac{d\check{h}}{d\eta} = 0. \qquad (2.2.130)$$

This is a second order linear differential equation for $\check{h}(\eta)$. It requires two boundary conditions. The first one,

$$\check{h} = h_w \quad \text{at} \quad \eta = -\infty, \qquad (2.2.131)$$

is obtained by substitution of the asymptotic expansion of h_0 in (2.2.128) into the boundary condition (2.2.116) for h_0 at the lower edge of the shear layer. To formulate the second condition, we need to perform the matching with the solution in region 6a (see Figure 2.15). It immediately follows from (2.2.114) that the leading-order term of the asymptotic expansion of the enthalpy in region 6a is written as

$$h_0 = h_{00}(N) + \cdots \quad \text{as} \quad s \to 0.$$

Taking into account that, according to (2.2.5),

$$h_{00}(N) = h_w + \cdots \quad \text{as} \quad N \to 0,$$

it is easily seen that at the outer edge of region 6b

$$\check{h} = h_w \quad \text{at} \quad \eta = \infty. \qquad (2.2.132)$$

Obviously, the solution of the boundary-value problem (2.2.130), (2.2.131), and (2.2.132) is given by

$$\check{h}(\eta) = h_w.$$

This confirms that the flow in region 6b is incompressible, and the density and the viscosity coefficient are constants:

$$\check{\rho}(\eta) = \rho_w, \qquad \check{\mu}(\eta) = \mu_w. \qquad (2.2.133)$$

To determine the velocity field in region 6b we need to consider the momentum equation (2.2.113a). The substitution of (2.2.129) together with (2.2.133) into (2.2.113a) results in the following equation for $\varphi(\eta)$:

$$\mu_w\frac{d^3\varphi}{d\eta^3} + \frac{2}{3}\varphi\frac{d^2\varphi}{d\eta^2} - \frac{1}{3}\left(\frac{d\varphi}{d\eta}\right)^2 = 0. \qquad (2.2.134)$$

Equation (2.2.134) requires three boundary conditions. The first of these is obtained from matching with the solution (2.2.118) in region 6a. To perform the matching we need to analyse the behaviour of this solution at the 'bottom' of region 6a, where equation (2.2.119) holds. The substitution of (2.2.119) into (2.2.118) yields

$$U_0' = \lambda N + \cdots . \tag{2.2.135}$$

Now, we need to express equation (2.2.135) in terms of the variables of region 6b. This is done by solving (2.2.127b) for N and substituting the result into (2.2.135). We have

$$U_0' = s^{1/3}\lambda\eta + \cdots . \tag{2.2.136}$$

It remains to compare (2.2.136) with the equation for U_0' in (2.2.129). Using (2.2.133) we can write this equation as

$$U_0' = s^{1/3}\frac{1}{\rho_w}\frac{d\varphi}{d\eta} + \cdots . \tag{2.2.137}$$

We see that the sought boundary condition is

$$\frac{d\varphi}{d\eta} = \lambda\rho_w\eta + \cdots \quad \text{as} \quad \eta \to \infty. \tag{2.2.138}$$

The boundary condition at the lower edge of region 6b is obtained by substituting (2.2.137) into the condition for U_0' in (2.2.116). We have

$$\frac{d\varphi}{d\eta} = 0 \quad \text{at} \quad \eta = -\infty. \tag{2.2.139}$$

Finally, by substituting the equation for V_0' in (2.2.129) into boundary condition (2.2.117), we find that

$$\varphi = 0 \quad \text{at} \quad \eta = 0. \tag{2.2.140}$$

Similar to the Chapman problem (1.4.12), the boundary-value problem (2.2.134), (2.2.138)–(2.2.140) requires a numerical solution. Before performing the calculations it is convenient to scale parameters μ_w, ρ_w, and λ out of the problem formulation. This is done with the help of the affine transformations

$$\varphi = \left(\mu_w^2\rho_w\lambda\right)^{1/3}\bar{\varphi}, \qquad \eta = \left(\frac{\mu_w}{\rho_w\lambda}\right)^{1/3}\bar{\eta}. \tag{2.2.141}$$

They turn the problem (2.2.134), (2.2.138)–(2.2.140) into

$$\left.\begin{aligned}
\frac{d^3\bar{\varphi}}{d\bar{\eta}^3} + \frac{2}{3}\bar{\varphi}\frac{d^2\bar{\varphi}}{d\bar{\eta}^2} - \frac{1}{3}\left(\frac{d\bar{\varphi}}{d\bar{\eta}}\right)^2 = 0, \\
\frac{d\bar{\varphi}}{d\bar{\eta}}\bigg|_{\bar{\eta}=-\infty} = 0, \quad \bar{\varphi}\big|_{\bar{\eta}=0} = 0, \quad \frac{d^2\bar{\varphi}}{d\bar{\eta}^2}\bigg|_{\bar{\eta}=\infty} = 1.
\end{aligned}\right\} \tag{2.2.142}$$

The results of a numerical solution of the boundary-value problem (2.2.142) are displayed in Figure 2.16, where in addition to function $\bar{\varphi}$ we also show the behaviour

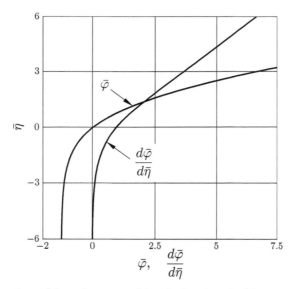

Fig. 2.16: Distribution of function $\bar{\varphi}$ and its derivative $d\bar{\varphi}/d\bar{\eta}$ across the shear layer.

of the derivative $d\bar{\varphi}/d\bar{\eta}$. The latter represents the longitudinal velocity profile in the shear layer. Indeed, substituting (2.2.141) into (2.2.137) and then into the asymptotic expansion for V_τ in (2.2.112) we find that

$$V_\tau = s^{1/3}\left(\frac{\mu_w \lambda^2}{\rho_w}\right)^{1/3}\frac{d\bar{\varphi}}{d\bar{\eta}} + \cdots . \tag{2.2.143}$$

Before turning to the flow analysis in the reverse flow region, shown as region 7 in Figure 2.15, we need to describe the entrainment effect of the shear layer. Figure 2.16 shows that function $\bar{\varphi}$ has a finite negative limit at the lower edge of the shear layer, that is[21]

$$\bar{\varphi} = -\Lambda + \cdots \quad \text{as} \quad \bar{\eta} \to -\infty. \tag{2.2.144}$$

The numerical value of constant Λ was found to be $\Lambda = 1.2565$. Combining (2.2.144) with (2.2.141) we have

$$\varphi = -\left(\mu_w^2 \rho_w \lambda\right)^{1/3}\Lambda + \cdots \quad \text{as} \quad \eta \to -\infty. \tag{2.2.145}$$

Now we can substitute (2.2.145) into (2.2.127), and then into (2.2.125). We find that at the lower edge of the shear layer (region 6) the stream function ψ assume a value

$$\psi\Big|_{N=-\infty} = -Re^{-1/2}s^{2/3}\left(\mu_w^2 \rho_w \lambda\right)^{1/3}\Lambda + \cdots . \tag{2.2.146}$$

Reverse flow (region 7)

Region 7 is situated below the shear layer (see Figure 2.15). Remember that the angle (2.2.111) made by the shear layer with the plate surface is an $O(Re^{-1/4})$ quantity.

---

[21]Next order terms in the asymptotic expansion (2.2.144) can be found in the same way as was done in Section 1.4.2, where the entrainment effect of the Chapman shear layer was analysed.

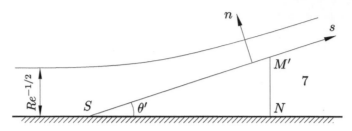

Fig. 2.17: Control volume $SM'N$.

This means that the asymptotic analysis of the Navier–Stokes equations in region 7 should be based on the limit

$$x = O(1), \quad \check{y} = Re^{1/4}y = O(1), \quad Re \to \infty. \tag{2.2.147}$$

Here, we shall use the Cartesian coordinate (x, y) with x measured along the plate surface from the leading edge of the plate, and y in the perpendicular direction. The separation point S is situated at $x = 1$.

To predict the form of the asymptotic expansions of the fluid-dynamic functions in region 7, we shall apply the mass conservation law to the control volume $SM'N$, where SM' lies along the upper boundary of region 7 and $M'N$ is perpendicular to the plate surface (see Figure 2.17). According to equation (3.3.13) in Part 1, for any two points M and M' in the (x, y)-plane, the mass flux Q through a line connecting these points is calculated as

$$Q = \psi(M') - \psi(M).$$

If we choose point M to coincide with the separation point S, and point M' to lie a distance s from point S, then the mass flux through the upper side SM' of the control volume will be given by (2.2.146):

$$Q = Re^{-1/2}s^{2/3}\left(\mu_w^2\rho_w\lambda\right)^{1/3}\Lambda. \tag{2.2.148}$$

This flux is produced by the entrainment effect of the shear layer and should be compensated by an equal flux through the rear side $M'N$ of the control volume:

$$Q = \int_N^{M'} \rho u \, dy. \tag{2.2.149}$$

Comparison of (2.2.148) and (2.2.149) leads to the following order-of-magnitude equation:

$$\rho u y \sim Re^{-1/2}.$$

Taking into account that the density ρ is an order one quantity, and the width of region 7 is estimated as $y \sim Re^{-1/4}$, we can conclude that the reverse flow velocity

$$u \sim Re^{-1/4}. \tag{2.2.150}$$

Now, by balancing the two terms in the continuity equation (2.2.2d), we find that

$$v \sim Re^{-1/2}. \tag{2.2.151}$$

To estimate the pressure variations in region 7, we balance the first convective term in the x-momentum equation (2.2.2a) with the pressure gradient. We have

$$\Delta p \sim u^2 \sim Re^{-1/2}. \tag{2.2.152}$$

Guided by (2.2.147), (2.2.150)–(2.2.152), we represent the fluid-dynamic functions in region 7 in the form of asymptotic expansions

$$\left.\begin{array}{ll}
u = Re^{-1/4}\check{u}_0(x,\check{y}) + \cdots, & v = Re^{-1/2}\check{v}_0(x,\check{y}) + \cdots, \\
\rho = \check{\rho}_0(x,\check{y}) + \cdots, & p = Re^{-1/4}P^*_{\text{pl}} + Re^{-1/2}\check{p}_1(x,\check{y}) + \cdots, \\
h = \check{h}_0(x,\check{y}) + \cdots, & \mu = \check{\mu}_0(x,\check{y}) + \cdots.
\end{array}\right\} \tag{2.2.153}$$

The substitution of (2.2.153) into the Navier–Stokes equations (2.2.2) yields

$$\check{\rho}_0\left(\check{u}_0\frac{\partial \check{u}_0}{\partial x} + \check{v}_0\frac{\partial \check{u}_0}{\partial \check{y}}\right) = -\frac{\partial \check{p}_1}{\partial x}, \tag{2.2.154a}$$

$$\frac{\partial \check{p}_1}{\partial \check{y}} = 0, \tag{2.2.154b}$$

$$\check{u}_0\frac{\partial \check{h}_0}{\partial x} + \check{v}_0\frac{\partial \check{h}_0}{\partial \check{y}} = 0, \tag{2.2.154c}$$

$$\frac{\partial(\check{\rho}_0\check{u}_0)}{\partial x} + \frac{\partial(\check{\rho}_0\check{v}_0)}{\partial \check{y}} = 0, \tag{2.2.154d}$$

$$\check{h}_0 = \frac{1}{(\gamma-1)M^2_\infty}\frac{1}{\check{\rho}_0}. \tag{2.2.154e}$$

Let us start with the energy equation (2.2.154c). It shows that the enthalpy $\check{h}_0$ remains constant along each streamline. This allows us to treat the flow in region 7 as incompressible. Indeed, let us assume, subject to subsequent confirmation, that all streamlines in region 7 originate from a stagnant fluid region situated sufficiently far downstream of the separation point S. Let us further assume that in the stagnation region the gas temperature coincides with the plate temperature, that is, $\check{h}_0 = h_w$. Then it follows from the energy equation (2.2.154c) that

$$\check{h}_0 = h_w \tag{2.2.155}$$

everywhere in region 7. Substituting (2.2.155) into the state equation (2.2.154e) and comparing the result with (2.2.6), we can see that the density $\check{\rho}$ is constant and equals ρ_w.

Now, we turn to the x-momentum equation (2.2.154a). The pressure gradient on the right-hand side of this equation is independent of $\check{y}$. Differentiating (2.2.154a) with respect to $\check{y}$ and using the continuity equation (2.2.154d), we find that

$$\breve{u}_0 \frac{\partial}{\partial \breve{x}}\left(\frac{\partial \breve{u}_0}{\partial \breve{y}}\right) + \breve{v}_0 \frac{\partial}{\partial \breve{y}}\left(\frac{\partial \breve{u}_0}{\partial \breve{y}}\right) = 0.$$

This means that the vorticity $\omega = \partial \breve{u}_0/\partial \breve{y}$ stays constant along each streamline. Since ω is zero in the stagnation region, we can conclude that

$$\frac{\partial \breve{u}_0}{\partial \breve{y}} = 0 \tag{2.2.156}$$

in the entire region 7.

It is convenient to express equation (2.2.156) in terms of the stream function. When dealing with the flow in region 6, we defined this function by equations (2.2.124) and (2.2.125). In the Cartesian coordinates used here, equations (2.2.124) are written as

$$u = \frac{1}{\rho}\frac{\partial \psi}{\partial y}, \qquad v = -\frac{1}{\rho}\frac{\partial \psi}{\partial x}. \tag{2.2.157}$$

By substituting (2.2.125) and the asymptotic expansion (2.2.153) for u in region 7 into the first equation in (2.2.157) we find that

$$\breve{u}_0 = \frac{1}{\rho_w}\frac{\partial \Psi}{\partial \breve{y}}. \tag{2.2.158}$$

Here it is taken into account that in region 7, the density $\rho = \rho_w$. The substitution of (2.2.158) into (2.2.156) results in

$$\frac{\partial^2 \Psi}{\partial \breve{y}^2} = 0. \tag{2.2.159}$$

When formulating the boundary conditions for (2.2.159) we have to keep in mind that equations (2.2.154) do not involve viscous terms. This means that the flow in region 7 is inviscid, and, therefore, instead of the no-slip conditions on the plate surface we must use the impermeability condition. It is written in terms of the stream function as

$$\Psi = 0 \quad \text{at} \quad \breve{y} = 0. \tag{2.2.160}$$

The condition of matching with the solution (2.2.146) in region 6 reads[22]

$$\Psi = -(x-1)^{2/3}\left(\mu_w^2 \rho_w \lambda\right)^{1/3}\Lambda \quad \text{at} \quad \breve{y} = \sqrt{M_\infty^2 - 1}\, P_{\text{pl}}^*\,(x-1). \tag{2.2.161}$$

The solution of the boundary-value problem (2.2.159)–(2.2.161) is written as

$$\Psi = -\frac{\left(\mu_w^2 \rho_w \lambda\right)^{1/3}\Lambda}{\sqrt{M_\infty^2 - 1}\, P_{\text{pl}}^*}\frac{\breve{y}}{(x-1)^{1/3}}. \tag{2.2.162}$$

It remains to substitute (2.2.162) into (2.2.158) and then into the asymptotic expansion for u in (2.2.153). We find that in region 7 the longitudinal velocity is given by

$$u = -Re^{-1/4}\frac{\mu_w^{2/3}\lambda^{1/3}\Lambda}{\rho_w^{2/3}\beta P_{\text{pl}}^*}\,(x-1)^{-1/3}, \tag{2.2.163}$$

where $\beta = \sqrt{M_\infty^2 - 1}$.

---

[22]Here it has been taken into account that in the leading-order approximation, s can be substituted with $x - 1$.

Near-wall boundary layer (region 8)

As expected, the solution (2.2.163) in region 7 does not satisfy the no-slip condition on the plate surface. Consequently, to complete the analysis of the flow behind the separation point, we also need to consider the boundary layer that forms on the plate surface in the reverse flow region. This boundary layer (shown as region 8 in Figure 2.15) belongs to the Falkner and Skan class (see Section 1.3). The solution for region 8 is discussed in detail in Problem 5 in Exercises 9. In particular, the longitudinal velocity is found to be

$$u = -Re^{-1/4}\frac{\mu_w^{2/3}\lambda^{1/3}\Lambda}{\rho_w^{2/3}\beta P_{\rm pl}^*}(x-1)^{-1/3}f'(\zeta) + \cdots \text{as} \quad x \to 0, \qquad (2.2.164a)$$

where

$$\zeta = Re^{3/8}\left(\frac{\rho_w\lambda}{\mu_w\beta^3}\right)^{1/6}\sqrt{\frac{\Lambda}{P_{\rm pl}^*}}\frac{y}{(x-1)^{2/3}}, \qquad (2.2.164b)$$

and function $f(\zeta)$ is the solution of the following boundary-value problem:

$$\left.\begin{array}{l} f''' - \dfrac{1}{3}ff'' + \dfrac{1}{3}(1-f'^2) = 0, \\[2mm] f(0) = f'(0) = 0, \quad f'(\infty) = 1. \end{array}\right\} \qquad (2.2.165)$$

Downstream boundary condition for the interaction region

It is convenient to express the results of the analysis of the flow behind the interaction region in the form of a composite asymptotic expansion.[23] We start with the observation that the solution (2.2.164) for region 8 is also valid in region 7. Indeed, in region 7, the argument ζ of function $f(\zeta)$ is large, and, therefore, $f'(\zeta) = 1$, which turns (2.2.164) into (2.2.163). Now, let us turn to region 6. The solution (2.2.143) for this region was obtained using the 'body-fitted' coordinates (s, n). However, since the angle θ' between the s- and x-axes is small, we can substitute V_τ in (2.2.143) by u and s by $x - 1$. We have

$$u = \left(\frac{\mu_w\lambda^2}{\rho_w}\right)^{1/3}(x-1)^{1/3}\frac{d\bar\varphi}{d\bar\eta} + \cdots. \qquad (2.2.166)$$

To construct a composite asymptotic expansion for u valid in regions 6, 7, and 8, we simply add (2.2.166) to (2.2.164a):

$$u = \left(\frac{\mu_w\lambda^2}{\rho_w}\right)^{1/3}(x-1)^{1/3}\bar\varphi'(\bar\eta)$$

$$- Re^{-1/4}\left(\frac{\mu_w^2\lambda}{\rho_w^2\beta^3}\right)^{1/3}\frac{\Lambda}{P_{\rm pl}^*}(x-1)^{-1/3}f'(\zeta) + \cdots. \qquad (2.2.167)$$

In region 6, the second term in (2.2.167) is much smaller than the first term and may be disregarded in the leading-order approximations. However, in regions 7 and 8, the derivative $d\bar\varphi/d\bar\eta$ becomes transcendentally small, and (2.2.167) reduces to (2.2.166).

---

[23]See page 49 in Part 2 of this book series.

The argument ζ of function $f(\zeta)$ is expressed in Cartesian coordinates (x, y) by equation (2.2.164b). To cast the argument $\bar{\eta}$ of function $\bar{\varphi}(\bar{\eta})$ in (x, y) coordinates, we use Prandtl's transposition theorem (see Section 1.4.3). It follows from (2.2.111) that the dividing streamline is given by the equation

$$y = Re^{-1/4} \beta P_{\text{pl}}^* (x - 1).$$

Therefore,

$$\eta = \frac{N}{s^{1/3}} = \frac{Re^{1/2} n}{s^{1/3}} = \frac{Re^{1/2} \left[y - Re^{-1/4} \beta P_{\text{pl}}^* (x - 1) \right]}{(x - 1)^{1/3}}.$$

It remains to apply the affine transformation (2.2.141), and we will have

$$\bar{\eta} = \left(\frac{\rho_w \lambda}{\mu_w} \right)^{1/3} \frac{Re^{1/2} \left[y - Re^{-1/4} \beta P_{\text{pl}}^* (x - 1) \right]}{(x - 1)^{1/3}}. \qquad (2.2.168)$$

Now, we are ready to perform the matching with the solution in the lower tier (region 3) of the interaction region. For this purpose we express (2.2.167), (2.2.168), and (2.2.164b) in terms of variables (2.2.30) of region 3. This yields

$$u = Re^{-1/8} \left[x_*^{1/3} \left(\frac{\mu_w \lambda^2}{\rho_w} \right)^{1/3} \bar{\varphi}'(\bar{\eta}) - x_*^{-1/3} \left(\frac{\mu_w^2 \lambda}{\rho_w^2 \beta^3} \right)^{1/3} \frac{\Lambda}{P_{\text{pl}}^*} f'(\zeta) \right] + \cdots, \qquad (2.2.169a)$$

$$\bar{\eta} = \left(\frac{\rho_w \lambda}{\mu_w} \right)^{1/3} \frac{Y_* - \beta P_{\text{pl}}^* x_*}{x_*^{1/3}}, \qquad \zeta = \left(\frac{\rho_w \lambda}{\mu_w \beta^3} \right)^{1/6} \sqrt{\frac{\Lambda}{P_{\text{pl}}^*} \frac{Y_*}{x_*^{2/3}}}. \qquad (2.2.169b)$$

The comparison of (2.2.169a) with the asymptotic expansion (2.2.31) for u in region 3 shows that the sought matching condition is written as

$$U^* = x_*^{1/3} \left(\frac{\mu_w \lambda^2}{\rho_w} \right)^{1/3} \bar{\varphi}'(\bar{\eta})$$

$$- x_*^{-1/3} \left(\frac{\mu_w^2 \lambda}{\rho_w^2 \beta^3} \right)^{1/3} \frac{\Lambda}{P_{\text{pl}}^*} f'(\zeta) + \cdots \quad \text{as} \quad x_* \to \infty. \qquad (2.2.170)$$

This completes the formulation of the viscous-inviscid interaction problem for the self-induced separation of the boundary layer in a supersonic flow.

2.2.7   Canonical form of the interaction problem

Boundary-value problem (2.2.87)–(2.2.92), (2.2.170) involves four parameters: μ_w, ρ_w, λ, and $\beta = \sqrt{M_\infty^2 - 1}$. These depend on the flow regime considered. Fortunately, we do not need to solve the interaction problem each time a new combination of μ_w, ρ_w, λ, and β is considered. Instead, we can use the affine transformations

$$\left. \begin{array}{ccc} x_* = \dfrac{\mu_w^{-1/4} \rho_w^{-1/2}}{\lambda^{5/4} \beta^{3/4}} \, \bar{X}, & Y_* = \dfrac{\mu_w^{1/4} \rho_w^{-1/2}}{\lambda^{3/4} \beta^{1/4}} \, \bar{Y}, & U^* = \dfrac{\mu_w^{1/4} \rho_w^{-1/2}}{\lambda^{-1/4} \beta^{1/4}} \, \bar{U}, \\[4mm] V^* = \dfrac{\mu_w^{3/4} \rho_w^{-1/2}}{\lambda^{-3/4} \beta^{-1/4}} \, \bar{V}, & A = \dfrac{\mu_w^{1/4} \rho_w^{-1/2}}{\lambda^{-1/4} \beta^{1/4}} \, \bar{A}, & P^* = \dfrac{\mu_w^{1/2}}{\lambda^{-1/2} \beta^{1/2}} \, \bar{P} \end{array} \right\} \qquad (2.2.171)$$

that allow us to exclude μ_w, ρ_w, λ, and β from the problem formulation, and cast the interaction problem in the following canonical form:

$$\left.\begin{aligned} \bar{U}\frac{\partial \bar{U}}{\partial \bar{X}} + \bar{V}\frac{\partial \bar{U}}{\partial \bar{Y}} &= -\frac{d\bar{P}}{d\bar{X}} + \frac{\partial^2 \bar{U}}{\partial \bar{Y}^2}, \\ \frac{\partial \bar{U}}{\partial \bar{X}} + \frac{\partial \bar{V}}{\partial \bar{Y}} &= 0, \\ \bar{P} &= -\frac{d\bar{A}}{d\bar{X}}, \\ \bar{U} = \bar{V} = 0 \qquad & \text{at} \quad \bar{Y} = 0, \\ \bar{U} = \bar{Y} + \cdots \qquad & \text{as} \quad \bar{X} \to -\infty, \\ \bar{U} = \bar{Y} + \bar{A}(\bar{X}) + \cdots \qquad & \text{as} \quad \bar{Y} \to \infty, \\ \bar{U} = \bar{X}^{1/3}\bar{\varphi}'(\bar{\eta}) - \bar{X}^{-1/3}\frac{\Lambda}{\bar{P}_{\text{pl}}}f'(\zeta) + \cdots \qquad & \text{as} \quad \bar{X} \to \infty. \end{aligned}\right\} \qquad (2.2.172)$$

Here $\bar{P}_{\text{pl}}$ is the transformed plateau pressure:

$$\bar{P}_{\text{pl}} = \left(\frac{\beta}{\mu_w\lambda}\right)^{1/2} P^*_{\text{pl}}. \qquad (2.2.173)$$

Functions $\bar{\varphi}(\bar{\eta})$ and $f(\zeta)$ are found by solving the boundary-value problems (2.2.142) and (2.2.165), respectively. The arguments (2.2.169b) of these functions are written in the new variables as

$$\bar{\eta} = \frac{\bar{Y} - \bar{P}_{\text{pl}}\bar{X}}{\bar{X}^{1/3}}, \qquad \zeta = \sqrt{\frac{\Lambda}{\bar{P}_{\text{pl}}}}\frac{\bar{Y}}{\bar{X}^{2/3}}.$$

The solution of the interaction problem (2.2.172) requires special numerical techniques. These will not be discussed here; the interested reader is referred to Chapter 7 of Sychev *et al.* (1998). The calculation results are displayed in Figure 2.18, where the pressure distribution along the interaction region is shown together with the skin friction. The latter is defined as

$$\bar{\tau}_w = \left.\frac{\partial \bar{U}}{\partial \bar{Y}}\right|_{\bar{Y}=0}.$$

In the variables used here, the skin friction in the boundary layer before the interaction region assumes a value of $\bar{\tau}_w = 1$. As the pressure starts to rise in the interaction region, the flow experiences a deceleration and the skin friction starts to decrease. It becomes zero at the separation point, and then turns negative, which signifies that fluid near the plate surface is moving in the direction opposite to the flow outside the boundary layer. While the pressure grows monotonically through the separation point and further downstream, the skin friction reaches a minimum some distance downstream of the separation point and then starts to recover tending to zero as $\bar{X} \to \infty$. This confirms that in the reverse flow region the fluid velocity decreases

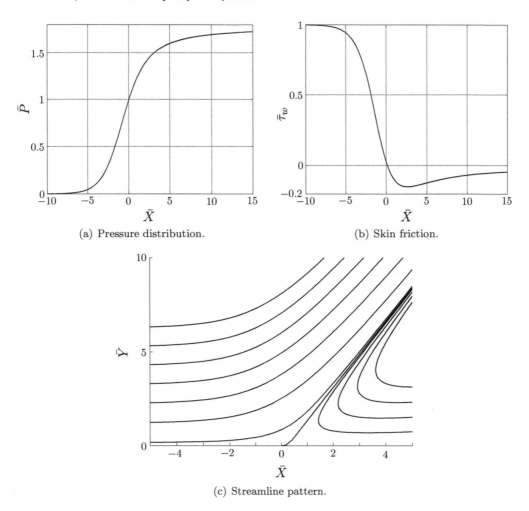

(a) Pressure distribution.

(b) Skin friction.

(c) Streamline pattern.

Fig. 2.18: Results of the numerical solution of the interaction problem (2.2.172).

with $\bar{X}$, which explains why the pressure develops a plateau. The numerical value of the plateau pressure is found to be

$$\bar{P}_{\text{pl}} = 1.723. \tag{2.2.174}$$

The streamline pattern in the viscous sublayer (region 3) is shown in Figure 2.18(c). For a comparison of these results with experimental data, the reader is referred to Neiland (1969*a*) and Stewartson and Williams (1969).

To conclude, we make the following important remark. We have demonstrated that the solution of the boundary-value problem (2.2.172) provides a universal description of the phenomenon of self-induced separation in supersonic flows. To apply this solution to a particular flow, one needs to calculate the boundary layer upstream of the interaction region, and find the skin friction parameter λ, the gas density near the body surface ρ_w, and the corresponding value of viscosity coefficient μ_w. Also, one needs to calculate parameter $\beta = \sqrt{M_\infty^2 - 1}$ using instead of the free-stream Mach number M_∞ the local Mach number at the outer edge of the boundary layer. Once the parameters λ, ρ_w, μ_w, and β are found, one can return to the original variables with the help of the affine transformations (2.2.171). Obviously, these transformations may also be used to relate one flow to another, thereby establishing the similarity rules for this class of flows. Unlike the classical similarity rules of fluid dynamics (see Section 1.7.4 in Part 1 of this book series), the similarity rules expressed by equations (2.2.171) involve a degree of approximation inherent to an asymptotic theory. Hence, we shall call them *asymptotic similarity rules*.

Exercises 9

1. Consider the flow in the middle tier (region 4) of the interaction region; see Figure 2.13 on page 157. You are given that the pressure p rises by a small amount Δp. By means of the inspection analysis, show that the variations of the gas density may be estimated as

$$\Delta\rho \sim \Delta p.$$

   You may use without proof the fact that the variations of the enthalpy are $\Delta h \sim \Delta p$.

   Suggestion: Write the state equation (2.2.2e) in the form

$$\rho = \left[\frac{1}{(\gamma - 1)M_\infty^2} + \frac{\gamma}{\gamma - 1}p \right] \frac{1}{h}, \qquad (2.2.175)$$

   and represent the variations of the density ρ using the Taylor expansion

$$\Delta\rho = \frac{\partial\rho}{\partial p}\Delta p + \frac{\partial\rho}{\partial h}\Delta h.$$

   By differentiating equation (2.2.175), calculate the derivatives $\partial\rho/\partial p$ and $\partial\rho/\partial h$, and argue that they are order one quantities in a supersonic flow.

2. The inspection analysis presented in Section 2.2.3 is applicable to any high-Reynolds-number flow where the solution for the boundary layer upstream of the interaction region may be represented by equations (2.2.4), (2.2.5). Written together, these show that near the body surface the longitudinal velocity

$$u = \lambda Y + \cdots \quad \text{as} \quad Y \to 0. \qquad (2.2.176)$$

   Here $Y = Re^{1/2}y$.

   However, there are various situations when, instead of being linear function of Y, the longitudinal velocity u behaves as

$$u = \lambda Y^{1/m} + \cdots \quad \text{as} \quad Y \to 0,$$

   with m and λ being positive constants.[24]

---

[24]An example of such flow is discussed in Problem 6 in Exercises 10; see also Ackerberg (1970), Ruban and Turkyilmaz (2000), and Ruban *et al.* (2006).

Your task is to find the value of m for which the displacement effect of the main part of the boundary layer (region 4 in Figure 2.13) is comparable with that of the nonlinear near-wall sublayer (region 3).

3. When analysing the flow in the middle tier (region 4), we represented the velocity components in the form of asymptotic expansions (2.2.60):

$$u = U_{00}(Y) + Re^{-1/8}\tilde{U}_1(x_*, Y) + \cdots, \qquad (2.2.177a)$$

$$v = Re^{-1/4}\tilde{V}_1(x_*, Y) + \cdots, \qquad (2.2.177b)$$

and we used expansion (2.2.64) for the pressure:

$$p = Re^{-1/4}\tilde{P}_1(x_*, Y) + \cdots. \qquad (2.2.178)$$

These appear to contradict the results of the inspection analysis, according to which the variations of the longitudinal velocity along a streamline are related to the pressure variations by the order-of-magnitude equation (2.2.9):

$$\Delta u \sim \Delta p. \qquad (2.2.179)$$

Your task is to resolve this 'contradiction'. You can perform this task in the following steps:

(a) Consider the longitudinal momentum equation (2.2.65a):

$$U_{00}(Y)\frac{\partial \tilde{U}_1}{\partial x_*} + \tilde{V}_1 U'_{00}(Y) = 0. \qquad (2.2.180)$$

Given that, according to (2.2.70)

$$\tilde{V}_1 = -\frac{1}{\lambda}\frac{dA}{dx_*}U_{00}(Y), \qquad (2.2.181)$$

show that the solution of equation (2.2.180) satisfying the upstream attenuation condition

$$\tilde{U}_1 \to 0 \quad \text{as} \quad x_* \to -\infty,$$

is written as

$$\tilde{U}_1 = \frac{1}{\lambda}A(x_*)\,U'_{00}(Y). \qquad (2.2.182)$$

(b) Show further that in region 4 the streamlines are represented by the equation

$$Y + Re^{-1/8}\frac{1}{\lambda}A(x_*) = const.$$

Suggestion: Remember that the streamlines are given by equation (1.4.13) in Part 1 of this book series. In a two-dimensional flow this equation is written as

$$\frac{dx}{u} = \frac{dy}{v}. \qquad (2.2.183)$$

When substituting (2.2.177) into (2.2.183) you may retain only the leading-order term on (2.2.177a). Remember that in the region considered

$$x = 1 + Re^{-3/8}x_*, \qquad y = Re^{-1/2}Y.$$

(c) Finally, substitute (2.2.182) into (2.2.177a) and argue that in region 4 the longitudinal velocity may be written as

$$u = U_{00}(Y') + O(Re^{-1/4}),$$

where

$$Y' = Y + Re^{-1/8}\frac{1}{\lambda}A(x_*).$$

What is the physical content of this result?

4. Consider the boundary layer on the surface of a flat plate in a hypersonic flow under conditions of weak interaction (see Section 1.11.2) when the hypersonic interaction parameter

$$\chi = M_\infty^2 Re_*^{-1/2}$$

is small.

Assume that the self-induced separation of the boundary layer takes place at a distance L from the leading edge of the plate. Your task is to analyse the region of viscous-inviscid interaction that forms in a vicinity of the separation point and show, using the inspection analysis, that the longitudinal extent of the interaction region

$$\Delta\hat{x} \sim L\chi^{3/4}.$$

Suggestion: To perform this task, introduce the dimensionless variables:

$$u = \frac{\hat{u}}{V_\infty}, \qquad p = \frac{\hat{p} - p_\infty}{\rho_\infty V_\infty^2}, \qquad x = \frac{\hat{x}}{L}, \qquad y = \frac{\hat{y}}{L},$$

and formulate four order-of-magnitude equations for the values of u, Δp, Δx, and y in the lower tier of the interaction region. You may proceed as follows:

(a) The first two order-of-magnitude equations may be obtained by balancing the first convective term with the pressure gradient and with the 'primary' viscous term in the $\hat{x}$-momentum equation (1.10.1a):

$$\hat{\rho}\hat{u}\frac{\partial\hat{u}}{\partial\hat{x}} \sim \frac{\partial\hat{p}}{\partial\hat{x}} \sim \frac{\partial}{\partial\hat{y}}\left(\hat{\mu}\frac{\partial\hat{u}}{\partial\hat{y}}\right).$$

When expressing the above equations in finite-difference form, keep in mind that in the hypersonic flow, the density $\hat{\rho} \sim \rho_\infty/M_\infty^2$ and the viscosity coefficient $\hat{\mu}$ is comparable with its value μ_* at the stagnation temperature.

(b) The third equation may be deduced by matching the longitudinal velocity u in the lower tier with that in the unperturbed boundary layer upstream of the interaction region. Remember that, according to (1.11.21b), in the regime of weak interaction, the thickness of the hypersonic boundary layer is estimated as

$$\hat{y} \sim L\frac{M_\infty}{\sqrt{Re_*}}.$$

(c) Finally, the fourth equation is deduced by applying the Ackeret formula

$$\hat{p} = p_\infty + \rho_\infty V_\infty^2 \frac{\theta}{\sqrt{M_\infty^2 - 1}}$$

to estimate the pressure perturbations in the viscous-inviscid interaction. When calculating the streamline slope angle θ, you may assume, without proof, that the displacement effect of the main part of the boundary layer is small compared to the displacement effect of the lower tier.

5. Consider the near-wall boundary layer that forms in the reverse flow downstream of the separation point. This boundary layer is shown as region 8 in Figure 2.15. When performing your analysis, use the non-dimensional variables defined by equations (2.2.1), and argue that in the boundary-layer approximation, the momentum (2.2.2a) and continuity (2.2.2d) equations reduce to

$$u \frac{\partial u}{\partial x} + v \frac{\partial u}{\partial y} = u_e \frac{du_e}{dx} + \frac{1}{Re} \frac{\mu_w}{\rho_w} \frac{\partial^2 u}{\partial y^2}, \tag{2.2.184}$$

$$\frac{\partial u}{\partial x} + \frac{\partial v}{\partial y} = 0. \tag{2.2.185}$$

You may assume without proof that the flow in region 8 is incompressible with the density $\rho = \rho_w$ and viscosity $\mu = \mu_w$. You may also use without derivation the following expression for the velocity at the outer edge of the boundary layer:

$$u_e = -Re^{-1/4} C x^{-1/3}, \qquad C = \frac{\mu_w^{2/3} \lambda^{1/3} \Lambda}{\rho_w^{2/3} \beta P_{\text{pl}}^*}, \tag{2.2.186}$$

given by the solution (2.2.163) in region 7.

Your task is to find the longitudinal velocity u in the near-wall boundary layer (region 8). You may perform this task in the following steps:

(a) Argue that in region 8

$$u \sim Re^{-1/4}.$$

Then, balancing the convective and viscous terms in the momentum equation (2.2.184), show that the thickness of this region is estimated as

$$y \sim Re^{-3/8}.$$

(b) Use the continuity equation (2.2.185) to introduce the stream function ψ such that

$$u = \frac{\partial \psi}{\partial y}, \qquad v = -\frac{\partial \psi}{\partial x}, \tag{2.2.187}$$

and deduce that in region 8,

$$\psi \sim Re^{-5/8}.$$

(c) Hence, seek the solution in the flow region 8 in the form

$$\psi = -Re^{-5/8}Ax^{\alpha}f(\zeta), \qquad \zeta = Re^{3/8}B\frac{y}{x^{\beta}}. \qquad (2.2.188)$$

Substitute (2.2.188) into the momentum equation (2.2.184), and show that

$$A^2B^2x^{2\alpha-2\beta-1}\big[(\alpha-\beta)f'^2 - \alpha ff''\big]$$
$$= -\frac{1}{3}C^2x^{-5/3} - AB^3\frac{\mu_w}{\rho_w}x^{\alpha-3\beta}f'''. \qquad (2.2.189)$$

Find the values of α and β that allow us to eliminate x from (2.2.189). Also, choose constants A and B such that equation (2.2.189) turns into

$$f''' - \frac{1}{3}ff'' + \frac{1}{3}\left(1 - f'^2\right) = 0.$$

Formulate the boundary conditions for this equation.

(d) Finally, confirm that in region 8 the longitudinal velocity component u is given by equation (2.2.164a).

2.3   Incompressible Flow Separation from a Smooth Body Surface

In this section we consider the subsonic flow separation from a body with a smooth surface. Asymptotic theory of this separation was developed by Sychev (1972). Following his original paper, we shall assume that the flow is incompressible, steady, and two-dimensional. The theory can be extended to the subsonic flow regime in the same way as this is done in Section 4.2.4 for the flow past a corner.

Remember that in Prandtl's (1904) classical description of the separation process, the boundary layer is assumed to undergo a gradual deceleration under the action of an adverse pressure gradient that is distributed over a finite interval of the body contour. However, in this formulation, the solution of the boundary-layer equations develops Goldstein's (1948) singularity at the point of zero skin friction, which precludes a rational theory of the boundary-layer separation to be formulated.[25] An alternative scenario is suggested by the theory of self-induced separation, presented in Section 2.2 for supersonic flows. In this theory, the appearance of Goldstein's singularity is prevented by the interaction of the boundary layer with the inviscid flow. However, the interaction can only take place if the boundary layer is exposed to a strong pressure gradient (2.2.29):[26]

$$\frac{\partial p}{\partial x} \sim Re^{1/8}. \qquad (2.3.1)$$

In these conditions, the boundary layer experiences a sharp deceleration over a short distance along the body contour, resulting in the displacement effect that is strong enough to support the pressure gradient (2.3.1).

---

[25] A detailed discussion of Goldstein's theory is given in Section 5.3.2.

[26] The inspection analysis, that led to equation (2.2.29), is easily extended to subsonic flows; see Problem 1 in Exercises 10.

In the supersonic flow, the required sharp pressure rise Δp is possible thanks to the compression waves,[27] with the Ackeret formula

$$\Delta p = \frac{\hat{p} - p_\infty}{\rho_\infty V_\infty^2} = \frac{\theta}{\sqrt{M_\infty^2 - 1}}$$

establishing a simple link between Δp and the streamline slope angle θ. For the viscous-inviscid interaction concept to be applicable to subsonic flows, the Euler equations, which describe the behaviour of the inviscid flow outside the interaction region, should allow for singular solutions that would support a large pressure gradient (2.3.1) in a vicinity of the separation point. At first sight this seems impossible, because of elliptic character of the subsonic flows. However, the free streamline theory (see Section 3.8 in Part 1), put forward by Kirchhoff (1869) for describing the separated flows, does produce the solution with singularities at the separation point.[28] We shall see that, remarkably, these singularities are consistent with the viscous-inviscid interaction theory.

The theory we deal with in this section is applicable to a wide variety of body shapes. The only restriction is that the separation takes place on a smooth segment of the body contour. For the purpose of presentation of the theory, it is convenient to think of a particular body shape; we have chosen the flow past a circular cylinder.

2.3.1   Problem formulation

Let a circular cylinder be placed in a uniform flow of an incompressible fluid with the free-stream velocity V_∞ and the pressure p_∞. The flow develops separation at point S on the upper surface of the cylinder and at point S' on the lower surface; see Figure 2.19. We shall assume that the flow considered is steady and two-dimensional. Then the Navier–Stokes equations can be written as[29]

$$\left.\begin{aligned}
\hat{u}\frac{\partial \hat{u}}{\partial \hat{x}} + \hat{v}\frac{\partial \hat{u}}{\partial \hat{y}} &= -\frac{1}{\rho}\frac{\partial \hat{p}}{\partial \hat{x}} + \nu\left(\frac{\partial^2 \hat{u}}{\partial \hat{x}^2} + \frac{\partial^2 \hat{u}}{\partial \hat{y}^2}\right), \\
\hat{u}\frac{\partial \hat{v}}{\partial \hat{x}} + \hat{v}\frac{\partial \hat{v}}{\partial \hat{y}} &= -\frac{1}{\rho}\frac{\partial \hat{p}}{\partial \hat{y}} + \nu\left(\frac{\partial^2 \hat{v}}{\partial \hat{x}^2} + \frac{\partial^2 \hat{v}}{\partial \hat{y}^2}\right), \\
\frac{\partial \hat{u}}{\partial \hat{x}} + \frac{\partial \hat{v}}{\partial \hat{y}} &= 0.
\end{aligned}\right\}$$

(2.3.2)

Here, ρ is the fluid density and ν is the kinematic viscosity coefficient. Both are assumed to remain constant in the entire flow field. As usual we use the 'hat' to denote the dimensional variables. We introduce the non-dimensional variables by means of the transformations

$$\left.\begin{aligned}
\hat{u} &= V_0 u, & \hat{v} &= V_0 v, & \hat{p} &= p_\infty + \rho V_0^2 p, \\
\hat{x} &= a x, & \hat{y} &= a y,
\end{aligned}\right\}$$

(2.3.3)

---

[27] See Figure 4.20 on page 273 in Part 1 of this book series.
[28] See Problem 6 in Exercises 12 in Part 1 and Problem 1 in Exercises 9 in Part 2.
[29] See equations (1.7.6) on page 62 in Part 1.

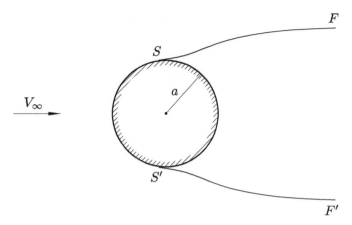

Fig. 2.19: Separated flow past a circular cylinder.

with a denoting the radius of the cylinder and V_0 the fluid velocity at the separation point S immediately outside the boundary layer. To find V_0 we need to solve the Euler equations for the inviscid flow past the cylinder. The corresponding solution in the framework of the *Kirchhoff model* was constructed by Levi-Civita (1907).

The substitution of (2.3.3) into (2.3.2) renders the Navier–Stokes equations in the dimensionless form:

$$u\frac{\partial u}{\partial x} + v\frac{\partial u}{\partial y} = -\frac{\partial p}{\partial x} + \frac{1}{Re}\left(\frac{\partial^2 u}{\partial x^2} + \frac{\partial^2 u}{\partial x^2}\right), \tag{2.3.4a}$$

$$u\frac{\partial v}{\partial x} + v\frac{\partial v}{\partial y} = -\frac{\partial p}{\partial y} + \frac{1}{Re}\left(\frac{\partial^2 v}{\partial x^2} + \frac{\partial^2 v}{\partial x^2}\right), \tag{2.3.4b}$$

$$\frac{\partial u}{\partial x} + \frac{\partial v}{\partial y} = 0. \tag{2.3.4c}$$

Here, the Reynolds number is defined as

$$Re = \frac{V_0\, a}{\nu}.$$

Our task will be to construct the solution of equations (2.3.4), assuming that $Re \to \infty$.

2.3.2 Flow outside the interaction region

We start our analysis with the main inviscid flow region, where

$$x = O(1), \quad y = O(1), \quad Re \to \infty. \tag{2.3.5}$$

We shall call it region 1. The asymptotic expansions of the velocity components (u, v) and the pressure p in this region are sought in the form

$$\left.\begin{array}{l} u(x, y; Re) = u_0(x, y) + \cdots, \qquad v(x, y; Re) = v_0(x, y) + \cdots, \\[2mm] p(x, y; Re) = p_0(x, y) + \cdots. \end{array}\right\} \tag{2.3.6}$$

Substituting (2.3.6) into the Navier–Stokes equations (2.3.4) and setting $Re \to \infty$, we find that

$$u_0 \frac{\partial u_0}{\partial x} + v_0 \frac{\partial u_0}{\partial y} = -\frac{\partial p_0}{\partial x},$$ (2.3.7a)

$$u_0 \frac{\partial v_0}{\partial x} + v_0 \frac{\partial v_0}{\partial y} = -\frac{\partial p_0}{\partial y},$$ (2.3.7b)

$$\frac{\partial u_0}{\partial x} + \frac{\partial v_0}{\partial y} = 0.$$ (2.3.7c)

These are the Euler equations describing the behaviour of incompressible inviscid flows. A detailed discussion of the properties of these flows is given in Chapter 3 in Part 1 of this book series. Here our task is to find the solution of (2.3.7) in a vicinity of the separation point S. To perform this task we will use the fact that the complex conjugate velocity $\overline{V} = u_0 - i v_0$ is an analytic function of the complex variable $z = x + iy$. We will also use the Bernoulli equation

$$\frac{u_0^2 + v_0^2}{2} + p_0 = \frac{1}{2} \left(\frac{V_\infty}{V_0} \right)^2.$$ (2.3.8)

The analyticity of $\overline{V}$ is proven by cross-differentiation of equations (2.3.7a) and (2.3.7b). Indeed, differentiating the x-momentum equation (2.3.7a) with respect to y, we have

$$u_0 \frac{\partial}{\partial x} \left(\frac{\partial u_0}{\partial y} \right) + \frac{\partial u_0}{\partial y} \frac{\partial u_0}{\partial x} + v_0 \frac{\partial}{\partial y} \left(\frac{\partial u_0}{\partial y} \right) + \frac{\partial v_0}{\partial y} \frac{\partial u_0}{\partial y} = -\frac{\partial^2 p_0}{\partial x \partial y},$$

which, in view of the continuity equation (2.3.7c), may be written as

$$u_0 \frac{\partial}{\partial x} \left(\frac{\partial u_0}{\partial y} \right) + v_0 \frac{\partial}{\partial y} \left(\frac{\partial u_0}{\partial y} \right) = -\frac{\partial^2 p_0}{\partial x \partial y}.$$ (2.3.9)

Similarly, differentiation of the y-momentum equation (2.3.7b) with respect to x yields

$$u_0 \frac{\partial}{\partial x} \left(\frac{\partial v_0}{\partial x} \right) + v_0 \frac{\partial}{\partial y} \left(\frac{\partial v_0}{\partial x} \right) = -\frac{\partial^2 p_0}{\partial x \partial y}.$$ (2.3.10)

The pressure p_0 can now be eliminated from (2.3.9) and (2.3.10), which leads to the following equation:

$$u_0 \frac{\partial \omega}{\partial x} + v_0 \frac{\partial \omega}{\partial y} = 0$$ (2.3.11)

for the vorticity

$$\omega = \frac{\partial v_0}{\partial x} - \frac{\partial u_0}{\partial y}.$$

Equation (2.3.11) signifies that the vorticity remains unchanged along each stream-line. In the uniform flow upstream of the cylinder, $\omega = 0$. Hence, everywhere outside the separation region

$$\frac{\partial v_0}{\partial x} - \frac{\partial u_0}{\partial y} = 0.$$ (2.3.12)

Equation (2.3.12) considered together with the continuity equation (2.3.7c) constitute the Cauchy–Riemann conditions that are necessary and sufficient for the complex conjugate velocity $\overline{V} = u_0 - iv_0$ to be an analytic function of the complex variable $z = x + iy$.

Now, let us turn to the Bernoulli equation (2.3.8). It can be deduced in various ways.[30] When dealing with an irrotational flow satisfying equation (2.3.12) we can substitute $\partial u_0/\partial y$ in the x-momentum equation (2.3.7a) by $\partial v_0/\partial x$. This leads to

$$\frac{\partial}{\partial x}\left(\frac{u_0^2 + v_0^2}{2} + p_0\right) = 0.$$

Similarly, combining (2.3.12) with the y-momentum equation (2.3.7b), we have

$$\frac{\partial}{\partial y}\left(\frac{u_0^2 + v_0^2}{2} + p_0\right) = 0.$$

Consequently,

$$\frac{u_0^2 + v_0^2}{2} + p_0 = const. \tag{2.3.13}$$

To find the constant on the right-hand side of (2.3.13), we need to return to equations (2.3.3). These show that in the free-stream flow before the cylinder, the non-dimensional velocity and pressure are

$$\sqrt{u_0^2 + v_0^2} = \frac{V_\infty}{V_0}, \qquad p_0 = 0.$$

We see that

$$const = \frac{1}{2}\left(\frac{V_\infty}{V_0}\right)^2,$$

which confirms the validity of the Bernoulli equation (2.3.8).

We are ready now to study the flow behaviour near the separation point. When performing this task, it is convenient to use a Cartesian coordinate system with the origin at the separation point S and the x-axis directed tangentially to the cylinder surface, as shown in Figure 2.20. We also show in this figure the dividing streamline SF that originates from the separation point S and separates the main stream from the recirculating flow region. Since the dividing streamline is tangent to the cylinder surface at point S, we can claim that at this point

$$v_0 = 0.$$

Remember that the dimensionless velocity was introduced in (2.3.3) by referring it to the inviscid velocity V_0 at the separation point. This means that at point S

$$u_0 = 1.$$

Therefore, in the vicinity of the separation point the coordinate expansion of the complex conjugate velocity should be sought in the form

$$\overline{V}(z) = 1 + Cz^\alpha + \cdots \quad \text{as} \quad z \to 0. \tag{2.3.14}$$

---

[30]See Section 3.1.1 in Part 1 of this book series.

Fig. 2.20: Inviscid flow near the point of separation.

Here, $C = C_r + iC_i$ is a complex constant, and parameter α is assumed positive:

$$\alpha > 0. \tag{2.3.15}$$

At any point different from the point of separation S, the complex argument z may be expressed as

$$z = re^{i\vartheta}. \tag{2.3.16}$$

Here, r is the distance from point z to the separation point S and ϑ is the angle made by the radius r with the x-axis (see Figure 2.20). By substituting (2.3.16) into (2.3.14) and separating the real and imaginary parts, we find that the velocity components

$$\left.\begin{aligned} u_0 &= 1 + \left[C_r\cos(\alpha\vartheta) - C_i\sin(\alpha\vartheta)\right]r^\alpha + \cdots, \\ v_0 &= -\left[C_r\sin(\alpha\vartheta) + C_i\cos(\alpha\vartheta)\right]r^\alpha + \cdots \end{aligned}\right\} \quad \text{as} \quad r \to 0. \tag{2.3.17}$$

The corresponding asymptotic expansion for the pressure

$$p_0 = \frac{1}{2}\left[\left(\frac{V_\infty}{V_0}\right)^2 - 1\right] - \left[C_r\cos(\alpha\vartheta) - C_i\sin(\alpha\vartheta)\right]r^\alpha + \cdots \tag{2.3.18}$$

is obtained by substituting (2.3.17) into the Bernoulli equation (2.3.8).

Now, we need to consider the boundary conditions on the cylinder surface before the separation and on the dividing streamline SF after the separation. Since the flow in region 1 is inviscid, we have to use the impermeability condition on the cylinder surface:

$$\left.\frac{v_0}{u_0}\right|_{y=y_w(x)} = \frac{dy_w}{dx}, \quad x < 0. \tag{2.3.19}$$

Here, $y_w(x)$ is the body shape function. Taking into account that in the dimensionless variables the radius of the cylinder equals unity, we write

$$y_w(x) = -\frac{1}{2}x^2 + \cdots \quad \text{as} \quad |x| \to 0.$$

When using (2.3.17) in (2.3.19), we have to set $\vartheta = \pi$ and substitute r by $(-x)$. As a result, the impermeability condition (2.3.19) takes the form

$$-\left[C_r \sin(\alpha\pi) + C_i \cos(\alpha\pi)\right](-x)^\alpha + \cdots = (-x) + \cdots .$$

We shall assume, subject to subsequent confirmation, that in the asymptotic expansion (2.3.14) there exists a term with

$$\alpha < 1. \tag{2.3.20}$$

Then we have to set

$$C_r \sin(\alpha\pi) + C_i \cos(\alpha\pi) = 0. \tag{2.3.21}$$

Turning to the dividing streamline SF, we shall first consider the boundary condition for the pressure. Remember that in the Kirchhoff model, the fluid in the separation region is assumed to be motionless, which makes the pressure constant on the free streamlines SF and $S'F'$; see Figure 2.19. Of course, in reality, the fluid in the separation region is involved in a recirculation motion, but since point S is a stagnation point for this motion, the derivative of the pressure p_0 along SF should be zero at point S.[31] Setting $\vartheta = 0$ (see Figure 2.20) and substituting r by x in (2.3.18), we have the following equation for the pressure on SF:

$$p_0 = \frac{1}{2}\left[\left(\frac{V_\infty}{V_0}\right)^2 - 1\right] - C_r x^\alpha + \cdots .$$

For α satisfying condition (2.3.20), the pressure gradient can only be zero if

$$C_r = 0.$$

Since we are interested in a nontrivial term $C z^\alpha$ in the asymptotic expansion (2.3.14), we have to assume that $C_i \neq 0$, and then it follows from (2.3.21) that

$$\cos(\alpha\pi) = 0. \tag{2.3.22}$$

The solution of equation (2.3.22), which satisfies conditions (2.3.15) and (2.3.20), is

$$\alpha = \frac{1}{2}.$$

Thus, we can conclude that the asymptotic expansion of the complex conjugate velocity (2.3.14) near the separation point S has the form

$$\overline{V}(z) = 1 - 2i\varkappa z^{1/2} + \cdots \quad \text{as} \quad z \to 0, \tag{2.3.23}$$

where $\varkappa = -C_i/2$.

---

[31] A detailed analysis of the flow in the recirculation region is given in Section 2.3.5.

The velocity components u_0 and v_0 are obtained by separating the real and imaginary parts in (2.3.23):

$$\left.\begin{aligned} u_0 &= 1 - \Re\{2i\varkappa z^{1/2}\} + \cdots, \\ v_0 &= \Im\{2i\varkappa z^{1/2}\} + \cdots \end{aligned}\right\} \quad \text{as} \quad z \to 0. \tag{2.3.24}$$

To find the pressure p_0, we substitute (2.3.24) into the Bernoulli equation (2.3.8). We have

$$p_0 = P_0 + \Re\{2i\varkappa z^{1/2}\} + \cdots \quad \text{as} \quad z \to 0, \tag{2.3.25}$$

where P_0 is the pressure in the separation region given by

$$P_0 = \frac{1}{2}\left[\left(\frac{V_\infty}{V_0}\right)^2 - 1\right].$$

In particular, the pressure distribution on the cylinder surface before the separation point S may be found by setting $\vartheta = \pi$ in (2.3.18):

$$p_0 = P_0 - 2\varkappa(-x)^{1/2} + \cdots \quad \text{as} \quad x \to 0-. \tag{2.3.26}$$

Finally, to find the shape $y = y_S(x)$ of the dividing streamline SF, we use the equation

$$\frac{dy_S}{dx} = \left.\frac{v_0}{u_0}\right|_{SF}. \tag{2.3.27}$$

Setting $\vartheta = 0$ in (2.3.17) and substituting the resulting equations for u_0 and v_0 into (2.3.27), we have

$$\frac{dy_S}{dx} = 2\varkappa x^{1/2} + \cdots. \tag{2.3.28}$$

The integration of (2.3.28) with the initial condition $y_S(0) = 0$ yields

$$y_S(x) = \frac{4}{3}\varkappa x^{3/2} + \cdots \quad \text{as} \quad x \to 0+. \tag{2.3.29}$$

It should be noted that the 'local solution' of the Euler equations (2.3.7), presented above, leaves parameter $\varkappa$ undetermined. The solution for the entire inviscid region 1 was constructed by Levi-Civita (1907) and Brodetsky (1923) using the Kirchhoff model.[32] They demonstrated that the Euler equations admit a family of solutions with the position of the separation point S on the cylinder surface being a free parameter (see Figure 2.21). For each member of the family, parameter $\varkappa$ was found to be uniquely defined. If the separation point S is situated close to the front stagnation point O, then $\varkappa$ appears to be negative. It grows monotonically as the separation point moves along the cylinder surface in the downstream direction, and turns zero when the separation point reaches point B situated at 55° from the front stagnation point. After this, $\varkappa$ becomes positive and continues to grow with the distance from point B.

---

[32]A detailed discussion of this model is presented in Section 3.8 in Part 1 of this book series.

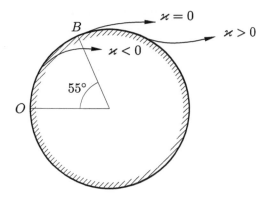

Fig. 2.21: Dependence of parameter κ on the position of the separation point.

Of course, the solutions with negative $\varkappa$ are physically impossible; it follows from (2.3.29) that, in this case, the free streamline would intersect the body surface. The case of positive $\varkappa$ leads to a contradiction of a different kind. It follows from (2.3.26) that the pressure gradient acting on the boundary layer,

$$\frac{dp_0}{dx} = \frac{\varkappa}{(-x)^{1/2}} + \cdots \quad \text{as} \quad x \to 0-, \tag{2.3.30}$$

is adverse and singular. Clearly, in these conditions, the boundary layer cannot reach the assumed position of the separation point S. Instead, it develops Goldstein's singularity some distance before point S. Keeping this in mind, Brillouin (1911) and Villat (1914) suggested that the 'correct solution' should be the one with $\varkappa = 0$. This requirement, known now as the *Brillouin–Villat condition*, allows us to find the position of the separation point on the body surface, and thereby makes the inviscid solution unique. However, a closer look at the solution with $\varkappa = 0$ reveals yet another contradiction. It appears that if the separation point is placed at point B (see Figure 2.21), then the pressure on the cylinder surface decays monotonically all the way from the front stagnation point O to the point of separation B, which makes the boundary-layer separation unexplained from a physical standpoint.

This paradox was resolved by Sychev (1972), who suggested that the separation point should lie a short distance downstream of point B, in which case the parameter $\varkappa$ assumes a small positive value. Remember that when dealing with the self-induced separation in a supersonic flow, we found the plateau pressure (2.2.174) uniquely defined. This suggests that the boundary-layer separation requires a certain pressure rise in the interaction region. In the subsonic flow considered here, the role of the plateau pressure is played by parameter $\varkappa$. Therefore, one can expect to find $\varkappa$ as a result of the flow analysis in the viscous-inviscid interaction region. Of course, an estimate for $\varkappa$ is easily obtained with the help of inspection analysis. According to (2.2.29), the pressure gradient in the interaction region is estimated as

$$\frac{\partial p}{\partial x} \sim Re^{1/8}. \tag{2.3.31}$$

Immediately outside the interaction region, $\partial p / \partial x$ is given by (2.3.30). We also know from (2.2.28) that the longitudinal extent of the interaction region

$$|x| \sim Re^{-3/8}. \tag{2.3.32}$$

Substituting (2.3.32) into (2.3.30) and comparing the result with (2.3.31), we see that

$$\varkappa = O(Re^{-1/16}).$$

Keeping this in mind, we seek parameter $\varkappa$ in the form of the asymptotic expansion

$$\varkappa = Re^{-1/16}\varkappa_0 + \cdots \quad \text{as} \quad Re \to \infty, \tag{2.3.33}$$

where $\varkappa_0$ is an order one constant.

2.3.3   Boundary layer before the interaction region

Now, let us consider the boundary layer that forms on the cylinder surface. It is shown as region 2 in Figure 2.22. To perform the flow analysis in this region, we use the body-fitted coordinates (s, n), with s measured along the cylinder surface from the separation point S, and n in the normal direction. The coordinates (s, n) and the corresponding velocity components (V_τ, V_n) are made dimensionless in the same way as the Cartesian coordinates (2.3.3).

It follows from (2.3.30) and (2.3.33) that the pressure gradient, acting on the boundary layer, may be expressed by the asymptotic expansion

$$\frac{dp}{ds} = \frac{dp_0}{ds} + Re^{-1/16}\frac{dp_1}{ds} + \cdots \quad \text{as} \quad Re \to \infty, \tag{2.3.34}$$

where dp_0/ds and dp_1/ds are such that

$$\left.\begin{array}{l} \dfrac{dp_0}{ds} \to 0, \\[2mm] \dfrac{dp_1}{ds} = \dfrac{\varkappa_0}{(-s)^{1/2}} + \cdots \end{array}\right\} \quad \text{as} \quad s \to 0-. \tag{2.3.35}$$

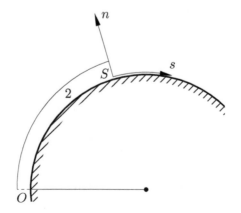

Fig. 2.22: Body-fitted coordinates (s, n).

Corresponding to this, we seek the solution in the boundary layer in the form

$$
\left.
\begin{aligned}
V_\tau(s, n; Re) &= U_0(s, N) + Re^{-1/16}U_1(s, N) + \cdots, \\
V_n(s, n; Re) &= Re^{-1/2}V_0(s, N) + Re^{-9/16}V_1(s, N) + \cdots,
\end{aligned}
\right\}
\tag{2.3.36}
$$

where $N = Re^{1/2}n$.

The substitution of (2.3.36) together with (2.3.34) into the Navier–Stokes equations (1.2.8) yields, in the leading-order approximation, the classical boundary-layer equations

$$
U_0 \frac{\partial U_0}{\partial s} + V_0 \frac{\partial U_0}{\partial N} = -\frac{dp_0}{ds} + \frac{\partial^2 U_0}{\partial N^2},
\tag{2.3.37a}
$$

$$
\frac{\partial U_0}{\partial s} + \frac{\partial V_0}{\partial N} = 0.
\tag{2.3.37b}
$$

We are interested in the behaviour of the solution to these equations as $s \to 0-$. We know that the leading-order pressure gradient dp_0/ds remains favourable all the way between the front stagnation point O and the point of separation S. It is also regular as $s \to 0-$. In these conditions, the solution to (2.3.37) may be expressed in the form of the Taylor expansions:

$$
\left.
\begin{aligned}
U_0(s, N) &= U_{00}(N) + sU_{01}(N) + \cdots, \\
V_0(s, N) &= V_{00}(N) + sV_{01}(N) + \cdots
\end{aligned}
\right\}
\quad \text{as} \quad s \to 0-.
\tag{2.3.38}
$$

Clearly, near the cylinder surface

$$
\left.
\begin{aligned}
U_{00}(N) &= \lambda N + \cdots, \\
U_{01}(N) &= \lambda' N + \cdots
\end{aligned}
\right\}
\quad \text{as} \quad N \to 0.
\tag{2.3.39}
$$

Here, λ is a positive constant representing the dimensionless skin friction immediately upstream of the interaction region. By substituting (2.3.38), (2.3.39) into the continuity equation (2.3.37b), and solving this equation for $V_{00}(N)$ with the initial condition $V_{00}(0) = 0$, we find that

$$
V_{00}(N) = -\frac{1}{2}\lambda' N^2 + \cdots \quad \text{as} \quad N \to 0.
\tag{2.3.40}
$$

The situation with the $O(Re^{-1/16})$ terms in (2.3.36) is more complex. Functions $U_1(s, N)$ and $V_1(s, N)$ satisfy the linearized boundary-layer equations:

$$
U_0 \frac{\partial U_1}{\partial s} + U_1 \frac{\partial U_0}{\partial s} + V_0 \frac{\partial U_1}{\partial N} + V_1 \frac{\partial U_0}{\partial N} = -\frac{dp_1}{ds} + \frac{\partial^2 U_1}{\partial N^2},
\tag{2.3.41a}
$$

$$
\frac{\partial U_1}{\partial s} + \frac{\partial V_1}{\partial N} = 0.
\tag{2.3.41b}
$$

To predict the form of the asymptotic solution of these equations, we shall use the following order-of-magnitude arguments. We start by showing that due to singular pressure gradient dp_1/ds, the boundary layer (region 2) splits into viscous sublayer $2a$ and 'locally inviscid' region $2b$; see Figure 2.23. Indeed, for the flow to be viscous, the first convective term on the left-hand side of (2.3.41a) should be in balance with the viscous term, that is

$$U_0 \frac{\partial U_1}{\partial s} \sim \frac{\partial^2 U_1}{\partial N^2}. \tag{2.3.42}$$

If we assume, subject to subsequent confirmation, that the viscous sublayer is thin ($N \ll 1$), then using (2.3.38) and (2.3.39), we can write

$$U_0 = \lambda N + \cdots . \tag{2.3.43}$$

Representing the derivatives in (2.3.42) by finite differences and using (2.3.43) for U_0, we have

$$N \frac{U_1}{\Delta s} \sim \frac{U_1}{N^2}.$$

We see that

$$N \sim (-s)^{1/3}, \tag{2.3.44}$$

which confirms that the thickness of the viscous sublayer $2a$ does tend to zero as $s \to 0-$.

To obtain an estimate for U_1 in region $2a$, we compare the convective term in the momentum equation (2.3.41a) with the pressure gradient:

$$U_0 \frac{\partial U_1}{\partial s} \sim \frac{dp_1}{ds}. \tag{2.3.45}$$

The right-hand side in (2.3.45) is given by (2.3.35). An estimate for U_0 is obtained by combining (2.3.44) with (2.3.43):

$$U_0 \sim (-s)^{1/3}. \tag{2.3.46}$$

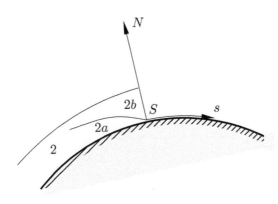

Fig. 2.23: Splitting of the boundary layer (region 2) into viscous sublayer $2a$ and 'locally inviscid' region $2b$.

Substituting (2.3.35) and (2.3.46) into (2.3.45), and approximating the derivative $\partial U_1/\partial s$ by finite differences, we have

$$(-s)^{1/3}\frac{U_1}{\Delta s} \sim (-s)^{-1/2}.$$

We see that in region $2a$

$$U_1 \sim (-s)^{1/6}. \tag{2.3.47}$$

The continuity equation (2.3.41b) allows us to introduce the stream function $\Psi_1(s, N)$ such that

$$\frac{\partial \Psi_1}{\partial s} = -V_1, \qquad \frac{\partial \Psi_1}{\partial N} = U_1. \tag{2.3.48}$$

An estimate for Ψ_1 in region $2a$ may be obtained with the help of the second equation in (2.3.48). We have

$$\Psi_1 \sim U_1 N. \tag{2.3.49}$$

The substitution of (2.3.47) and (2.3.44) into (2.3.49) yields

$$\Psi_1 \sim (-s)^{1/2}. \tag{2.3.50}$$

Now, we shall return to equations (2.3.41) and perform their analysis on a rigorous mathematical basis.

Viscous sublayer (region $2a$)

Guided by (2.3.50) and (2.3.44), we seek the solution for the stream function $\Psi_1(s, N)$ in region $2a$ in the form of asymptotic expansion

$$\Psi_1(s, N) = (-s)^{1/2}g(\xi) + \cdots \quad \text{as} \quad s \to 0-, \tag{2.3.51a}$$

where

$$\xi = \frac{N}{(-s)^{1/3}} \tag{2.3.51b}$$

is an order-one quantity in region $2a$. The substitution of (2.3.51) into (2.3.48) yields

$$U_1 = (-s)^{1/6}g'(\xi) + \cdots, \qquad V_1 = (-s)^{-1/2}\left[\frac{1}{2}g(\xi) - \frac{1}{3}\xi g'(\xi)\right] + \cdots. \tag{2.3.52}$$

It follows from (2.3.43) and (2.3.51b) that the leading-order velocity U_0 is represented in region $2a$ by

$$U_0 = (-s)^{1/3}\lambda\xi + \cdots. \tag{2.3.53}$$

The other coefficients in the momentum equation (2.3.41a) are

$$\frac{\partial U_0}{\partial s} = (-s)^{1/3}\lambda'\xi + \cdots, \qquad V_0 = -(-s)^{2/3}\frac{1}{2}\lambda'\xi^2 + \cdots, \qquad \frac{\partial U_0}{\partial N} = \lambda. \tag{2.3.54}$$

It remains to substitute (2.3.52)–(2.3.54), together with the equation for dp_1/ds in (2.3.35), into the linearized momentum equation (2.3.41a). Setting $s \to 0-$, and working with the leading-order terms, we arrive at the conclusion that function $g(\xi)$ satisfies the following equation:

$$g''' - \frac{1}{3}\lambda\xi^2 g'' + \frac{1}{2}\lambda\xi g' - \frac{1}{2}\lambda g = \varkappa_0. \qquad (2.3.55)$$

Now, we need to formulate the boundary conditions for this (2.3.55). Since region $2a$ is adjacent to the cylinder surface, the solution in this region should satisfy the no-slip conditions

$$U_1 = V_1 = 0 \quad \text{at} \quad N = 0. \qquad (2.3.56)$$

The substitution of (2.3.52) into (2.3.56) yields

$$g(0) = g'(0) = 0. \qquad (2.3.57)$$

As a third boundary condition we shall use the requirement that function $g(\xi)$ does not grow exponentially as $\xi \to \infty$. This is because an exponential growth of $g(\xi)$ would preclude the matching of solutions in regions $2a$ and $2b$.[33]

The solution to equation (2.3.55) may be found in an analytic form. We start by setting $\xi = 0$ in (2.3.55). We see that

$$g'''(0) = \varkappa_0. \qquad (2.3.58)$$

Now, we differentiate (2.3.55) with respect to ξ. This leads to the following equation:

$$h'' - \frac{1}{3}\lambda\xi^2 h' - \frac{1}{6}\lambda\xi h = 0 \qquad (2.3.59)$$

for function $h = g''$. The substitution of the independent variable, $z = \frac{1}{9}\lambda\xi^3$, turns (2.3.59) into a confluent hypergeometric equation

$$z\frac{d^2 h}{dz^2} + \left(\frac{2}{3} - z\right)\frac{dh}{dz} - \frac{1}{6}h = 0, \qquad (2.3.60a)$$

with the boundary condition (2.3.58) taking the form

$$\frac{dh}{dz} = \frac{\varkappa_0}{\sqrt[3]{3\lambda}} z^{-2/3} + \cdots \quad \text{as} \quad z \to 0. \qquad (2.3.60b)$$

In general, the confluent hypergeometric equation (also known as Kummer's equation) is written as

$$z\frac{d^2 h}{dz^2} + (b - z)\frac{dh}{dz} - ah = 0.$$

The general solution to this equation may be expressed as a linear superposition of Kummer's function $M(a, b; z)$ and Tricomi's function $U(a, b; z)$:[34]

$$h = C_1 M(a, b; z) + C_2 U(a, b; z). \qquad (2.3.61)$$

---

[33]See Section 1.4.2 in Part 2 of this book series.
[34]See, for example, Abramowitz and Stegun (1965).

Kummer's function is known to be regular at $z = 0$, but grows exponentially as $z \to \infty$, and thus we have to set $C_1 = 0$. Tricomi's function displays an algebraic behaviour at large values of z:

$$U(a, b; z) = z^{-a}\left[1 + O(z^{-1})\right] \quad \text{as} \quad z \to \infty. \tag{2.3.62}$$

Near $z = 0$ it can be represented by the asymptotic expansion

$$U(a, b; z) = \frac{\Gamma(1-b)}{\Gamma(a-b+1)} + \frac{\Gamma(b-1)}{\Gamma(a)}z^{1-b} + O(z) \quad \text{as} \quad z \to 0, \tag{2.3.63}$$

where Γ stands for Euler's gamma function.

Taking into account that the parameters of equation (2.3.60a) are

$$a = \frac{1}{6}, \qquad b = \frac{2}{3},$$

and using (2.3.63) in (2.3.61), we have

$$h = C_2\left[\frac{\Gamma(1/3)}{\Gamma(1/2)} + \frac{\Gamma(-1/3)}{\Gamma(1/6)}z^{1/3} + O(z)\right] \quad \text{as} \quad z \to 0.$$

Now it is easily found from boundary condition (2.3.60b) that

$$C_2 = \frac{3\varkappa_0}{\sqrt[3]{3\lambda}}\frac{\Gamma(1/6)}{\Gamma(-1/3)} = -\frac{\varkappa_0}{\sqrt[3]{3\lambda}}\frac{\Gamma(1/6)}{\Gamma(2/3)}. \tag{2.3.64}$$

It remains to substitute (2.3.64) into (2.3.61), and we will have

$$h(\xi) = g''(\xi) = -\frac{\varkappa_0}{\sqrt[3]{3\lambda}}\frac{\Gamma(1/6)}{\Gamma(2/3)}U\left(\frac{1}{6}, \frac{2}{3}; \frac{\lambda}{9}\xi^3\right). \tag{2.3.65}$$

Function $g(\xi)$ may now be found by integrating (2.3.65) twice with initial conditions (2.3.57).

To determine the behaviour of $g(\xi)$ at large values of ξ, we use (2.3.62) in (2.3.65), which yields

$$g''(\xi) = -\frac{\varkappa_0}{\sqrt[3]{3\lambda}}\frac{\Gamma(1/6)}{\Gamma(2/3)}\left(\frac{9}{\lambda}\right)^{1/6}\xi^{-1/2} + \cdots \quad \text{as} \quad \xi \to \infty.$$

Keeping this in mind, we express the result of the first integration of (2.3.65) in the form

$$g'(\xi) = -\frac{\varkappa_0}{\sqrt[3]{3\lambda}}\frac{\Gamma(1/6)}{\Gamma(2/3)}\int_0^\xi \left[U\left(\frac{1}{6}, \frac{2}{3}; \frac{\lambda}{9}\xi^3\right) - \left(\frac{9}{\lambda}\right)^{1/6}\xi^{-1/2}\right]d\xi$$

$$-\frac{2\varkappa_0}{\sqrt{\lambda}}\frac{\Gamma(1/6)}{\Gamma(2/3)}\xi^{1/2}. \tag{2.3.66}$$

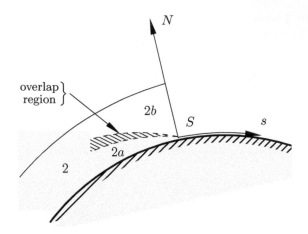

Fig. 2.24: Matching through the overlap region.

We see that the asymptotic expansion of $g'(\xi)$ at large values of ξ is written as

$$g'(\xi) = A_0\xi^{1/2} + A_1 + \cdots \quad \text{as} \quad \xi \to \infty, \qquad (2.3.67)$$

where

$$A_0 = -\frac{2\varkappa_0}{\sqrt{\lambda}}\frac{\Gamma(1/6)}{\Gamma(2/3)}, \qquad (2.3.68)$$

$$A_1 = -\frac{\varkappa_0}{\sqrt[3]{3\lambda}}\frac{\Gamma(1/6)}{\Gamma(2/3)}\int_0^\infty \left[U\left(\frac{1}{6}, \frac{2}{3}; \frac{\lambda}{9}\xi^3\right) - \left(\frac{9}{\lambda}\right)^{1/6}\xi^{-1/2}\right]d\xi.$$

Integrating (2.3.67) again, we have

$$g(\xi) = \frac{2}{3}A_0\xi^{3/2} + A_1\xi + \cdots \quad \text{as} \quad \xi \to \infty. \qquad (2.3.69)$$

Now, we shall consider the velocity components (U_1, V_1). By substituting (2.3.69) into (2.3.52), we find that near the outer edge of region 2a

$$U_1 = (-s)^{1/6}A_0\xi^{1/2} + (-s)^{1/6}A_1 + \cdots, \qquad V_1 = (-s)^{-1/2}\frac{1}{6}A_1\xi + \cdots. \qquad (2.3.70)$$

Equations (2.3.70) are valid in the overlap region that lies between regions 2a and 2b; see Figure 2.24. The substitution of (2.3.51b) into equations (2.3.70) casts these equations in the form

$$U_1 = A_0 N^{1/2} + (-s)^{1/6}A_1 + \cdots, \qquad V_1 = (-s)^{-5/6}\frac{1}{6}A_1 N + \cdots, \qquad (2.3.71)$$

which is suitable for describing the behaviour of U_1 and V_1 at the 'bottom' of region 2b.

Main part of the boundary layer (region 2*b*)

In region 2*b*, the asymptotic analysis of the linearized boundary-layer equations (2.3.41) is based on the limit

$$N = O(1), \quad s \to 0-.$$

Guided by (2.3.71), we seek the solution of (2.3.41) in this region in the form

$$\left. \begin{aligned} U_1(s, N) &= U_{10}(N) + (-s)^{1/6} U_{11}(N) + \cdots, \\ V_1(s, N) &= (-s)^{-5/6} V_{11}(N) + \cdots \end{aligned} \right\} \quad \text{as} \quad s \to 0-. \tag{2.3.72}$$

It further follows from (2.3.71) that the conditions of matching with the solution in region 2*a* are written as

$$\left. \begin{aligned} U_{10}(N) &= A_0 N^{1/2} + \cdots, \\ U_{11}(N) &= A_1 + \cdots, \\ V_{11}(N) &= \frac{1}{6} A_1 N + \cdots \end{aligned} \right\} \quad \text{as} \quad N \to 0. \tag{2.3.73}$$

The substitution of (2.3.72) together with (2.3.38) and (2.3.35) into the momentum (2.3.41a) results in

$$-(-s)^{-5/6} \frac{1}{6} U_{00} U_{11} + (-s)^{-5/6} V_{11} \frac{dU_{00}}{dN} = -\frac{\varkappa_0}{(-s)^{1/2}} + \frac{d^2 U_{10}}{dN^2}.$$

Clearly, the viscous term and the pressure gradient can be disregarded in the leading-order approximation, and we are left with

$$\frac{1}{6} U_{00} U_{11} - \frac{dU_{00}}{dN} V_{11} = 0. \tag{2.3.74a}$$

Similarly, by substituting (2.3.72) into the continuity equation (2.3.41b), we find that

$$\frac{1}{6} U_{11} - \frac{dV_{11}}{dN} = 0. \tag{2.3.74b}$$

Equations (2.3.74) may be solved as follows. We first eliminate U_{11}, which results in

$$U_{00} \frac{dV_{11}}{dN} - \frac{dU_{00}}{dN} V_{11} = 0. \tag{2.3.75}$$

Now, we divide both terms in (2.3.75) by U_{00}^2. We have

$$\frac{1}{U_{00}} \frac{dV_{11}}{dN} - \frac{1}{U_{00}^2} \frac{dU_{00}}{dN} V_{11} = 0,$$

or, equivalently,

$$\frac{d}{dN} \left(\frac{V_{11}}{U_{00}} \right) = 0.$$

Consequently, we can write

$$\frac{V_{11}}{U_{00}} = const, \tag{2.3.76}$$

which means that the streamline slope angle, produced by the displacement effect of the viscous sublayer (region 2*a*), remains unchanged across the main part of the boundary layer (region 2*b*). The constant on the right-hand side of (2.3.76) is easily found by taking into account that according to (2.3.39) and (2.3.73)

$$U_{00} = \lambda N + \cdots, \quad V_{11} = \frac{1}{6}A_1 N + \cdots \quad \text{as} \quad N \to 0.$$

We have

$$const = \frac{A_1}{6\lambda},$$

and we can conclude that

$$V_{11} = \frac{A_1}{6\lambda} U_{00}(N). \tag{2.3.77}$$

Now, we can return to the inviscid flow in region 1, and study the perturbations produced in this region by the displacement effect of the boundary layer. The results of this analysis (see Problem 2 in Exercises 10) may be expressed by an additional $O(Re^{-9/16})$ term in the asymptotic expansion (2.3.23) the complex conjugate velocity $\overline{V} = u - iv$:

$$\overline{V} = 1 - Re^{-1/16}2i\varkappa_0 z^{1/2} + Re^{-9/16}\frac{iA_1}{3\sqrt{3}\lambda}z^{-5/6} + \cdots \quad \text{as} \quad z \to 0. \tag{2.3.78}$$

Corresponding to this, the asymptotic expansion for the pressure (2.3.25) becomes

$$p = P_0 + Re^{-1/16}\Re\{2i\varkappa_0 z^{1/2}\} - Re^{-9/16}\Re\left\{\frac{iA_1}{3\sqrt{3}\lambda}z^{-5/6}\right\} + \cdots. \tag{2.3.79}$$

2.3.4   Viscous-inviscid interaction

Until now, the flow analysis near the point of separation was conducted based on Prandtl's hierarchical strategy. We began in Section 2.3.2 with the inviscid flow in region 1. To describe this flow, the Euler equations were solved subject to the impermeability condition on the cylinder surface. At this stage, the existence of the boundary layer was completely ignored. As a part of the solution, we found that the pressure gradient on the cylinder surface before the separation was given by equation (2.3.30), which, when combined with (2.3.33), is written as

$$\frac{dp}{dx} = Re^{-1/16}\varkappa_0(-x)^{-1/2} + \cdots \quad \text{as} \quad x \to 0-.$$

Then, we turned to the second step of the hierarchical strategy. In Section 2.3.3 we considered the flow in the boundary layer (region 2) that formed on the cylinder surface. In this region, the classical boundary-layer equations were used. After solving these equations, the displacement effect of the boundary layer on the flow in region 1 were evaluated, which represented the third step of the hierarchical strategy, and leads to equations (2.3.78), (2.3.79). In particular, the pressure gradient in the body surface upstream of the separation appears to be

$$\frac{dp}{dx} = Re^{-1/16}\varkappa_0(-x)^{-1/2}$$

$$- Re^{-9/16}\frac{5A_1}{36\sqrt{3}\lambda}(-x)^{-11/6} + \cdots \quad \text{as} \quad x \to 0-. \tag{2.3.80}$$

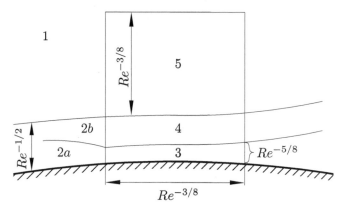

Fig. 2.25: Viscous-inviscid interaction region.

In the hierarchical approach, the second term in (2.3.80) is presumed to be much smaller than the first one. This assumption does hold for $|x| = O(1)$, thanks to the fact that the coefficient in front of the second term is $Re^{-1/2}$ times smaller than that in front of the first term. However, the solution displays a singular behaviour as $x \to 0-$, with the second term growing faster than the first one. They become same order quantities in a region where

$$|x| \sim Re^{-3/8}. \tag{2.3.81}$$

We can say that in the vicinity of the separation point defined by (2.3.81), the pressure acting on the boundary layer and the pressure induced by the displacement effect of the boundary layer become indistinguishable from one another, and, therefore, the flow should be reexamined, now based on the *viscous-inviscid interaction* concept.

The interaction region has the three-tiered structure (see Figure 2.25) that can be already recognized in the solution outside the interaction region. Indeed, we have seen that the boundary layer on the cylinder surface (region 2) splits up into two subregions: viscous sublayer (region 2a), and the main part of the boundary layer (region 2b). When continued into the vicinity (2.3.81) of the separation point, they form the lower tier (region 3) and the middle tier (region 4) of the interaction region. These should be supplemented with region 5, which serves to 'convert' the displacement effect of the boundary layer into the pressure perturbations.

When dealing with the interaction region we use the body-fitted coordinates (s, n), with s measured along the cylinder surface from the separation point S, and n in the normal direction; see Figure 2.22. We start with region 3.

Viscous sublayer (region 3)

We know from (2.3.44) that the thickness of the viscous sublayer 2a is estimated as

$$n = Re^{-1/2}N \sim Re^{-1/2}(-x)^{1/3}. \tag{2.3.82}$$

To find the characteristic thickness of region 3, we need to use (2.3.81) in (2.3.82), which yields

$$n \sim Re^{-5/8}. \tag{2.3.83}$$

It follows from (2.3.81) and (2.3.83) that the asymptotic analysis of the Navier–Stokes equations in region 3 should be conducted using the limit

$$x_* = \frac{s}{Re^{-3/8}} = O(1), \quad Y_* = \frac{n}{Re^{-5/8}} = O(1), \quad Re \to \infty. \tag{2.3.84}$$

The form of the asymptotic expansions of the velocity components V_τ, V_n and the pressure p in this region may be predicted either by using the inspection analysis (Problem 1 in Exercises 10) or with the help of the following simple arguments. According to (2.3.36), the tangential velocity is represented in region 2 by the asymptotic expansion

$$V_\tau(s, n; Re) = U_0(s, N) + Re^{-1/16}U_1(s, N) + \cdots . \tag{2.3.85}$$

In the viscous sublayer (region 2a), functions U_0 and U_1 are given by equations (2.3.53) and (2.3.52), respectively:

$$U_0 = (-s)^{1/3}\lambda\xi + \cdots , \qquad U_1 = (-s)^{1/6}g'(\xi) + \cdots , \tag{2.3.86}$$

where

$$\xi = \frac{N}{(-s)^{1/3}}.$$

Substituting (2.3.86) into (2.3.85) and expressing the resulting equation in terms of variables (2.3.84) of region 3, we have

$$V_\tau = Re^{-1/8}\left[(-x_*)^{1/3}\lambda\xi + (-x_*)^{1/6}g'(\xi) + \cdots\right], \tag{2.3.87a}$$

where

$$\xi = \frac{Y_*}{(-x_*)^{1/3}}. \tag{2.3.87b}$$

Equation (2.3.87a) suggests that in region 3, the tangential velocity should be sought in the form

$$V_\tau(s, n; Re) = Re^{-1/8}U^*(x_*, Y_*) + \cdots , \tag{2.3.88}$$

and it also follows from (2.3.87a) that function U^* in (2.3.88) should satisfy the following condition of matching with the solution in region 2a:

$$U^*(x_*, Y_*) = (-x_*)^{1/3}\lambda\xi + (-x_*)^{1/6}g'(\xi) + \cdots \quad \text{as} \quad x_* \to -\infty. \tag{2.3.89}$$

The form of the solution for the normal velocity component may be predicted in the same way, that is, by matching with the solution for V_n in region 2a. Alternatively, we can use the continuity equation (1.2.8c). Written in the order-of-magnitude form

$$\frac{V_\tau}{\Delta s} \sim \frac{V_n}{n},$$

it yields

$$V_n \sim \frac{n}{\Delta s}V_\tau \sim Re^{-3/8},$$

which suggests that in region 3 the normal velocity component should be represented in the form the asymptotic expansion

$$V_n(s,n;Re) = Re^{-3/8}V^*(x_*,Y_*) + \cdots .$$

It remains to determine the form of the asymptotic expansion of the pressure. For this purpose we can consider the pressure (2.3.26) acting on the boundary layer before the interaction region. Using (2.3.33) in (2.3.26), and substituting the independent variable $x = Re^{-3/8}x_*$, we have[35]

$$p = P_0 - Re^{-1/4}2\varkappa_0(-x_*)^{1/2} + \cdots .$$

This suggests that in region 3 the pressure should be sought in the form

$$p = P_0 + Re^{-1/4}P^*(x_*,Y_*) + \cdots , \qquad (2.3.90)$$

where function P^* is such that

$$P^* = -2\varkappa_0(-x_*)^{1/2} + \cdots \quad \text{as} \quad x_* \to -\infty.$$

Consequently, we can conclude that in region 3

$$\left.\begin{aligned} V_\tau(s,n;Re) &= Re^{-1/8}U^*(x_*,Y_*) + \cdots , \\ V_n(s,n;Re) &= Re^{-3/8}V^*(x_*,Y_*) + \cdots , \\ p(s,n;Re) &= P_0 + Re^{-1/4}P^*(x_*,Y_*) + \cdots \end{aligned}\right\} \quad \text{as} \quad Re \to \infty. \qquad (2.3.91)$$

By substituting (2.3.91) into the Navier–Stokes equations (1.2.8), and taking into account that the curvature of the cylinder surface $\kappa = 1$, we find that the flow in region 3 is described by the boundary-layer equations:

$$U^*\frac{\partial U^*}{\partial x_*} + V^*\frac{\partial U^*}{\partial Y_*} = -\frac{\partial P^*}{\partial x_*} + \frac{\partial^2 U^*}{\partial Y_*^2}, \qquad (2.3.92a)$$

$$\frac{\partial P^*}{\partial Y_*} = 0, \qquad (2.3.92b)$$

$$\frac{\partial U^*}{\partial x_*} + \frac{\partial V^*}{\partial Y_*} = 0. \qquad (2.3.92c)$$

These have to be solved with the no-slip conditions on the cylinder surface

$$U^* = V^* = 0 \quad \text{at} \quad Y_* = 0, \qquad (2.3.93)$$

and the condition of matching with the solution in region 2a. The latter is expressed by equations (2.3.89), (2.3.87b):

$$U^* = (-x_*)^{1/3}\lambda\xi + (-x_*)^{1/6}g'(\xi) + \cdots \quad \text{as} \quad x_* \to -\infty, \qquad (2.3.94a)$$

where

$$\xi = \frac{Y_*}{(-x_*)^{1/3}} = O(1). \qquad (2.3.94b)$$

---

[35]It is apparent that near the separation point, the curvilinear coordinate s is indistinguishable from the Cartesian coordinate x.

Finally, since the boundary-layer equations (2.3.92) are parabolic, we also need a condition for U^* at the outer edge of region 3. It will be formulated after the solution in the middle tier (region 4) is found.

Middle tier (region 4)

The asymptotic analysis of the Navier–Stokes equations (1.2.8) in region 4 (see Figure 2.25) is based on the limit

$$x_* = \frac{s}{Re^{-3/8}} = O(1), \quad Y = \frac{n}{Re^{-1/2}} = O(1), \quad Re \to \infty. \tag{2.3.95}$$

The form of the solution in this region can be predicted in the same way as it was for region 3, but now we need to consider the matching with the solution in region 2b. With this goal in mind we return to the asymptotic expansions (2.3.36) of the velocity components in region 2:

$$\left.\begin{aligned} V_\tau(s,n;Re) &= U_0(s,N) + Re^{-1/16}U_1(s,N) + \cdots, \\ V_n(s,n;Re) &= Re^{-1/2}V_0(s,N) + Re^{-9/16}V_1(s,N) + \cdots. \end{aligned}\right\} \tag{2.3.96}$$

These are valid in the entire region 2. Also in region 2, the leading-order terms in (2.3.96) are given by (2.3.38):

$$\left.\begin{aligned} U_0(s,N) &= U_{00}(N) + sU_{01}(N) + \cdots, \\ V_0(s,N) &= V_{00}(N) + sV_{01}(N) + \cdots \end{aligned}\right\} \quad \text{as} \quad s \to 0-. \tag{2.3.97}$$

As far as the perturbation functions U_1 and V_1 are concerned, they are represented in region 2b by equations (2.3.72):

$$\left.\begin{aligned} U_1(s,N) &= U_{10}(N) + (-s)^{1/6}U_{11}(N) + \cdots, \\ V_1(s,N) &= (-s)^{-5/6}V_{11}(N) + \cdots \end{aligned}\right\} \quad \text{as} \quad s \to 0-. \tag{2.3.98}$$

Substituting (2.3.97) and (2.3.98) into (2.3.96), and expressing the resulting equations in terms of the variables (2.3.95) of region 4, we have

$$\begin{aligned} V_\tau &= U_{00}(Y) + Re^{-1/16}U_{10}(Y) + Re^{-1/8}(-x_*)^{1/6}U_{11}(Y) + \cdots, \\ V_n &= Re^{-1/4}(-x_*)^{-5/6}V_{11}(Y) + \cdots. \end{aligned}$$

This suggests that the solution in region 4 should be sought in the form

$$\left.\begin{aligned} V_\tau &= U_{00}(Y) + Re^{-1/16}U_{10}(Y) + Re^{-1/8}\widetilde{U}_1(x_*,Y) + \cdots, \\ V_n &= Re^{-1/4}\widetilde{V}_1(x_*,Y) + \cdots, \\ p &= P_0 + Re^{-1/4}\widetilde{P}_1(x_*,Y) + \cdots. \end{aligned}\right\} \tag{2.3.99}$$

Notice that since the pressure p does not change across the boundary layer, the asymptotic expansion for p has been written in (2.3.99) in the same form as that in the lower tier; see equation (2.3.90).

The substitution of (2.3.99) into the Navier–Stokes equations (1.2.8) yields

$$U_{00}\frac{\partial \tilde{U}_1}{\partial x_*} + \tilde{V}_1\frac{dU_{00}}{dY} = 0, \tag{2.3.100a}$$

$$\frac{\partial \tilde{P}_1}{\partial Y} = 0, \tag{2.3.100b}$$

$$\frac{\partial \tilde{U}_1}{\partial x_*} + \frac{\partial \tilde{V}_1}{\partial Y} = 0. \tag{2.3.100c}$$

Equations (2.3.100a) and (2.3.100c) coincide with equations (2.2.65a) and (2.2.66) for the middle tier in the corresponding supersonic flow. The general solution to these equations is written as

$$\tilde{U}_1 = \frac{1}{\lambda}A(x_*)\frac{dU_{00}}{dY}, \qquad \tilde{V}_1 = -\frac{1}{\lambda}\frac{dA}{dx_*}U_{00}(Y), \tag{2.3.101}$$

where $A(x_*)$ is referred to as the *displacement function*. It remains unknown at this stage of the flow analysis.

Let us now perform the matching of the tangential velocity component V_τ in regions 3 and 4 (see Figure 2.25). For this purpose, we substitute the first equation from (2.3.101) into the asymptotic expansion for V_τ in (2.3.99). We have

$$V_\tau = U_{00}(Y) + Re^{-1/16}U_{10}(Y) + Re^{-1/8}\frac{1}{\lambda}A(x_*)\frac{dU_{00}}{dY} + \cdots . \tag{2.3.102}$$

For small values of Y, the behaviour of functions U_{00} and U_{10} is given by (2.3.39) and (2.3.73), respectively:

$$U_{00}(Y) = \lambda Y + \cdots , \qquad U_{10}(Y) = A_0 Y^{1/2} + \cdots .$$

These turn (2.3.102) into

$$V_\tau = \lambda Y + Re^{-1/16}A_0 Y^{1/2} + Re^{-1/8}A(x_*) + \cdots . \tag{2.3.103}$$

Now, we need to express (2.3.103) in terms of variables of region 3. Of course, the longitudinal coordinate x_* is common for regions 3 and 4, while the normal coordinates are related as

$$Y = Re^{-1/8}Y_*. \tag{2.3.104}$$

The substitution of (2.3.104) into equation (2.3.103) renders it in the form

$$V_\tau = Re^{-1/8}\left[\lambda Y_* + A_0 Y_*^{1/2} + A(x_*)\right] + O(Re^{-3/16}). \tag{2.3.105}$$

It remains to compare (2.3.105) with the asymptotic expansion for V_τ in (2.3.91), and we can see that the sought boundary condition for U^* at the outer edge of region 3 is written as

$$U^* = \lambda Y_* - \frac{2\varkappa_0}{\sqrt{\lambda}}\frac{\Gamma(1/6)}{\Gamma(2/3)}Y_*^{1/2} + A(x_*) + \cdots \quad \text{as} \quad Y_* \to \infty. \tag{2.3.106}$$

Here, we have used equation (2.3.68) for A_0.

Upper tier (region 5)

Region 5 lies outside the boundary layer and represents the upper tier of the triple-deck structure (see Figure 2.25). To start the flow analysis in this region, we first need to determine its dimensions. Of course, the longitudinal extent of region 5 coincides with that of the interaction region as a whole: $s \sim Re^{-3/8}$. To determine the lateral size, we use the fact that any inviscid potential flow of an incompressible fluid is described by the Laplace equation

$$\frac{\partial^2 \varphi}{\partial x^2} + \frac{\partial^2 \varphi}{\partial y^2} = 0$$

for the velocity potential φ.[36] The principle of least degeneration, applied to this equation, suggests that $y \sim x$. Consequently, the asymptotic analysis of the Navier–Stokes equations (1.2.8) in region 5 should be performed based on the limit

$$x_* = \frac{s}{Re^{-3/8}} = O(1), \quad y_* = \frac{n}{Re^{-3/8}} = O(1), \quad Re \to \infty. \qquad (2.3.107)$$

To predict the form of the asymptotic expansions of the velocity components (V_τ, V_n) and the pressure p in region 5, we note that there is an apparent analogy with the flow past a thin aerofoil.[37] Remember that the perturbations produced by a thin aerofoil in inviscid flow are proportional to the aerofoil thickness. For the flow considered here, a more suitable measure is the slope of the streamlines at the outer edge of region 4. The angle ϑ made by the streamlines with the cylinder surface is calculated as

$$\tan \vartheta = \frac{V_n}{V_\tau}.$$

Using the asymptotic expansions (2.3.99) for V_τ and V_n in region 4, and the solution for $\widetilde{V}_1$ given by (2.3.101), we can see that

$$\vartheta = Re^{-1/4}\left(-\frac{1}{\lambda}\frac{dA}{dx_*} \right) + \cdots . \qquad (2.3.108)$$

This suggests that the solution in region 5 should be sought in the form

$$\left.\begin{aligned} V_\tau(s, n; Re) &= 1 + Re^{-1/4}u_*(x_*, y_*) + \cdots , \\ V_n(s, n; Re) &= Re^{-1/4}v_*(x_*, y_*) + \cdots , \\ p(s, n; Re) &= P_0 + Re^{-1/4}p_*(x_*, y_*) + \cdots \end{aligned}\right\} \quad \text{as} \quad Re \to \infty. \qquad (2.3.109)$$

The substitution of (2.3.109) into the Navier–Stokes equations (1.2.8) results in

$$\frac{\partial u_*}{\partial x_*} = -\frac{\partial p_*}{\partial x_*}, \quad \frac{\partial v_*}{\partial x_*} = -\frac{\partial p_*}{\partial y_*}, \quad \frac{\partial u_*}{\partial x_*} + \frac{\partial v_*}{\partial y_*} = 0. \qquad (2.3.110)$$

Equations (2.3.110) coincide with the corresponding equations of the thin aerofoil theory.[38] When analysing the inviscid subsonic flow past a thin aerofoil, the solution

---

[36]See Section 3.2 in Part 1 of this book series.

[37]See Section 2.1 in Part 2.

[38]See equations (2.1.10) on page 85 in Part 2.

to these equations was sought subject to the perturbations' attenuation condition far from the aerofoil and the impermeability condition on the aerofoil surface. Now, instead of the perturbations' attenuation condition, we need to use the condition of matching with the solution in region 1; see Figure 2.25. To perform the matching it is convenient to introduce the complex variable $z_* = x_* + iy_*$, which is related to the complex variable $z = x + iy$ of region 1 as

$$z = Re^{-3/8}z_*. \tag{2.3.111}$$

Using (2.3.111) in (2.3.78) and (2.3.79), we have

$$\left.\begin{aligned}
\overline{V} &= 1 + Re^{-1/4}\left(-2i\varkappa_0 z_*^{1/2} + \frac{iA_1}{3\sqrt{3}\lambda}z_*^{-5/6}\right) + \cdots, \\
p &= P_0 + Re^{-1/4}\Re\left\{2i\varkappa_0 z_*^{1/2} - \frac{iA_1}{3\sqrt{3}\lambda}z_*^{-5/6}\right\} + \cdots.
\end{aligned}\right\} \tag{2.3.112}$$

Comparing (2.3.112) with (2.3.109), and taking into account that in the interaction region the body-fitted coordinates are indistinguishable from the Cartesian coordinates, we see that the matching condition is written as

$$\left.\begin{aligned}
\overline{V}_* &= -2i\varkappa_0 z_*^{1/2} + \frac{iA_1}{3\sqrt{3}\lambda}z_*^{-5/6} + \cdots, \\
p_* &= \Re\left\{2i\varkappa_0 z_*^{1/2} - \frac{iA_1}{3\sqrt{3}\lambda}z_*^{-5/6}\right\} + \cdots
\end{aligned}\right\} \quad \text{as} \quad z_* \to \infty, \tag{2.3.113}$$

where $\overline{V}_* = u_* - iv_*$.

An analogue of the impermeability condition on the aerofoil surface is the condition of matching of the streamline slope angle ϑ in regions 4 and 5. It follows from (2.3.109) that in region 5

$$\vartheta = Re^{-1/4}v_* + \cdots. \tag{2.3.114}$$

Comparing (2.3.114) with the solution (2.3.108) for ϑ in region 4, we see that the sought boundary condition is written as

$$v_*\Big|_{y_*=0} = -\frac{1}{\lambda}\frac{dA}{dx_*}. \tag{2.3.115}$$

Equations (2.3.110) considered together with boundary conditions (2.3.113) and (2.3.115) constitute a boundary-value problem for region 5 that may be solved as follows. We start by integrating the first equation in (2.3.110) with respect to x_*. We have

$$u_* + p_* = \Phi(y_*). \tag{2.3.116}$$

To find function $\Phi(y_*)$ we choose a value of y_* and, keeping it fixed, assume that $x_* \to -\infty$. Then, the right-hand side of (2.3.116) will remain constant, while the left-hand side will tend to zero, according to (2.3.113). Thus, we can conclude that $\Phi(y_*) \equiv 0$, which reduces equation (2.3.116) to the linearized Bernoulli equation:

$$p_* = -u_*. \tag{2.3.117}$$

We use it to eliminate p_* from (2.3.110), which leads to the equations

$$\frac{\partial u_*}{\partial x_*} = -\frac{\partial v_*}{\partial y_*}, \qquad \frac{\partial u_*}{\partial y_*} = \frac{\partial v_*}{\partial x_*}.$$

These are the Cauchy–Riemann equations; they represent the necessary and sufficient conditions for the function $\overline{V}_* = u_* - iv_*$ to be an analytic function of the complex variable $z_* = x_* + iy_*$.

According to (2.3.113), the complex conjugate velocity $\overline{V}_*$ is growing at large values of z. We shall consider instead the function

$$F(z_*) = u_* - iv_* + 2i\varkappa_0 z_*^{1/2} \tag{2.3.118}$$

that tends to zero as $z_* \to \infty$. Here, $z_*^{1/2}$ is defined as

$$z_*^{1/2} = (re^{i\vartheta})^{1/2} = r^{1/2}e^{i\vartheta/2},$$

with r and ϑ being the modulus and the argument of z_*, respectively. In particular, on the real axis

$$\left. z_*^{1/2} \right|_{y_*=0} = \begin{cases} x_*^{1/2} & \text{if } x_* > 0, \\ i\,(-x_*)^{1/2} & \text{if } x_* < 0. \end{cases} \tag{2.3.119}$$

Keeping this in mind, we can formulate the following boundary-value problem for function $F(z_*)$:

Problem 2.1 *Find function $F(z_*)$ such that*

1. *$F(z_*)$ is analytic in the upper half of complex plane z_*,*
2. *the imaginary part of $F(z_*)$ is given on the real axis:*

$$\left. \Im\{F\} \right|_{y_*=0} = \frac{1}{\lambda}\frac{dA}{dx_*} + 2\varkappa_0 x_*^{1/2}\mathcal{H}(x_*), \tag{2.3.120}$$

3. *and it is know that*

$$F(z_*) \to 0 \quad as \quad z_* \to \infty. \tag{2.3.121}$$

Here, $\mathcal{H}(x_*)$ is the Heaviside step function given by

$$\mathcal{H}(x_*) = \begin{cases} 1 & \text{if } x_* > 0, \\ 0 & \text{if } x_* < 0. \end{cases} \tag{2.3.122}$$

To solve this problem, consider an arbitrary point z_* in the upper half of the complex plane, and surround it with a close contour C that is composed of a segment

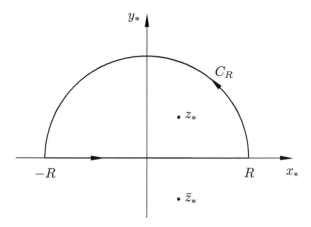

Fig. 2.26: Integration contour for the Cauchy formulae (2.3.123) and (2.3.125).

$[-R, R]$ of the real axis and a semi-circle C_R of radius R as shown in Figure 2.26. Since function $F(z_*)$ is analytic inside C, we can use the Cauchy formula:

$$F(z_*) = \frac{1}{2\pi i} \oint_C \frac{F(\zeta)}{\zeta - z_*} d\zeta$$

$$= \frac{1}{2\pi i} \int_{-R}^{R} \frac{F(\zeta)}{\zeta - z_*} d\zeta + \frac{1}{2\pi i} \int_{C_R} \frac{F(\zeta)}{\zeta - z_*} d\zeta, \qquad (2.3.123)$$

where the integration point ζ is assumed to follow contour C in the anti-clockwise direction. Since function $F(z_*)$ satisfies condition (2.3.121), Jordan's lemma is applicable to the second integral in (2.3.123). According to the lemma,[39]

$$\lim_{R \to \infty} \int_{C_R} \frac{F(\zeta)}{\zeta - z_*} d\zeta = 0.$$

Consequently, setting $R \to \infty$ in equation (2.3.123) reduces it to

$$F(z_*) = \frac{1}{2\pi i} \int_{-\infty}^{\infty} \frac{F(\zeta)}{\zeta - z_*} d\zeta. \qquad (2.3.124)$$

Here, the integration is performed along the real axis.

Now, let us repeat the above procedure using, instead of z_*, the conjugate point $\bar{z}_*$. Since this point is situated outside the integration contour (see Figure 2.26), we can write

$$\oint_C \frac{F(\zeta)}{\zeta - \bar{z}_*} d\zeta = 0. \qquad (2.3.125)$$

[39]See, for example, Dettman (1965).

The integral in (2.3.125) may be treated in the same way as that in (2.3.123). We again split it into two integrals,

$$\oint_C \frac{F(\zeta)}{\zeta - \bar{z}_*}\, d\zeta = \int_{-R}^{R} \frac{F(\zeta)}{\zeta - \bar{z}_*}\, d\zeta + \int_{C_R} \frac{F(\zeta)}{\zeta - \bar{z}_*}\, d\zeta,$$

and note that in the limit $R \to \infty$, the integral along C_R tends to zero. This turns equation (2.3.125) into

$$\int_{-\infty}^{\infty} \frac{F(\zeta)}{\zeta - \bar{z}_*}\, d\zeta = 0. \tag{2.3.126}$$

Now, we take the complex conjugates on both sides of (2.3.126), which results in[40]

$$\int_{-\infty}^{\infty} \frac{\overline{F(\zeta)}}{\zeta - z_*}\, d\zeta = 0. \tag{2.3.127}$$

Equations (2.3.127) and (2.3.124) can be combined together as follows:

$$F(z_*) = \frac{1}{2\pi i} \int_{-\infty}^{\infty} \frac{F(\zeta) - \overline{F(\zeta)}}{\zeta - z_*}\, d\zeta. \tag{2.3.128}$$

The numerator of the integrand in (2.3.128) is calculated using boundary condition (2.3.120). We have

$$F(\zeta) - \overline{F(\zeta)} = 2i\, \Im\{F(\zeta)\} = 2i\left[\frac{1}{\lambda}\frac{dA}{d\zeta} + 2\varkappa_0\zeta^{1/2}\mathcal{H}(\zeta)\right],$$

which, when substituted into (2.3.128), renders the solution of Problem 2.1 in the form

$$F(z_*) = \frac{1}{2\pi i} \int_{-\infty}^{\infty} 2i\left[\frac{1}{\lambda}\frac{dA}{d\zeta} + 2\varkappa_0\zeta^{1/2}\mathcal{H}(\zeta)\right]\frac{d\zeta}{\zeta - z_*}. \tag{2.3.129}$$

The integral on the right-hand side of (2.3.129) is a Cauchy type integral. A detailed discussion of the properties of integrals of this kind is given in Section 2.1.5 in Part 2 of this book series. In general case, the Cauchy type integral is written as

$$F(z) = \frac{1}{2\pi i} \int_C \frac{f(\zeta)}{\zeta - z}\, d\zeta, \tag{2.3.130}$$

where the integration is performed along an open contour C connecting points a and b in the complex z-plane (see Figure 2.27). If point z finds itself on contour C, then

---

[40]Remember that an integral can be treated as a sum to which the following well-known rule applies: $\overline{z_1 + z_2} = \bar{z}_1 + \bar{z}_2$. Also remember that $\overline{z_1 \cdot z_2} = \bar{z}_1 \cdot \bar{z}_2$ and $\overline{z_1/z_2} = \bar{z}_1/\bar{z}_2$.

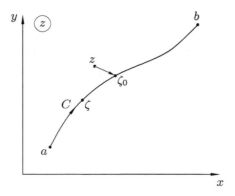

Fig. 2.27: Integration contour C.

we denote it as ζ. We assume that function $f(\zeta)$ is piecewise continuous on C. Then $F(z)$ proves to be analytic everywhere outside C. When point z approaches a point ζ_0 that lies on C, as shown in Figure 2.27, then the limiting value of function $F(z)$ is given by the Sokhotsky–Plemelj formula:[41]

$$\lim_{z \to \zeta_0} F(z) = \frac{1}{2} f(\zeta_0) + F(\zeta_0). \tag{2.3.131}$$

Here, $F(\zeta_0)$ is the principal value of the Cauchy type integral:

$$F(\zeta_0) = \frac{1}{2\pi i} \int_C \frac{f(\zeta)}{\zeta - \zeta_0}\, d\zeta.$$

Let us now return to equation (2.3.129), which represents the solution for region 5. In the integral on the right-hand side of this equation, the role of integration contour C is played by the real axis. Equation (2.3.129) allows us to find function $F(z_*)$ at any point z_* in the upper half-plane. Our main interest is in the behaviour of the pressure p_* at the 'bottom' of region 5. Therefore, we assume that z_* tends to a point $\zeta_0 = (x_*, 0)$ lying on the real axis, and apply the Sokhotsky–Plemelj formula (2.3.131) to the integral in (2.3.129). We see that the limiting value of function $F(z_*)$ is given by

$$F\Big|_{y_*=0} = i\left[\frac{1}{\lambda}\frac{dA}{dx_*} + 2\varkappa_0 x_*^{1/2}\mathcal{H}(x_*)\right] + \frac{1}{\pi}\int_{-\infty}^{\infty}\left[\frac{1}{\lambda}\frac{dA}{d\zeta} + 2\varkappa_0\zeta^{1/2}\mathcal{H}(\zeta)\right]\frac{d\zeta}{\zeta - x_*}.$$

Separating the real and imaginary parts in the above equation, we have

$$\Re\{F\}\Big|_{y_*=0} = \frac{1}{\pi}\int_{-\infty}^{\infty}\left[\frac{1}{\lambda}\frac{dA}{d\zeta} + 2\varkappa_0\zeta^{1/2}\mathcal{H}(\zeta)\right]\frac{d\zeta}{\zeta - x_*}, \tag{2.3.132a}$$

$$\Im\{F\}\Big|_{y_*=0} = \frac{dA}{dx_*} + 2\varkappa_0 x_*^{1/2}\mathcal{H}(x_*). \tag{2.3.132b}$$

---

[41]See equation (2.1.79) on page 104 in Part 2 of this book series.

Equation (2.3.132b) confirms that solution (2.3.129) does satisfy boundary condition (2.3.120). To find the pressure distribution at the 'bottom' of region 5, we need to consider equation (2.3.132a). It follows from (2.3.118) and (2.3.119) that the left-hand side of (2.3.132a) may be written as

$$\Re\{F\}\Big|_{y_*=0} = u_*(x_*,0) - 2\varkappa_0(-x_*)^{1/2}\mathcal{H}(-x_*). \tag{2.3.133}$$

It further follows from the Bernoulli equation (2.3.117) that

$$u_*(x_*,0) = -p_*(x_*,0). \tag{2.3.134}$$

By substituting (2.3.134) into (2.3.133), and then into (2.3.132a), we arrive at a conclusion that

$$p_*\Big|_{y_*=0} = -2\varkappa_0(-x_*)^{1/2}\mathcal{H}(-x_*) - \frac{1}{\pi}\int\limits_{-\infty}^{\infty}\left[\frac{1}{\lambda}\frac{dA}{d\zeta} + 2\varkappa_0\zeta^{1/2}\mathcal{H}(\zeta)\right]\frac{d\zeta}{\zeta - x_*}. \tag{2.3.135}$$

Formulation of the interaction problem

Let us summarize the results of the flow analysis in the viscous-inviscid interaction region, obtained thus far. We found that the flow in region 5 is described by the boundary-layer equations (2.3.92a), (2.3.92c):

$$U^*\frac{\partial U^*}{\partial x_*} + V^*\frac{\partial U^*}{\partial Y_*} = -\frac{dP^*}{dx_*} + \frac{\partial^2 U^*}{\partial Y_*^2}, \tag{2.3.136}$$

$$\frac{\partial U^*}{\partial x_*} + \frac{\partial V^*}{\partial Y_*} = 0. \tag{2.3.137}$$

These have to be solved with the no-slip conditions (2.3.93):

$$U^* = V^* = 0 \quad\text{at}\quad Y_* = 0, \tag{2.3.138}$$

the matching condition (2.3.94) with the solution in region 2a:

$$U^* = (-x_*)^{1/3}\lambda\xi + (-x_*)^{1/6}g'(\xi) + \cdots \quad\text{as}\quad x_* \to -\infty, \tag{2.3.139}$$

and the matching condition (2.3.106) with the solution in region 4:

$$U^* = \lambda Y_* - \frac{2\varkappa_0}{\sqrt{\lambda}}\frac{\Gamma(1/6)}{\Gamma(2/3)}Y_*^{1/2} + A(x_*) + \cdots \quad\text{as}\quad Y_* \to \infty. \tag{2.3.140}$$

It should be noted that neither the pressure P^* nor the displacement function $A(x_*)$ is known in advance. Instead, it is known that these functions are related to one another by means of the *interaction law* (2.3.135):

$$P^*(x_*) = -2\varkappa_0(-x_*)^{1/2}\mathcal{H}(-x_*) - \frac{1}{\pi}\int\limits_{-\infty}^{\infty}\left[\frac{1}{\lambda}\frac{dA}{d\zeta} + 2\varkappa_0\zeta^{1/2}\mathcal{H}(\zeta)\right]\frac{d\zeta}{\zeta - x_*}. \tag{2.3.141}$$

Here, it is taken into account that the pressure does not change across regions 4 and 3, and, therefore,

$$p_*\Big|_{y_*=0} = P^*(x_*).$$

2.3.5 Flow behind the interaction region

To complete the formulation of the interaction problem (2.3.136)–(2.3.141), we also need to pose a downstream boundary condition for the longitudinal velocity U^*. This is to account for the upstream influence in the reverse flow region that forms as a result of the boundary-layer separation. To perform this task we need to study the flow behind the viscous-inviscid interaction region; see Figure 2.28. The physical processes taking place in this flow are the same as in the correspondent supersonic flow (Section 2.2.6) and consist of the following. As the boundary layer separates from the cylinder surface, it turns into the shear layer, shown as region 6 in Figure 2.28. The entrainment effect of the shear layer produces the reverse flow in region 7, which is inviscid, and, therefore, we also have to consider the near-wall boundary layer, shown as region 8 in Figure 2.28. We shall start our analysis with region 6.

Shear layer (region 6)

Since the shear layer is, in essence, the boundary layer after its separation from the body surface, the asymptotic analysis of the Navier–Stokes equations in region 6 should be performed based on the limit

$$s' = O(1), \quad N' = \frac{n'}{Re^{-1/2}} = O(1), \quad Re \to \infty.$$

Here, we use a new set of 'body-fitted' coordinates (s', n'), with s' measured along the dividing streamline, and n' in the perpendicular direction. The shape of the dividing streamline is given by equation (2.3.29). Using (2.3.33) in (2.3.29), we can express it in the form

$$y_S = Re^{-1/16} \frac{4}{3} \varkappa_0 x^{3/2} + \cdots . \tag{2.3.142}$$

We shall seek the solution of the Navier–Stokes equations (1.2.8) in region 6 in the form

$$\left. \begin{aligned} V_\tau(s', n'; Re) &= U_0'(s', N') + \cdots , \\ V_n(s', n'; Re) &= Re^{-1/2} V_0'(s', N') + \cdots , \\ p(s', n'; Re) &= P_0 + \cdots . \end{aligned} \right\} \tag{2.3.143}$$

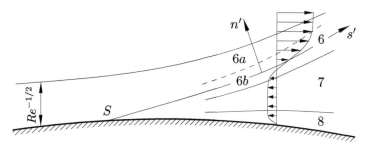

Fig. 2.28: Flow behind the viscous-inviscid interaction region.

By substituting (2.3.143) into (1.2.8) and setting $Re \to \infty$, we find that the flow in the shear layer is described by the boundary-layer equations:

$$U_0' \frac{\partial U_0'}{\partial s'} + V_0' \frac{\partial U_0'}{\partial N'} = \frac{\partial^2 U_0'}{\partial N'^2}, \tag{2.3.144a}$$

$$\frac{\partial U_0'}{\partial s'} + \frac{\partial V_0'}{\partial N'} = 0. \tag{2.3.144b}$$

The boundary conditions for these equations are written as

$$
\begin{aligned}
U_0' &= 1 & &\text{at} \quad N' = \infty, & &(2.3.145a)\\
U_0' &= 0 & &\text{at} \quad N' = -\infty, & &(2.3.145b)\\
U_0' &= U_{00}(N') & &\text{at} \quad s' = 0, & &(2.3.145c)\\
V_0' &= 0 & &\text{at} \quad N' = 0. & &(2.3.145d)
\end{aligned}
$$

When formulating the boundary condition (2.3.145a) at the upper edge of the shear layer, it is taken into account that the velocity components have been made dimensionless in (2.3.3) by referring them to the velocity V_0 at the separated streamline as it is defined by the solution of the Euler equations in region 1. In condition (2.3.145b) we assume, subject to subsequent confirmation, that the fluid velocity in region 7 is small compared to that in region 6 (see Figure 2.28). The initial condition (2.3.145c) is formulated by matching with the solution (2.3.96), (2.3.97) for V_τ in the middle tier (region 4) of the triple-deck structure; see Figure 2.25. Finally, (2.3.145d) is a kinematic condition that follows directly from the fact that the s'-axis is chosen to be aligned with the dividing streamline.

We know that in the interaction region the boundary layer is composed of the middle tier (region 4), where the flow is inviscid, and the viscous nonlinear sublayer (region 3). Correspondingly, we expect the shear layer to split, as $s' \to 0^+$, into two regions: viscous region 6b, and 'locally inviscid' region 6a. It follows from (2.3.145c) that the leading-order term of the asymptotic expansion of U_0' in region 6a is given by

$$U_0' = U_{00}(N') + \cdots \quad \text{as} \quad s' \to 0. \tag{2.3.146}$$

To find the form of the solution in region 6b, we note that since the flow in this region is viscous, the following balance should hold in the momentum equation (2.3.144a):

$$U_0' \frac{\partial U_0'}{\partial s'} \sim \frac{\partial^2 U_0'}{\partial N'^2}. \tag{2.3.147}$$

The factor U_0' on the left-hand side of (2.3.147) can be easily estimated through matching with the solution (2.3.146) in region 6a. Assuming that the thickness of region 6b tends to zero as $s' \to 0^+$, we can use for $U_{00}(N')$ the first equation in (2.3.39):

$$U_{00}(N') = \lambda N' + \cdots . \tag{2.3.148}$$

We see that in region 6b

$$U_0' \sim N', \tag{2.3.149}$$

which allows us to express the order-of-magnitude equation (2.3.147) in the form

$$N'\frac{U_0'}{s'} \sim \frac{U_0'}{N'^2}.$$

Solving this equation for N', we see the thickness of region $6b$ does tend to zero as

$$N' \sim (s')^{1/3}. \tag{2.3.150}$$

An estimate for the longitudinal velocity

$$U_0' \sim (s')^{1/3} \tag{2.3.151}$$

can now be obtained by substituting (2.3.150) into (2.3.149).

The continuity equation (2.3.144b) allows us to introduce the stream function $\Psi(s', N')$ such that

$$U_0' = \frac{\partial \Psi}{\partial N'}, \qquad V_0' = -\frac{\partial \Psi}{\partial s'}. \tag{2.3.152}$$

Using the first equation in (2.3.152) and equations (2.3.150) and (2.3.151), we find that

$$\Psi \sim U_0' N' \sim (s')^{2/3}. \tag{2.3.153}$$

Guided by (2.3.153) and (2.3.150), we seek the solution in region $6b$ in the form

$$\Psi(s', N') = (s')^{2/3}\varphi(\eta) + \cdots \quad \text{as} \quad s' \to 0, \tag{2.3.154a}$$

where

$$\eta = \frac{N'}{(s')^{1/3}}. \tag{2.3.154b}$$

If we substitute (2.3.154) into (2.3.152), then we will have

$$U_0' = (s')^{1/3}\frac{d\varphi}{d\eta} + \cdots, \qquad V_0' = -(s')^{-1/3}\left(\frac{2}{3}\varphi - \frac{1}{3}\eta\frac{d\varphi}{d\eta}\right) + \cdots. \tag{2.3.155}$$

The equation for function $\varphi(\eta)$

$$\frac{d^3\varphi}{d\eta^3} + \frac{2}{3}\varphi\frac{d^2\varphi}{d\eta^2} - \frac{1}{3}\left(\frac{d\varphi}{d\eta}\right)^2 = 0 \tag{2.3.156}$$

is obtained by substituting (2.3.155) into the momentum equation (2.3.144a).

Equation (2.3.156) requires three boundary conditions. Two of these are obtained by substituting (2.3.155) into (2.3.145b) and (2.3.145d). We have

$$\frac{d\varphi}{d\eta} = 0 \quad \text{at} \quad \eta = -\infty, \tag{2.3.157}$$

$$\varphi = 0 \quad \text{at} \quad \eta = 0. \tag{2.3.158}$$

To formulate the third boundary condition, we need to perform the matching with the solution (2.3.146) in region $6a$. We substitute (2.3.148) into (2.3.146), and using

(2.3.154b), express the resulting equation for U_0' in terms of variables of region 6b. We have

$$U_0' = (s')^{1/3}\lambda\eta + \cdots. \tag{2.3.159}$$

It remains to compare (2.3.159) with the asymptotic expansion for U_0' in (2.3.155), and we can conclude that

$$\frac{d\varphi}{d\eta} = \lambda\eta + \cdots \quad \text{as} \quad \eta \to \infty. \tag{2.3.160}$$

The affine transformations

$$\varphi = \lambda^{1/3}\bar{\varphi}, \qquad \eta = \lambda^{-1/3}\bar{\eta} \tag{2.3.161}$$

allow us to cast the boundary-value problem (2.3.156)–(2.3.158), (2.3.160) in the following canonical form:

$$
\left.
\begin{aligned}
&\frac{d^3\bar{\varphi}}{d\bar{\eta}^3} + \frac{2}{3}\bar{\varphi}\frac{d^2\bar{\varphi}}{d\bar{\eta}^2} - \frac{1}{3}\left(\frac{d\bar{\varphi}}{d\bar{\eta}}\right)^2 = 0, \\
&\frac{d\bar{\varphi}}{d\bar{\eta}}\bigg|_{\bar{\eta}=-\infty} = 0, \quad \bar{\varphi}\big|_{\bar{\eta}=0} = 0, \quad \frac{d^2\bar{\varphi}}{d\bar{\eta}^2}\bigg|_{\bar{\eta}=\infty} = 1.
\end{aligned}
\right\} \tag{2.3.162}
$$

The problem (2.3.162) is identical to the correspondent problem (2.2.142) in the supersonic theory of self-induced separation. The results of numerical solution of (2.3.162) are displayed in Figure 2.16.

To study the flow in region 7 (see Figure 2.28), we need to evaluate the entrainment effect of the shear layer (region 6).[42] The latter is defined by the asymptotic behaviour of the solution to (2.3.162) at the lower edge of the shear layer, given by (2.2.144):

$$\bar{\varphi} = -\Lambda + \cdots \quad \text{as} \quad \bar{\eta} \to -\infty. \tag{2.3.163}$$

Here, Λ is a positive constant. Remember that its numerical value was found to be $\Lambda = 1.2565$. Substitution of (2.3.163) into (2.3.161) and then into the equation for V_0' in (2.3.155) yields

$$V_0' = (s')^{-1/3}\frac{2}{3}\lambda^{1/3}\Lambda + \cdots \quad \text{as} \quad s' \to 0. \tag{2.3.164}$$

It remains to substitute (2.3.164) into the asymptotic expansion for V_n in (2.3.143), and we can conclude that at the lower edge of region 6,

$$V_n\bigg|_{N'=-\infty} = Re^{-1/2}(s')^{-1/3}\frac{2}{3}\lambda^{1/3}\Lambda + \cdots. \tag{2.3.165}$$

Reverse flow (region 7)

Region 7 is situated between the shear layer (region 6) and the cylinder surface (see Figure 2.28). Remember that the dividing streamline is given by equation (2.3.142).

---

[42]For a detailed discussion of this phenomenon, see Section 1.4.2.

This means that in region 7 the asymptotic analysis of the Navier–Stokes equations (2.3.4) should be performed based on the limit

$$x = O(1), \quad \check{y} = Re^{1/16}y = O(1), \quad Re \to \infty. \tag{2.3.166}$$

When performing this task, we shall use the dimensionless Cartesian coordinates (x, y) with the origin situated at the separation point S and the x-axis directed tangentially to the cylinder surface at point S.

It follows from (2.3.165) that in region 7 the y-component of the velocity vector is estimated as

$$v \sim Re^{-1/2}. \tag{2.3.167}$$

To find an estimate for the x-component of the velocity vector, we use the continuity equation (2.3.4c). Balancing the two terms in this equation

$$\frac{\partial u}{\partial x} \sim \frac{\partial v}{\partial y}, \tag{2.3.168}$$

and using (2.3.166) and (2.3.167), we find that

$$u \sim Re^{-7/16}. \tag{2.3.169}$$

Now, the pressure perturbations can be estimated by balancing the first convective term on the left-hand side of the x-momentum equation (2.3.4a) with the pressure gradient on the right-hand side:

$$u\frac{\partial u}{\partial x} \sim \frac{\partial p}{\partial x}. \tag{2.3.170}$$

Representing the derivatives in (2.3.170) by finite differences, and using (2.3.169), we can see that

$$\Delta p \sim u^2 \sim Re^{-7/8}. \tag{2.3.171}$$

Guided by (2.3.167), (2.3.169), and (2.3.171) we seek the solution in region 7 in the form

$$\left. \begin{aligned} u = Re^{-7/16}\check{u}_0(x, \check{y}) + \cdots, \qquad v = Re^{-1/2}\check{v}_0(x, \check{y}) + \cdots, \\ p = P_0 + Re^{-7/8}\check{p}_1(x, \check{y}) + \cdots. \end{aligned} \right\} \tag{2.3.172}$$

The substitution of (2.3.172) together with (2.3.166) into the Navier–Stokes equations (2.3.4) yields

$$\check{u}_0\frac{\partial \check{u}_0}{\partial x} + \check{v}_0\frac{\partial \check{u}_0}{\partial \check{y}} = -\frac{\partial \check{p}_1}{\partial x}, \tag{2.3.173a}$$

$$\frac{\partial \check{p}_1}{\partial \check{y}} = 0, \tag{2.3.173b}$$

$$\frac{\partial \check{u}_0}{\partial x} + \frac{\partial \check{v}_0}{\partial \check{y}} = 0. \tag{2.3.173c}$$

Since equations (2.3.173) do not involve viscous terms, we have to pose the impermeability condition of the cylinder surface:

$$\check{v}_0 = 0 \quad \text{at} \quad \check{y} = 0. \tag{2.3.174}$$

The condition of matching with the solution (2.3.165) in region 6 is written as[43]

$$\check{v}_0 - 2\varkappa_0 x^{1/2}\check{u}_0 = \frac{2}{3}\lambda^{1/3}\Lambda x^{-1/3} + \cdots \quad \text{at} \quad \check{y} = \frac{4}{3}\varkappa_0 x^{3/2} + \cdots . \tag{2.3.175}$$

The boundary-value problem (2.3.173)–(2.3.175) may be solved as follows. We start by differentiating the x-momentum equation (2.3.173a) with respect to $\check{y}$. This results in

$$\check{u}_0 \frac{\partial}{\partial x}\left(\frac{\partial\check{u}_0}{\partial\check{y}}\right) + \frac{\partial\check{u}_0}{\partial\check{y}}\frac{\partial\check{u}_0}{\partial x} + \check{v}_0\frac{\partial}{\partial\check{y}}\left(\frac{\partial\check{u}_0}{\partial\check{y}}\right) + \frac{\partial\check{v}_0}{\partial\check{y}}\frac{\partial\check{u}_0}{\partial\check{y}} = 0. \tag{2.3.176}$$

Here, it has been taken into account that the pressure $\check{p}_1$ is independent of $\check{y}$. Using further the continuity equation (2.3.173c), we can see that the second and fourth terms in (2.3.176) cancel each other, which leads to

$$\check{u}_0 \frac{\partial}{\partial x}\left(\frac{\partial\check{u}_0}{\partial\check{y}}\right) + \check{v}_0\frac{\partial}{\partial\check{y}}\left(\frac{\partial\check{u}_0}{\partial\check{y}}\right) = 0. \tag{2.3.177}$$

Equation (2.3.177) signifies that the vorticity $\omega = \partial\check{u}_0/\partial\check{y}$ stays constant along each streamline. Since ω is zero in the stagnation region, we can conclude that

$$\frac{\partial\check{u}_0}{\partial\check{y}} = 0 \tag{2.3.178}$$

everywhere in region 7.

It follows from (2.3.178) that the first term $\partial\check{u}_0/\partial x$ in the continuity equation (2.3.173c) is independent of $\check{y}$. This allows us to integrate (2.3.173c) with respect to $\check{y}$ using (2.3.174) as an initial condition. We have

$$\check{v}_0 = -\frac{d\check{u}_0}{dx}\check{y}. \tag{2.3.179}$$

The substitution of (2.3.179) into the condition (2.3.175) of matching with the solution in region 6 leads to the following equation for $\check{u}_0(x)$:

$$\frac{4}{3}\varkappa_0 x^{3/2}\frac{d\check{u}_0}{dx} + 2\varkappa_0 x^{1/2}\check{u}_0 = -\frac{2}{3}\lambda^{1/3}\Lambda x^{-1/3} + \cdots \quad \text{as} \quad x \to 0.$$

The general solution of this equation is

$$\check{u}_0 = -\frac{3}{4}\frac{\lambda^{1/3}\Lambda}{\varkappa_0}x^{-5/6} + Cx^{-3/2}. \tag{2.3.180}$$

The first term on the right-hand side of (2.3.180) is due to the entrainment effect of the shear layer. The second term represents the flow from a source situated at the separation point S. Since no such source is present in the flow considered here, we can conclude that the solution in region 7 is given by

$$\check{u}_0 = -\frac{3}{4}\frac{\lambda^{1/3}\Lambda}{\varkappa_0}x^{-5/6} + \cdots \quad \text{as} \quad x \to 0. \tag{2.3.181}$$

---

[43]See Problem 4 in Exercises 10.

Near-wall boundary layer (region 8)

Since the solution (2.3.181) in region 7 does not satisfy the no-slip condition on the cylinder surface, to complete the flow description we also need to consider the near-wall boundary layer, shown as region 8 in Figure 2.28. This boundary layer belongs to the Falkner and Skan class.[44] It may be studied in the same way as in Section 2.2.6 for the correspondent supersonic flow. In particular, we find that the longitudinal velocity in region 8 is given by

$$u = -Re^{-7/16} \frac{3\lambda^{1/3}\Lambda}{4\varkappa_0} x^{-5/6} f'(\zeta) + \cdots \quad \text{as} \quad x \to 0, \tag{2.3.182a}$$

where

$$\zeta = Re^{9/32} \sqrt{\frac{3\lambda^{1/3}\Lambda}{4\varkappa_0}} \frac{y}{x^{11/12}}, \tag{2.3.182b}$$

and function $f(\zeta)$ is the solution of the following boundary-value problem:

$$\left. \begin{array}{l} f''' - \dfrac{1}{12} f f'' + \dfrac{5}{6}(1 - f'^2) = 0, \\[2mm] f(0) = f'(0) = 0, \quad f'(\infty) = 1. \end{array} \right\} \tag{2.3.183}$$

Downstream boundary condition for the interaction region

We are ready now to perform the matching of the solution in the lower tier (region 3) of the interaction region with the solution behind the interaction region. For this purpose, it is convenient to express the latter in the form of the composite asymptotic expansion valid in regions 6b, 7, and 8.[45] We start with the observation that the solution (2.3.182) for region 8 is also valid in region 7. Indeed, in region 7, the argument ζ of function $f(\zeta)$ is large, and, therefore, $f'(\zeta) = 1$, which turns (2.3.182a) into (2.3.181).

Now, let us turn to region 6b. The longitudinal velocity in this region is given by the first equation in (2.3.155):

$$U_0' = (s')^{1/3} \frac{d\varphi}{d\eta} + \cdots . \tag{2.3.184}$$

The substitution of (2.3.161) into (2.3.184) yields

$$U_0' = (s')^{1/3} \lambda^{2/3} \frac{d\bar{\varphi}}{d\bar{\eta}} + \cdots . \tag{2.3.185}$$

It remains to substitute (2.3.185) into the asymptotic expansion (2.3.143) of V_τ in region 6, which yields

$$u = x^{1/3} \lambda^{2/3} \bar{\varphi}'(\bar{\eta}) + \cdots . \tag{2.3.186}$$

Here, we use again the fact that in the limit as $s' \to 0+$, the curvilinear coordinates (s', n') become indistinguishable from the Cartesian coordinate (x, y), and V_τ tends to u.

---

[44]See Section 1.3.

[45]For a definition of the composite asymptotic expansion see page 49 in Part 2 of this book series.

To construct a composite asymptotic expansion for u valid in regions 6b, 7, and 8, we simply add (2.3.182a) to (2.3.186):

$$u = x^{1/3}\lambda^{2/3}\bar{\varphi}'(\bar{\eta}) - Re^{-7/16}\frac{3\lambda^{1/3}\Lambda}{4\varkappa_0}x^{-5/6}f'(\zeta) + \cdots. \tag{2.3.187}$$

Indeed, in region 6b, the second term in (2.3.187) is much smaller than the first term and may be disregarded in the leading-order approximations. However, in regions 7 and 8, the derivative $d\bar{\varphi}/d\bar{\eta}$ becomes transcendentally small, and (2.3.187) reduces to (2.3.182a).

The argument ζ of function $f(\zeta)$ is given by (2.3.182b). We also need to express the argument $\bar{\eta}$ of $\bar{\varphi}(\bar{\eta})$ in terms of Cartesian coordinates (x, y). We will use for this purpose Prandtl's transposition theorem (see Section 1.4.3). Keeping in mind that the dividing streamline is given by (2.3.142), we can write equation (2.3.154b) as

$$\eta = \frac{N'}{(s')^{1/3}} = \frac{Re^{1/2}n'}{(s')^{1/3}} = \frac{Re^{1/2}\big[y - Re^{-1/16}\frac{4}{3}\varkappa_0 x^{3/2}\big]}{x^{1/3}}.$$

It remains to apply the affine transformation (2.3.161), and we will have

$$\bar{\eta} = \lambda^{1/3}\frac{Re^{1/2}\big[y - Re^{-1/16}\frac{4}{3}\varkappa_0 x^{3/2}\big]}{x^{1/3}}. \tag{2.3.188}$$

We are ready now to perform the matching with the viscous sublayer (region 3) of the interaction region. For this purpose, we express (2.3.187), (2.3.188), and (2.3.182b) in terms of the variables (2.3.84) of region 3, which results in

$$u = Re^{-1/8}\left[\lambda^{2/3}x_*^{1/3}\bar{\varphi}'(\bar{\eta}) - \frac{3\lambda^{1/3}\Lambda}{4\varkappa_0}x_*^{-5/6}f'(\zeta)\right] + \cdots, \tag{2.3.189a}$$

where

$$\bar{\eta} = \lambda^{1/3}\frac{Y_* - \frac{4}{3}\varkappa_0 x_*^{3/2}}{x_*^{1/3}}, \qquad \zeta = \sqrt{\frac{3\lambda^{1/3}\Lambda}{4\varkappa_0}}\frac{Y_*}{x_*^{11/12}}. \tag{2.3.189b}$$

By comparing (2.3.189a) with the asymptotic expansion (2.3.91) of the longitudinal velocity V_τ in region 3, we can conclude that the sought matching condition is given by

$$U^* = \lambda^{2/3}x_*^{1/3}\bar{\varphi}'(\bar{\eta}) - \frac{3\lambda^{1/3}\Lambda}{4\varkappa_0}x_*^{-5/6}f'(\zeta) + \cdots \quad \text{as} \quad x_* \to \infty. \tag{2.3.190}$$

2.3.6   Canonical form of the interaction problem

The viscous-inviscid interaction problem (2.3.136)–(2.3.141), (2.3.190) involves the skin friction parameter λ. To find λ, we need to solve the classical Prandtl's equations for the boundary layer upstream of the interaction region, and then use equation (2.3.39) for U_{00}. Of course, the behaviour of the boundary layer depends on the shape of a body placed in the flow. Fortunately, we do not need to solve the interaction problem each time a new body shape is considered, because λ can be excluded from the problem formulation by means of the affine transformations

$$
\left.
\begin{aligned}
x_* &= \lambda^{-5/4}\bar{X}, & Y_* &= \lambda^{-3/4}\bar{Y}, & U^* &= \lambda^{1/4}\bar{U}, \\
V^* &= \lambda^{3/4}\bar{V}, & A &= \lambda^{1/4}\bar{A}, & P^* &= \lambda^{1/2}\bar{P}.
\end{aligned}
\right\}
\tag{2.3.191}
$$

These transformations establish the similarity rules for the class of flows considered, and allow us to cast the interaction problem in the following canonical form:[46]

$$
\left.
\begin{aligned}
&\bar{U}\frac{\partial \bar{U}}{\partial \bar{X}} + \bar{V}\frac{\partial \bar{U}}{\partial \bar{Y}} = -\frac{d\bar{P}}{d\bar{X}} + \frac{\partial^2 \bar{U}}{\partial \bar{Y}^2}, \\[4pt]
&\frac{\partial \bar{U}}{\partial \bar{X}} + \frac{\partial \bar{V}}{\partial \bar{Y}} = 0, \\[4pt]
&\bar{P} = -2\bar{\varkappa}_0(-\bar{X})^{1/2}\mathcal{H}(-\bar{X}) - \frac{1}{\pi}\int\limits_{-\infty}^{\infty}\left[\frac{d\bar{A}}{ds} + 2\bar{\varkappa}_0 s^{1/2}\mathcal{H}(s)\right]\frac{ds}{s-\bar{X}}, \\[4pt]
&\bar{U} = \bar{V} = 0 && \text{at} && \bar{Y} = 0, \\[4pt]
&\bar{U} = (-\bar{X})^{1/3}\bar{\xi} + (-\bar{X})^{1/6}\bar{\varkappa}_0\,\bar{g}'(\bar{\xi}) + \cdots && \text{as} && \bar{X} \to -\infty, \\[4pt]
&\bar{U} = \bar{Y} - 2\bar{\varkappa}_0\frac{\Gamma(1/6)}{\Gamma(2/3)}\bar{Y}^{1/2} + \bar{A}(\bar{X}) + \cdots && \text{as} && \bar{Y} \to \infty, \\[4pt]
&\bar{U} = \bar{X}^{1/3}\bar{\varphi}'(\bar{\eta}) - \bar{X}^{-5/6}\frac{3\Lambda}{4\bar{\varkappa}_0}f'(\zeta) + \cdots && \text{as} && \bar{X} \to \infty.
\end{aligned}
\right\}
\tag{2.3.192}
$$

Here

$$
\bar{\varkappa}_0 = \lambda^{-9/8}\varkappa_0.
\tag{2.3.193}
$$

Function $\bar{g}(\bar{\xi})$ satisfies the equation

$$
\bar{g}''' - \frac{1}{3}\bar{\xi}^2\bar{g}'' + \frac{1}{2}\bar{\xi}\bar{g}' - \frac{1}{2}\bar{g} = 1,
\tag{2.3.194a}
$$

which has to be solved with the no-slip conditions

$$
\bar{g}(0) = \bar{g}'(0) = 0,
\tag{2.3.194b}
$$

as well as the requirement that $\bar{g}(\bar{\xi})$ does not grow exponentially as $\bar{\xi} \to \infty$. The boundary-value problem (2.3.194) is obtained by applying the affine transformations

$$
g = \frac{\varkappa_0}{\lambda}\bar{g}, \qquad \xi = \lambda^{-1/3}\bar{\xi}
\tag{2.3.195}
$$

to (2.3.55) and (2.3.57). Using (2.3.195) and (2.3.191) in (2.3.94b), we can see that the argument $\bar{\xi}$ of function $\bar{g}(\bar{\xi})$ can be expressed in terms of the transformed coordinates $(\bar{X}, \bar{Y})$ as $\bar{\xi} = \bar{Y}/(-\bar{X})^{1/3}$.

---

[46]For a more accurate presentation of the boundary condition for $\bar{U}$ at the outer edge of the viscous sublayer, see Problem 5 in Exercises 10.

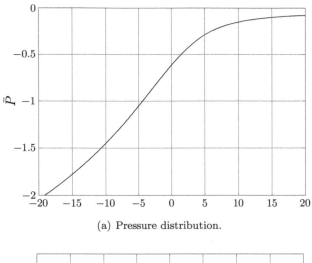

(a) Pressure distribution.

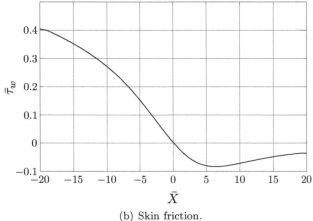

(b) Skin friction.

Fig. 2.29: Results of the numerical solution of the interaction problem (2.3.192).

Functions $\bar{\varphi}(\bar{\eta})$ and $f(\zeta)$ are found by solving boundary-value problems (2.3.162) and (2.3.183), respectively. The arguments (2.3.189b) of these functions are written in terms of the transformed coordinates $(\bar{X}, \bar{Y})$ as

$$\bar{\eta} = \frac{\bar{Y} - \frac{4}{3}\bar{\varkappa}_0 \bar{X}^{3/2}}{\bar{X}^{1/3}}, \qquad \zeta = \sqrt{\frac{3\Lambda}{4\bar{\varkappa}_0} \frac{\bar{Y}}{\bar{X}^{11/12}}}.$$

The results of the numerical solution of the interaction problem (2.3.192) are displayed in Figure 2.29. Figures 2.29(a) and 2.29(b) show the pressure $\bar{P}$ and the skin friction $\bar{\tau}_w$, respectively. The latter is calculated as

$$\bar{\tau}_w = \frac{\partial \bar{U}}{\partial \bar{Y}}\bigg|_{\bar{Y}=0}.$$

Notice that unlike in the corresponding supersonic flow (Figure 2.18) the pressure perturbations do not disappear upstream of the interaction region. Instead, the asymptotic behaviour of $\bar{P}$ is given by

$$\bar{P} = -2\bar{\varkappa}_0(-\bar{X})^{1/2} + \cdots \quad \text{as} \quad \bar{X} \to -\infty.$$

The calculation results show that the pressure grows monotonically in the interaction region, and develops a 'plateau' as $\bar{X} \to \infty$.

As far as the skin friction is concerned, it follows from the upstream boundary condition for $\bar{U}$ in (2.3.192) that

$$\bar{\tau}_w = 1 + (-\bar{X})^{-1/6}\bar{\varkappa}_0\bar{g}''(0) + \cdots \quad \text{as} \quad \bar{X} \to -\infty.$$

This means that $\bar{\tau}_w$ tends to its unperturbed value $\bar{\tau}_w = 1$ in the boundary layer before the interaction region, but does this rather slowly. Inside the interaction region, $\bar{\tau}_w$ first decreases, passing through zero at the separation point S. Behind this point, $\bar{\tau}_w$ becomes negative, which signifies that the fluid near the cylinder surface moves upstream. Some distance downstream of the separation point, $\bar{\tau}_w$ reaches a minimum, and starts to recover, approaching zero as $\bar{X} \to \infty$. This means that the recirculating motion of the fluid becomes progressively slower, which confirms the formation of the stagnant region behind the separation point S.

We see that there are obvious similarities in the behaviour of subsonic and supersonic flows near the separation point. However, there are also some differences. In particular, we know that in the supersonic flow, the dividing streamline tends to a straight line as the distance $\bar{X}$ from the separation point increases:

$$\bar{Y} = \bar{P}_{\mathrm{pl}}\bar{X} + \cdots \quad \text{as} \quad \bar{X} \to \infty.$$

Meanwhile, in the incompressible flow, this equation assumes the form

$$\bar{Y} = \frac{4}{3}\bar{\varkappa}_0\bar{X}^{3/2} + \cdots \quad \text{as} \quad \bar{X} \to \infty.$$

Parameter $\bar{\varkappa}_0$ is found as a part of the solution of the interaction problem (2.3.192). Its numerical value was calculated by a number of authors starting with Smith (1977). Our calculations show that

$$\bar{\varkappa}_0 = 0.2285. \tag{2.3.196}$$

To complete the analysis, it remains to return to the inviscid flow in region 1. Remember that the Euler equations describing the flow in this region admit a family of solutions with the position of the separation point on the cylinder surface being a free parameter (see Figure 2.21). The members of the family may also be identified by parameter $\varkappa$ in (2.3.26) and (2.3.29), because $\varkappa$ is uniquely defined by the position of the separation point. At the same time, $\varkappa$ is determined by the solution in the viscous-inviscid interaction region. Substituting (2.3.196) into (2.3.193) and then into (2.3.33), we have

$$\varkappa = 0.2285 \cdot Re^{-1/16}\lambda^{9/8}.$$

Since $\varkappa$ is positive and small, the separation point is situated a small distance downstream of the Brillouin–Villat point, shown as B in Figure 2.21.

Exercises 10

1. Adjust the inspection analysis of Section 2.2.3 to the case of subsonic flow in the viscous-inviscid interaction region, and show that estimates (2.2.28) for the fluid-dynamic functions and the dimensions of the lower tier remain unchanged.

    Suggestion: Notice that the main difference between the supersonic viscous-inviscid interaction and its subsonic counterpart is in the behaviour of the flow in the upper tier where, instead of the Ackeret formula (2.2.17), one has to use the subsonic version of the thin aerofoil theory; see Sections 2.1 and 2.2 in Part 2 of this book series.

2. Consider the inviscid flow in region 1 outside the viscous-inviscid interaction region. You may use without proof the fact that the leading-order solution in region 1 is expressed in terms of the complex conjugate velocity $\overline{V} = u_0 - iv_0$ by equation (2.3.23) with parameter $\varkappa$ given by (2.3.33):

$$\overline{V}(z) = 1 - Re^{-1/16}2i\varkappa_0 z^{1/2} + \cdots \quad \text{as} \quad z \to 0. \tag{2.3.197}$$

The corresponding equation for the pressure is given by (2.3.25):

$$p = P_0 + Re^{-1/16}\Re\{2i\varkappa_0 z^{1/2}\} + \cdots \quad \text{as} \quad z \to 0. \tag{2.3.198}$$

You may also use without proof the fact that the solution in the boundary layer (region 2) is expressed by the asymptotic expansions (2.3.36):

$$\left.\begin{array}{l} V_\tau(s, n; Re) = U_0(s, N) + Re^{-1/16}U_1(s, N) + \cdots, \\ V_n(s, n; Re) = Re^{-1/2}V_0(s, N) + Re^{-9/16}V_1(s, N) + \cdots. \end{array}\right\} \tag{2.3.199}$$

Here, the leading-order terms (U_0, V_0) are regular near the separation point, and may be represented by the Taylor expansions (2.3.38):

$$\left.\begin{array}{l} U_0(s, N) = U_{00}(N) + sU_{01}(N) + \cdots, \\ V_0(s, N) = V_{00}(N) + sV_{01}(N) + \cdots \end{array}\right\} \quad \text{as} \quad s \to 0-.$$

The second-order terms (U_1, V_1) in (2.3.199) are singular. In particular, the lateral velocity component V_1 is given in the main part of the boundary layer (region 2b in Figure 2.23) by equations (2.3.72) and (2.3.77). Combining these together we can write

$$V_1(s, N) = (-s)^{-5/6}\frac{A_1}{6\lambda}U_{00}(N) + \cdots \quad \text{as} \quad s \to 0-.$$

Your task is to study the displacement effect of the boundary layer on the flow in region 1, and to find the corresponding correction terms in the asymptotic expansions (2.3.197) and (2.3.198). You may perform this task in the following steps:

(a) Seek the solution in region 1 in the form

$$\left.\begin{array}{l} u = u_0(x, y) + Re^{-1/2}u_1(x, y) + Re^{-9/16}u_2(x, y) + \cdots, \\ v = v_0(x, y) + Re^{-1/2}v_1(x, y) + Re^{-9/16}v_2(x, y) + \cdots, \\ p = p_0(x, y) + Re^{-1/2}p_1(x, y) + Re^{-9/16}p_2(x, y) + \cdots. \end{array}\right\} \tag{2.3.200}$$

Substitute (2.3.200) into the Navier–Stokes equations (2.3.4) and deduce that functions u_2, v_2, and p_2 satisfy the equations

$$u_0\frac{\partial u_2}{\partial x} + u_2\frac{\partial u_0}{\partial x} + v_0\frac{\partial u_2}{\partial y} + v_2\frac{\partial u_0}{\partial y} = -\frac{\partial p_2}{\partial x}, \tag{2.3.201a}$$

$$u_0\frac{\partial v_2}{\partial x} + u_2\frac{\partial v_0}{\partial x} + v_0\frac{\partial v_2}{\partial y} + v_2\frac{\partial v_0}{\partial y} = -\frac{\partial p_2}{\partial y}, \tag{2.3.201b}$$

$$\frac{\partial u_2}{\partial x} + \frac{\partial v_2}{\partial y} = 0. \tag{2.3.201c}$$

(b) By cross-differentiating equations (2.3.201a) and (2.3.201b), prove that the flow considered is irrotational, that is,

$$\frac{\partial u_2}{\partial y} - \frac{\partial v_2}{\partial x} = 0. \tag{2.3.202}$$

Hint: Remember that in the leading-order approximation, the velocity components satisfy equations (2.3.12) and (2.3.7c):

$$\frac{\partial u_0}{\partial y} - \frac{\partial v_0}{\partial x} = 0, \tag{2.3.203}$$

$$\frac{\partial u_0}{\partial x} + \frac{\partial v_0}{\partial y} = 0.$$

(c) Using the zero-vorticity equations (2.3.202) and (2.3.203) in the momentum equations (2.3.201a) and (2.3.201b), show that the pressure perturbation function p_2 is related to the velocity components by means of the linearized Bernoulli equation

$$u_0 u_2 + v_0 v_2 + p_2 = const. \tag{2.3.204}$$

(d) Argue that function $\overline{V}_2 = u_2 - iv_2$ is an analytic function of the complex variable $z = x + iy$. Prove that this function satisfies the following boundary conditions on the real axis:

$$\left.\Im\{\overline{V}_2\}\right|_{y=0} = -(-x)^{-5/6}\frac{A_1}{6\lambda} + \cdots \quad \text{if} \quad x < 0, \atop \left.\Re\{\overline{V}_2\}\right|_{y=0} = 0 \qquad\qquad\qquad \text{if} \quad x > 0. \right\} \tag{2.3.205}$$

(e) Seek the solution for $\overline{V}_2(z)$ in the form

$$\overline{V}_2(z) = Cz^{-5/6} + \cdots \quad \text{as} \quad z \to 0,$$

and deduce from (2.3.205) that the complex constant $C = C_r + iC_i$ is given by

$$C = \frac{iA_1}{3\sqrt{3\lambda}}.$$

(f) Finally, use the Bernoulli equation (2.3.204) to show that

$$p_2 = -\Re\left\{\frac{iA_1}{3\sqrt{3}\lambda}z^{-5/6}\right\} + \cdots \quad \text{as} \quad z \to 0.$$

3. In equation (2.3.135), known as the interaction law, neither the pressure p_* nor the displacement function A is known in advance, and should be found as a part of the solution of the viscous-inviscid interaction problem as a whole. For numerical purposes some authors use the inverted form of this equation:

$$\frac{1}{\lambda}\frac{dA}{dx_*} + 2\varkappa_0 x_*^{1/2}\mathcal{H}(x_*) = \frac{1}{\pi}\int_{-\infty}^{\infty}\frac{p_*(\zeta,0) + 2\varkappa_0(-\zeta)^{1/2}\mathcal{H}(-\zeta)}{\zeta - x_*}\,d\zeta. \qquad (2.3.206)$$

Your task is to confirm the validity of equation (2.3.206). To perform this task, you need to manipulate equations (2.3.124) and (2.3.127):

$$F(z_*) = \frac{1}{2\pi i}\int_{-\infty}^{\infty}\frac{F(\zeta)}{\zeta - z_*}\,d\zeta, \qquad \int_{-\infty}^{\infty}\frac{\overline{F(\zeta)}}{\zeta - z_*}\,d\zeta = 0,$$

where $F(z_*) = u_* - iv_* + 2i\varkappa_0 z_*^{1/2}$ with $z_* = x_* + iy_*$. You may use these equations without derivation. You may also use without proof the Bernoulli equation (2.3.117)

$$p_* = -u_*$$

and the boundary condition (2.3.115)

$$v_*\Big|_{y_*=0} = -\frac{1}{\lambda}\frac{dA}{dx_*}.$$

4. The fluid motion in the reverse flow region (shown as region 7 in Figure 2.28) is caused by the entrainment effect of the shear layer (region 6). Mathematically, the entrainment effect is expressed by equation (2.3.165) for the normal velocity component at the lower edge of the shear layer:

$$V_n\Big|_{N'=-\infty} = Re^{-1/2}x^{-1/3}\frac{2}{3}\lambda^{1/3}\Lambda + \cdots. \qquad (2.3.207)$$

Taking into account that the dividing streamline is given by equation (2.3.142),

$$y = Re^{-1/16}\frac{4}{3}\varkappa_0 x^{3/2}, \qquad (2.3.208)$$

and that the velocity components in region 7 are expressed by the asymptotic expansions (2.3.172),

$$u = Re^{-7/16}\breve{u}_0(x,\breve{y}) + \cdots, \qquad v = Re^{-1/2}\breve{v}_0(x,\breve{y}) + \cdots, \qquad (2.3.209)$$

show that equation (2.3.207) may be expressed in terms of the variables (2.3.209) of region 7 as

$$\breve{v}_0 - 2\varkappa_0 x^{1/2}\breve{u}_0 = x^{-1/3}\frac{2}{3}\lambda^{1/3}\Lambda + \cdots.$$

Suggestion: Show that the unit vector **n** normal to (2.3.208) is given by

$$n_x = -Re^{-1/16}2\varkappa x^{1/2}, \qquad n_1 = 1,$$

and calculate normal velocity as $V_n = un_x + vn_y$.

5. Return to the viscous-inviscid interaction problem (2.3.192), and find the next term in the 'upstream' asymptotic expansion

$$\bar{U} = (-\bar{X})^{1/3}\bar{\xi} + (-\bar{X})^{1/6}\varkappa_0\,\bar{g}'(\bar{\xi}) + \cdots \quad \text{as} \quad \bar{X} \to -\infty.$$

Also, find the corresponding corrections in the asymptotic representation of $\bar{U}$ at the outer edge of the viscous sublayer:

$$\bar{U} = \bar{Y} - 2\varkappa_0\frac{\Gamma(1/6)}{\Gamma(2/3)}\,\bar{Y}^{1/2} + \bar{A}(\bar{X}) + \cdots \quad \text{as} \quad \bar{Y} \to \infty.$$

You may perform these tasks in the following steps:
(a) Introduce the stream function $\Psi(\bar{X},\bar{Y})$ such that

$$\frac{\partial\Psi}{\partial\bar{X}} = -\bar{V}, \qquad \frac{\partial\Psi}{\partial\bar{Y}} = \bar{U},$$

and seek the 'upstream' asymptotic expansion of Ψ in the form

$$\Psi = (-\bar{X})^{2/3}\frac{1}{2}\bar{\xi}^2 + (-\bar{X})^{1/2}\varkappa_0\,\bar{g}_1(\bar{\xi})$$
$$+ (-\bar{X})^{1/3}\varkappa_0^2\,\bar{g}_2(\bar{\xi}) + \cdots \quad \text{as} \quad \bar{X} \to -\infty,$$

where

$$\bar{\xi} = \frac{\bar{Y}}{(-\bar{X})^{1/3}}.$$

Show that the corresponding expansions of the velocity components are written as

$$\bar{U} = (-\bar{X})^{1/3}\bar{\xi} + (-\bar{X})^{1/6}\varkappa_0\bar{g}_1'(\bar{\xi}) + \varkappa_0^2\bar{g}_2'(\bar{\xi}) + \cdots, \tag{2.3.210a}$$

$$\bar{V} = (-\bar{X})^{-1/2}\varkappa_0\left(\frac{1}{2}\bar{g}_1 - \frac{1}{3}\bar{\xi}\bar{g}_1'\right)$$
$$+ (-\bar{X})^{-2/3}\varkappa_0^2\left(\frac{1}{3}\bar{g}_2 - \frac{1}{3}\bar{\xi}\bar{g}_2'\right) + \cdots. \tag{2.3.210b}$$

Substitute (2.3.210) into the momentum equation

$$\bar{U}\frac{\partial\bar{U}}{\partial\bar{X}} + \bar{V}\frac{\partial\bar{U}}{\partial\bar{Y}} = -\frac{d\bar{P}}{d\bar{X}} + \frac{\partial^2\bar{U}}{\partial\bar{Y}^2},$$

and confirm that function $\bar{g}_1(\bar{\xi})$ satisfies equation (2.3.194a). Also show that the equation for $\bar{g}_2(\bar{\xi})$ is written as

$$\bar{g}_2''' - \frac{1}{3}\bar{\xi}^2\bar{g}_2'' + \frac{1}{3}\bar{\xi}\,\bar{g}_2' - \frac{1}{3}\bar{g}_2 = \frac{1}{2}\bar{g}_1\bar{g}_1'' - \frac{1}{6}\left(\bar{g}_1'\right)^2. \tag{2.3.211}$$

Suggestion: When performing the substitution, remember that

$$\bar{P} = -2\varkappa_0(-\bar{X})^{1/2} + \cdots \quad \text{as} \quad \bar{X} \to -\infty.$$

(b) Taking into account that[47]

$$\bar{g}_1 = \frac{2}{3}\bar{A}_0\bar{\xi}^{3/2} + \bar{A}_1\bar{\xi} + \cdots \quad \text{as} \quad \bar{\xi} \to \infty, \tag{2.3.212}$$

show that at the outer edge of the viscous sublayer the right-hand side of equation (2.3.211) behaves as

$$RHS = -\frac{1}{12}\bar{A}_0\bar{A}_1\bar{\xi}^{1/2} + \cdots .$$

Hence, argue that a particular solution $\bar{g}_{2p}$ to (2.3.211) may be chosen such that

$$\bar{g}_{2p} = \bar{A}_0\bar{A}_1\bar{\xi}^{1/2} + \cdots \quad \text{as} \quad \bar{\xi} \to \infty.$$

(c) Now consider the homogeneous part of (2.3.211):

$$\bar{g}_2''' - \frac{1}{3}\bar{\xi}^2\bar{g}_2'' + \frac{1}{3}\bar{\xi}\bar{g}_2' - \frac{1}{3}\bar{g}_2 = 0. \tag{2.3.213}$$

Differentiate all the terms in (2.3.213) with respect to $\bar{\xi}$. Then, introduce a new independent variable $z = \frac{1}{9}\bar{\xi}^3$, and show that function $h_2 = d^2\bar{g}_2/d\bar{\xi}^2$ satisfies the confluent hypergeometric equation

$$z\frac{d^2 h_2}{dz^2} + \left(\frac{2}{3} - z\right)\frac{dh_2}{dz} - \frac{1}{3}h_2 = 0 \tag{2.3.214}$$

with parameters

$$a = \frac{1}{3}, \qquad b = \frac{2}{3}.$$

Remember that the solution of (2.3.214), which does not grow exponentially as $z \to 0$, is given by equation (2.3.61) with $C_1 = 0$:

$$h_2 = C_2 U(a, b; z).$$

Using the fact that at large values of the argument z, the Tricomi function behaves as

$$U(a, b; z) = z^{-a}\left[1 + O(z^{-1})\right] \quad \text{as} \quad z \to \infty,$$

show that

$$h_2 = \frac{d^2\bar{g}_2}{d\bar{\xi}^2} = \frac{B_0}{\bar{\xi}} + \cdots \quad \text{as} \quad \bar{\xi} \to \infty, \tag{2.3.215}$$

where $B_0 = 3^{2/3}C_2$.

---

[47]Equation (2.3.212) is obtained by performing affine transformations (2.3.195) in (2.3.69).

(d) Integrate equation (2.3.215) with respect to $\bar{\xi}$:

$$\frac{d\bar{g}_2}{d\bar{\xi}} = B_0 \ln \bar{\xi} + \cdots \quad \text{as} \quad \bar{\xi} \to \infty, \tag{2.3.216}$$

and substitute (2.3.216) together with (2.3.212) into (2.3.210a) to show that at large negative values of $\bar{X}$, the longitudinal velocity is represented at the outer edge of the viscous sublayer by

$$\bar{U} = \bar{Y} + \bar{\varkappa}_0 \bar{A}_0 \bar{Y}^{1/2} + \bar{\varkappa}_0^2 B_0 \ln \bar{Y}$$
$$+ \bar{\varkappa}_0 \bar{A}_1 (-\bar{X})^{1/6} - \frac{1}{3} \bar{\varkappa}_0^2 B_0 \ln(-\bar{X}) + \cdots.$$

Hence, argue that for $\bar{X} = O(1)$, the asymptotic expansion of $\bar{U}$ at the outer edge of the viscous sublayer has the form

$$\bar{U} = \bar{Y} + \bar{\varkappa}_0 \bar{A}_0 \bar{Y}^{1/2} + \bar{\varkappa}_0^2 B_0 \ln \bar{Y} + \bar{A}(\bar{X}) + \cdots \quad \text{as} \quad \bar{Y} \to \infty,$$

where $\bar{A}(\bar{X})$ is the displacement function that exhibits the following behaviour at large negative values of $\bar{X}$:

$$\bar{A}(\bar{X}) = \bar{\varkappa}_0 \bar{A}_1 (-\bar{X})^{1/6} - \frac{1}{3} \bar{\varkappa}_0^2 B_0 \ln(-\bar{X}) + \cdots \quad \text{as} \quad \bar{X} \to -\infty.$$

6. It is known that in transonic flow separating from a corner point O (see Figure 2.30) the pressure gradient, acting upon the boundary layer upstream of the separation, behaves as[48]

$$\frac{dP}{dx} = -k(-x)^{-2/3} + \cdots \quad \text{as} \quad x \to 0-,$$

where k is an order one positive constant, and x is the coordinate measured from the corner point O parallel to the body surface before the corner.

Your task is to analyse the flow in the boundary layer. You may use for this purpose the classical boundary-layer equations. These are written in suitably chosen non-dimensional variables as

$$\rho U \frac{\partial U}{\partial x} + \rho V \frac{\partial U}{\partial Y} = -\frac{dP}{dx} + \frac{\partial}{\partial Y}\left(\mu \frac{\partial U}{\partial Y}\right), \tag{2.3.217a}$$

$$\rho U \frac{\partial h}{\partial x} + \rho V \frac{\partial h}{\partial Y} = U \frac{dP}{dx} + \frac{1}{Pr} \frac{\partial}{\partial Y}\left(\mu \frac{\partial h}{\partial Y}\right) + \mu \left(\frac{\partial U}{\partial Y}\right)^2, \tag{2.3.217b}$$

$$\frac{\partial(\rho U)}{\partial x} + \frac{\partial(\rho V)}{\partial Y} = 0, \tag{2.3.217c}$$

$$h = \frac{\gamma}{\gamma - 1} \frac{P}{\rho}. \tag{2.3.217d}$$

---

[48]See equation (4.4.37) on page 233 in Part 1 of this book series.

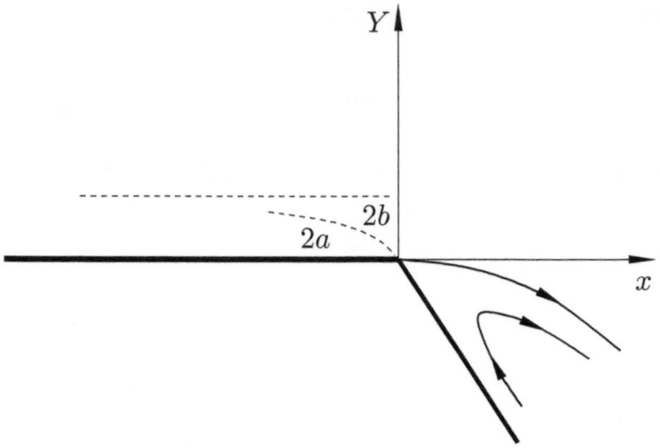

Fig. 2.30: Transonic flow separating from a corner.

You may perform the analysis in the following steps:

(a) First consider the near-wall viscous sublayer (region 2a). Seek the solution in this region in the form

$$\Psi = (-x)^{7/12} g(\xi) + \cdots \quad \text{as} \quad x \to 0-,$$

with the similarity variable

$$\xi = \frac{Y}{(-x)^{5/12}}.$$

Using the fact that the stream function Ψ is defined by the equations

$$\frac{\partial \Psi}{\partial x} = -\rho V, \qquad \frac{\partial \Psi}{\partial Y} = \rho U,$$

deduce that

$$U = (-x)^{1/6} \frac{1}{\rho_w} g'(\xi) + \cdots , \tag{2.3.218}$$

$$V = (-x)^{-5/12} \frac{1}{\rho_w} \left(\frac{7}{12} g - \frac{5}{12} \xi g' \right) + \cdots ,$$

and show that function $g(\xi)$ satisfies the following equation:

$$\mu_w g''' - \frac{7}{12} g g'' + \frac{1}{6} (g')^2 + k \rho_w = 0. \tag{2.3.219}$$

Suggestion: You may accept without proof the fact that flow in region 2a may be treated as incompressible with the density ρ and the viscosity coefficient μ assuming the values ρ_w and μ_w, respectively.

(b) Now study the flow behaviour at the outer edge of region $2a$. Represent function $g(\xi)$ in the form of the asymptotic expansion

$$g(\xi) = A\xi^\alpha + \cdots \quad \text{as} \quad \xi \to \infty. \tag{2.3.220}$$

Substitute (2.3.220) into (2.3.219), and show that if $\alpha > 1$, then

$$A^2\alpha(5\alpha - 7) = 0.$$

Hence, conclude that

$$\alpha = \frac{7}{5}. \tag{2.3.221}$$

(c) Substitute (2.3.220) with (2.3.221) into (2.3.218) to show that at the 'bottom' of region $2b$

$$U = \frac{7A}{5\rho_w} Y^{2/5} + \cdots .$$

3
Trailing-Edge Flow

With the introduction of the concept of viscous–inviscid interaction, the boundary-layer theory entered a new era. It took many years of experimentation and reasoning to achieve proper understanding of the physical processes involved in the viscous–inviscid interaction process,[1] but when the triple-deck theory was eventually put forward, it was done simultaneously by Neiland (1969a) and Stewartson and Williams (1969), who were concerned with the self-induced separation of the boundary layer in supersonic flow. Remarkably, at the same time, Messiter (1970) formulated the triple-deck theory in his analysis of a different problem: the high-Reynolds-number flow of an incompressible fluid near the trailing edge of a flat plate. This problem was also considered by Stewartson (1969) who, of course, was in a position to see an analogy between the trailing-edge flow and the self-induced separation of the boundary layer. Soon after that, it was recognized that the process of viscous–inviscid interaction plays a key role in a wide variety of flows, and since then the number of publications in the field has grown rapidly. In this chapter we consider the flow near the trailing edge of a flat plate. Chapter 4 discusses the boundary-layer separation at corner points. Finally, Chapter 5 presents the Marginal Separation theory in application to the boundary-layer separation at the leading edge of a thin aerofoil. For other applications of the viscous–inviscid interaction theory, the interested reader is referred to Sychev *et al.* (1998) and Neiland *et al.* (2008).

3.1 Problem Formulation

In this section we return to the high-Reynolds-number flow of an incompressible fluid past a flat plate (see Figure 3.1). We shall assume, as in Section 1.1, that the plate is parallel to the oncoming flow. We denote the plate length by L, the free-stream velocity by V_∞, the pressure in the oncoming flow by p_∞, and the fluid density and the kinematic viscosity coefficient by ρ and ν, respectively. Using these quantities, we introduce the non-dimensional variables:

$$\left.\begin{aligned} \hat{x} &= L\,x, \qquad \hat{y} = L\,y, \\ \hat{u} &= V_\infty u, \qquad \hat{v} = V_\infty v, \qquad \hat{p} = p_\infty + \rho\,V_\infty^2 p. \end{aligned}\right\} \tag{3.1.1}$$

Here, $(\hat{x}, \hat{y})$ are Cartesian coordinates, with $\hat{x}$ measured along the plate from its leading edge O and $\hat{y}$ in the perpendicular direction, $(\hat{u}, \hat{v})$ are the velocity components in these

---

[1] See Section 2.1.

Fluid Dynamics: Part 3: Boundary Layers. © Anatoly I. Ruban, 2018. Published 2018 by Oxford University Press. 10.1093/oso/9780199681754.001.0001

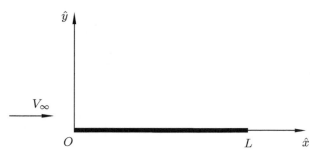

Fig. 3.1: Flow layout.

coordinates, and $\hat{p}$ is the pressure. As usual, we use the 'hat' to denote the dimensional variables.

The substitution of (3.1.1) into the Navier–Stokes equations (1.1.1) renders them in the form

$$u\frac{\partial u}{\partial x} + v\frac{\partial u}{\partial y} = -\frac{\partial p}{\partial x} + \frac{1}{Re}\left(\frac{\partial^2 u}{\partial x^2} + \frac{\partial^2 u}{\partial y^2}\right), \tag{3.1.2a}$$

$$u\frac{\partial v}{\partial x} + v\frac{\partial v}{\partial y} = -\frac{\partial p}{\partial y} + \frac{1}{Re}\left(\frac{\partial^2 v}{\partial x^2} + \frac{\partial^2 v}{\partial y^2}\right), \tag{3.1.2b}$$

$$\frac{\partial u}{\partial x} + \frac{\partial v}{\partial y} = 0. \tag{3.1.2c}$$

Here, Re is the Reynolds number defined as

$$Re = \frac{V_\infty L}{\nu}.$$

We seek the solution of equations (3.1.2) assuming that $Re \to \infty$.

To perform this task, we use Prandtl's hierarchical strategy. We start with the main inviscid part of the flow (region 1), where the asymptotic analysis of the Navier–Stokes equations (3.1.2) is based on the limit

$$x = O(1), \quad y = O(1), \quad Re \to \infty.$$

In the leading-order approximation, the solution in region 1 is represented by the unperturbed uniform flow (1.1.12):

$$u = 1, \quad v = 0, \quad p = 0. \tag{3.1.3}$$

As the second step of the hierarchical strategy, we have to consider the boundary layer shown as region 2 in Figure 1.3. In this region we have to use the limit

$$x = O(1), \quad Y = Re^{1/2}y = O(1), \quad Re \to \infty. \tag{3.1.4}$$

The velocity components in region 2 are sought in the form of asymptotic expansions

$$u = U_0(x, Y) + \cdots, \qquad v = Re^{-1/2}V_0(x, Y) + \cdots. \tag{3.1.5}$$

The substitution of (3.1.5) into the Navier–Stokes equations (3.1.2) reduces the latter to the boundary-layer equations

$$U_0 \frac{\partial U_0}{\partial x} + V_0 \frac{\partial U_0}{\partial Y} = \frac{\partial^2 U_0}{\partial Y^2}, \tag{3.1.6a}$$

$$\frac{\partial U_0}{\partial x} + \frac{\partial V_0}{\partial Y} = 0. \tag{3.1.6b}$$

To describe the boundary layer on the plate surface upstream of the trailing edge, we have to solve equations (3.1.6a), (3.1.6b) with the following initial condition at the leading edge of the plate:

$$U_0 = 1 \quad \text{at} \quad x = 0, \ Y \in (0, \infty). \tag{3.1.6c}$$

On the plate surface the no-slip conditions are used:

$$U_0 = V_0 = 0 \quad \text{at} \quad Y = 0, \ x \in [0, 1]. \tag{3.1.6d}$$

Finally, at the outer edge of the boundary layer, we impose the matching condition with the solution (3.1.3) in region 1:

$$U_0 = 1 \quad \text{at} \quad Y = \infty. \tag{3.1.6e}$$

In Section 1.1.2 we found that the boundary-value problem (3.1.6) admits a self-similar solution

$$U_0 = \varphi'(\eta), \qquad V_0 = \frac{1}{2\sqrt{x}} (\eta \varphi' - \varphi), \qquad \eta = \frac{Y}{\sqrt{x}}, \tag{3.1.7}$$

where function $\varphi(\eta)$ satisfies the Blasius equation

$$\varphi''' + \frac{1}{2} \varphi \varphi'' = 0. \tag{3.1.8a}$$

It has to be solved with the boundary conditions

$$\varphi(0) = \varphi'(0) = 0, \quad \varphi'(\infty) = 1. \tag{3.1.8b}$$

The results of the numerical solution of the boundary-value problem (3.1.8) are shown in Figure 1.5. It was also found that near the bottom of the boundary layer equation (1.1.49) holds:

$$\varphi(\eta) = \frac{1}{2} \lambda \eta^2 + \cdots \quad \text{as} \quad \eta \to 0, \tag{3.1.9}$$

while at the outer edge of the boundary layer the behaviour of function $\varphi(\eta)$ is described by equation (1.1.52):

$$\varphi(\eta) = \eta - A_- + \cdots \quad \text{as} \quad \eta \to \infty. \tag{3.1.10}$$

Now, our task will be to extend the Blasius solution downstream of the trailing edge. In Section 1.6 we studied the viscous wake at large distance from a trailing edge. Here, our interest is in the behaviour of the wake close to the trailing edge. This problem was first considered by Goldstein (1930), hence the name, *Goldstein's wake*.

3.2 Goldstein's Wake

Remember that the boundary-layer equations (3.1.6a), (3.1.6b) possess the following property: if the longitudinal velocity component U_0 remains positive, then at any point (x, Y) in the boundary layer, the solution of (3.1.6a), (3.1.6b) only depends on the boundary conditions upstream of this point, and cannot be influenced by a change of the boundary conditions downstream of the point considered. This means that the Blasius solution (3.1.7) holds everywhere in the boundary layer up to the trailing edge of the plate. Setting $x = 1$ in the first equation in (3.1.7), we find that the longitudinal velocity at the trailing edge is given by

$$U_0\Big|_{x=1} = \varphi'(Y). \tag{3.2.1}$$

Graphically this is illustrated in Figure 3.2(a), where it is taken into account that the boundary layer forms on both sides of the plate.

The deficit of the velocity, which forms in the boundary layer due to the no-slip condition on the plate surface, cannot disappear at once as the fluid leaves the plate at the trailing edge and moves further downstream. Hence, the boundary layer does not cease to exist behind the plate. Instead, it turns into a viscous wake (see Figure 3.2b). The thickness of the wake is, obviously, an order $Re^{-1/2}$ quantity, which means that the boundary-layer equations (3.1.6a), (3.1.6b) remain valid in the wake. Also, the condition of matching (3.1.6e) with the solution (3.1.3) in region 1 holds, but the no-slip conditions (3.1.6d) are no longer applicable.

To see what happens behind the plate, consider the axis of symmetry of the wake. Since at any point on the axis of symmetry $V_0 = 0$, equation (3.1.6a) reduces to

$$U_0 \frac{\partial U_0}{\partial x} = \frac{\partial^2 U_0}{\partial Y^2} \quad \text{at} \quad Y = 0, \ x > 1. \tag{3.2.2}$$

The left-hand side of (3.2.2) represents the acceleration of the fluid,[2] which is caused by the viscous force on the right-hand side of (3.2.2). Clearly, $\partial^2 U_0/\partial Y^2$ is infinite at the trailing edge, which means that the fluid particles experience a sharp acceleration as they leave the plate surface at the trailing edge. As a result, the velocity profile in the wake assumes the form depicted in Figure 3.2(b), which tells us that, in the wake, the no-slip conditions (3.1.6d) have to be substituted by

$$V_0 = \frac{\partial U_0}{\partial Y} = 0 \quad \text{at} \quad Y = 0, \ x > 1. \tag{3.2.3}$$

Thanks to the flow symmetry, we can restrict our attention to the upper half of the (x, Y)-plane. Our task will be to find the asymptotic solution of equations (3.1.6) as $s = x - 1 \rightarrow 0+$.

When solving the boundary-layer equations, the longitudinal velocity component U_0 is assumed to be a continuous function. Therefore, it immediately follows from

---

[2]See Section 1.4.1 in Part 1 of this book series.

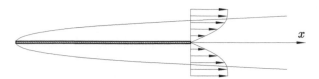

(a) Blasius boundary layer.

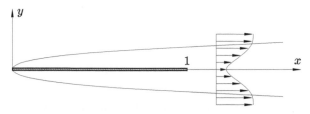

(b) Formation of the wake.

Fig. 3.2: Formation of the viscous wake in the flow past a flat plate.

(3.2.1) that the leading-order term of the asymptotic expansion of U_0 in the wake is given by

$$U_0(x, Y) = U_{00}(Y) + o(1) \quad \text{as} \quad s \to 0+, \tag{3.2.4}$$

where $U_{00}(Y) = \varphi'(Y)$. However, it follows from (3.1.9) that

$$U_{00}(Y) = \lambda Y + \cdots \quad \text{as} \quad Y \to 0, \tag{3.2.5}$$

which shows that the solution given by (3.2.4) does not satisfy the symmetry condition

$$\frac{\partial U_0}{\partial Y} = 0 \quad \text{at} \quad Y = 0,$$

imposed upon U_0 by (3.2.3). This suggests that we are dealing here with a singular perturbation problem, where (3.2.4) represents the outer solution. According to Goldstein (1930), in addition to the outer region, we have to introduce a viscous sublayer whose thickness becomes progressively smaller as $s \to 0+$. Indeed, for the flow to be viscous the following order-of-magnitude equation should hold:

$$U_0 \frac{\partial U_0}{\partial x} \sim \frac{\partial^2 U_0}{\partial Y^2}. \tag{3.2.6}$$

The factor U_0 on the left-hand side of (3.2.6) may be estimated by matching it with the solution (3.2.4), (3.2.5) in the outer region. We have

$$U_0 \sim Y, \tag{3.2.7}$$

which, being substituted into (3.2.6), yields

$$Y \frac{\partial U_0}{\partial x} \sim \frac{\partial^2 U_0}{\partial Y^2}. \tag{3.2.8}$$

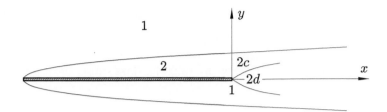

Fig. 3.3: Splitting of the wake into viscous and locally inviscid parts.

Representing the derivatives in (3.2.8) by finite differences

$$Y\frac{\Delta U_0}{\Delta x} \sim \frac{\Delta U_0}{Y^2}, \tag{3.2.9}$$

and solving (3.2.9) for Y, we see that the thickness of the viscous sublayer

$$Y \sim s^{1/3}, \tag{3.2.10}$$

does tend to zero as $s \to 0+$, leaving the rest of the boundary layer *locally inviscid*. Figure 3.3 shows the viscous sublayer as region 2d and the main part of the boundary layer as region 2c.

3.2.1 Viscous sublayer (region 2d)

It follows from (3.2.10) that in region 2d the solution of the boundary-layer equations (3.1.6) should be constructed based on the limit

$$\zeta = \frac{Y}{s^{1/3}} = O(1), \quad s \to 0+. \tag{3.2.11}$$

Instead of dealing with two velocity components, U_0 and V_0, it is easier to deal with the stream function $\Psi_0(x, Y)$. Using the continuity equation (3.1.6b) we define it such that

$$U_0 = \frac{\partial \Psi_0}{\partial Y}, \qquad V_0 = -\frac{\partial \Psi_0}{\partial x}. \tag{3.2.12}$$

An estimate for Ψ_0 may be obtained with the help of the first equation in (3.2.13). We see that

$$\Psi_0 \sim U_0 Y. \tag{3.2.13}$$

By substituting (3.2.10) into (3.2.7), we find that in region 2d

$$U_0 \sim s^{1/3}. \tag{3.2.14}$$

It remains to substitute (3.2.14) together with (3.2.10) into (3.2.13), and thus we can conclude that

$$\Psi_0 \sim s^{2/3}.$$

This suggests that the solution in region 2d should be sought in the form

$$\Psi_0(x, Y) = s^{2/3} f_0(\zeta) + \cdots \quad \text{as} \quad s \to 0+, \tag{3.2.15}$$

with the independent variable ζ given by (3.2.11). The substitution of (3.2.15) into (3.2.12) results in

$$\left.\begin{aligned} U_0(x, Y) &= s^{1/3} f_0'(\zeta) + \cdots, \\ V_0(x, Y) &= \frac{1}{3} s^{-1/3} \left(\zeta f_0' - 2 f_0 \right) + \cdots \end{aligned}\right\} \quad \text{as} \quad s \to 0+. \tag{3.2.16}$$

Here, function $f_0(\zeta)$ satisfies the equation

$$f_0''' + \frac{2}{3} f_0 f_0'' - \frac{1}{3} \left(f_0' \right)^2 = 0, \tag{3.2.17}$$

which is deduced by substituting (3.2.16) into the momentum equation (3.1.6a).

To formulate the boundary conditions for (3.2.17), we substitute (3.2.16) into (3.2.3), which yields

$$f_0(0) = f_0''(0) = 0. \tag{3.2.18}$$

One more boundary condition for equation (3.2.17) is deduced by matching with the solution (3.2.4) in region 2c. Substituting (3.2.5) into (3.2.4) and expressing the resulting equation for U_0 in terms of the inner variable (3.2.11), we have

$$U_0 = s^{1/3} \lambda \zeta + \cdots. \tag{3.2.19}$$

By comparing (3.2.19) with the asymptotic expansion for U_0 in (3.2.16), we can conclude that

$$f_0'(\zeta) = \lambda \zeta + \cdots \quad \text{as} \quad \zeta \to \infty, \tag{3.2.20}$$

where $\lambda = 0.3321$. The results of the numerical solution of equation (3.2.17) subject to boundary conditions (3.2.18) and (3.2.20) are displayed in Figure 3.4.

Before returning to the main part of the boundary layer, shown as region 2c in Figure 3.3, we shall study the behaviour of function $f_0(\zeta)$ at the outer edge of region 2d. It follows from (3.2.20) that the leading-order term of the asymptotic expansion of $f_0(\zeta)$ is given by

$$f_0 = \frac{1}{2} \lambda \zeta^2 + \cdots \quad \text{as} \quad \zeta \to \infty.$$

We seek the next term in the form of a power function:

$$f_0 = \frac{1}{2} \lambda \zeta^2 + A \zeta^\alpha + \cdots \quad \text{as} \quad \zeta \to \infty. \tag{3.2.21}$$

Here, A is a constant, and to ensure that the second term in (3.2.21) is much smaller than the first one (as required by the definition of an asymptotic expansion), we have to impose the following restriction on α:

$$\alpha < 2. \tag{3.2.22}$$

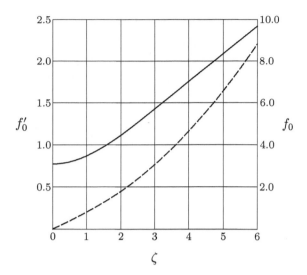

Fig. 3.4: Solution of the Goldstein problem (3.2.17), (3.2.18), and (3.2.20). The solid line shows the longitudinal velocity profile $f_0'(\zeta)$; the dashed line represents function $f_0(\zeta)$.

The substitution of (3.2.21) into (3.2.17) results in

$$A\alpha(\alpha - 1)(\alpha - 2)\zeta^{\alpha-3} + \frac{1}{3}A\lambda(\alpha^2 - 3\alpha + 2)\zeta^\alpha + \cdots = 0. \qquad (3.2.23)$$

Keeping in mind that $\zeta^{\alpha-3} \ll \zeta^\alpha$ for large ζ, we disregard the first term in (3.2.23). We have

$$A\lambda(\alpha^2 - 3\alpha + 2)\zeta^\alpha = 0.$$

Since our objective is to find a non-trivial second term in (3.2.21), we must suppose that $A \neq 0$, which leads to the following quadratic equation for α:

$$\alpha^2 - 3\alpha + 2 = 0.$$

It has two solutions: $\alpha_1 = 2$ and $\alpha_2 = 1$, but only α_2 satisfies condition (3.2.22). Thus, we take $\alpha = 1$, and write asymptotic expansion (3.2.21) in the form[3]

$$f_0 = \frac{1}{2}\lambda\zeta^2 + A_+\zeta + \cdots \quad \text{as} \quad \zeta \to \infty. \qquad (3.2.24)$$

Constant A_+ is found from the numerical solution of the boundary-value problem (3.2.17), (3.2.18), (3.2.20) to be $A_+ = 0.4278$.

[3]Here, we use for A a new notation A_+.

3.2.2 Main part of the boundary layer (region 2c)

The asymptotic analysis of the boundary-layer equations (3.1.6) in region 2c is based on the limit

$$Y = O(1), \quad s \to 0+.$$

To find the form of the asymptotic expansions of the velocity components U_0 and V_0 in region 2c, we shall use the following procedure. We substitute (3.2.24) into (3.2.16), which results in

$$U_0 = s^{1/3}\lambda\zeta + s^{1/3}A_+ + \cdots, \qquad V_0 = -\frac{1}{3}s^{-1/3}A_+\zeta + \cdots. \tag{3.2.25}$$

Equations (3.2.25) are valid in the overlap region that lies between regions 2c and 2d (see Figure 3.5). Setting $\zeta = Y/s^{1/3}$ in (3.2.25), we have

$$U_0 = \lambda Y + s^{1/3}A_+ + \cdots, \qquad V_0 = -\frac{1}{3}s^{-2/3}A_+Y + \cdots, \tag{3.2.26}$$

which suggests that in region 2c, where $Y = O(1)$, the asymptotic expansions of the velocity components should be sought in the form

$$\left.\begin{aligned}U_0(x,Y) &= U_{00}(Y) + s^{1/3}U_{01}(Y) + \cdots, \\ V_0(x,Y) &= s^{-2/3}V_{01}(Y) + \cdots\end{aligned}\right\} \quad \text{as} \quad s \to 0+. \tag{3.2.27}$$

It further follows from (3.2.26) that the conditions of matching with the solution in region 2d are written as

$$\left.\begin{aligned}U_{01}(Y) &= A_+ + \cdots, \\ V_{01}(Y) &= -\frac{1}{3}A_+Y + \cdots\end{aligned}\right\} \quad \text{as} \quad Y \to 0. \tag{3.2.28}$$

The leading-order term in the expansion (3.2.27) for longitudinal velocity component U_0 represents the velocity profile at the trailing edge of the flat plate, which is known from the Blasius solution: $U_{00}(Y) = \varphi'(Y)$. To find functions $U_{01}(Y)$ and

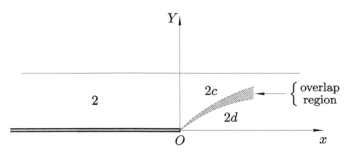

Fig. 3.5: Matching through the overlap region.

$V_{01}(Y)$, we substitute (3.2.27) into the boundary-layer equations (3.1.6a) and (3.1.6b). This results in

$$\frac{1}{3}s^{-2/3}U_{00}U_{01} + s^{-2/3}V_{01}\frac{U_{00}}{dY} = U_{00}'', \tag{3.2.29a}$$

$$\frac{1}{3}s^{-2/3}U_{01} + s^{-2/3}\frac{dV_{01}}{dY} = 0. \tag{3.2.29b}$$

We can see that the viscous term on the right-hand side of the momentum equation (3.2.29a) is much smaller than the convective terms on the left-hand side. This confirms that in the limit as $s = x - 1 \to 0$, the flow in region $2c$ can be treated as inviscid, which is why region $2c$ is called the *locally inviscid region*.

Disregarding the viscous term in (3.2.29a) we can write equations (3.2.29) as

$$\frac{1}{3}U_{00}U_{01} + V_{01}\frac{U_{00}}{dY} = 0, \qquad \frac{1}{3}U_{01} + \frac{dV_{01}}{dY} = 0. \tag{3.2.30}$$

To solve these equations we first eliminate U_{01} from (3.2.30). We find that

$$U_{00}\frac{dV_{01}}{dY} - V_{01}\frac{dU_{00}}{dY} = 0. \tag{3.2.31}$$

We then divide both terms in (3.2.31) by U_{00}^2, which results in

$$\frac{1}{U_{00}}\frac{dV_{01}}{dY} - \frac{V_{01}}{U_{00}^2}\frac{dU_{00}}{dY} = 0,$$

or, equivalently,

$$\frac{d}{dY}\left(\frac{V_{01}}{U_{00}}\right) = 0. \tag{3.2.32}$$

Thus, V_{01}/U_{00} does not change across region $2c$, and we can write

$$\frac{V_{01}}{U_{00}} = const. \tag{3.2.33}$$

To find the constant on the right-hand side of (3.2.33) we need to perform the matching with the solution in region $2d$ (see Figure 3.5). We shall consider for this purpose the streamline slope angle θ. It follows from (3.1.5) that in the boundary layer on the plate surface and in the wake

$$\tan\theta = \frac{v}{u} = Re^{-1/2}\frac{V_0}{U_0} + \cdots. \tag{3.2.34}$$

In region $2d$, the velocity components U_0, V_0 are given by (3.2.16). Hence,

$$\tan\theta = Re^{-1/2}s^{-2/3}\frac{(\zeta f_0' - 2f_0)}{3f_0'} + \cdots. \tag{3.2.35}$$

To determine θ at the outer edge of region $2d$, we have to substitute (3.2.24) into (3.2.35), which gives

$$\tan\theta = Re^{-1/2}s^{-2/3}\left(-\frac{A_+}{3\lambda}\right) + \cdots. \tag{3.2.36}$$

Fig. 3.6: The shape of a streamline near the trailing edge of the plate.

In the main part of the boundary layer (region $2c$), functions U_0 and V_0 are expressed by equations (3.2.27). Substituting these into (3.2.34), we have

$$\tan\theta = Re^{-1/2}s^{-2/3}\frac{V_{01}}{U_{00}} + \cdots . \tag{3.2.37}$$

By comparing (3.2.37) with (3.2.36), we can see that the constant in (3.2.33) is given by

$$const = -\frac{A_+}{3\lambda}. \tag{3.2.38}$$

It remains to substitute (3.2.38) into (3.2.33) and then into (3.2.37), and we can conclude that in region $2c$

$$\tan\theta = Re^{-1/2}s^{-2/3}\left(-\frac{A_+}{3\lambda}\right) + \cdots , \quad s > 0. \tag{3.2.39}$$

It is interesting to notice that it is the viscous sublayer (region $2d$) that produces the displacement effect of the wake. Indeed, setting $\zeta = 0$ in (3.2.35) yields $\theta = 0$, which confirms that the streamline slope angle is zero on the axis of symmetry of the flow; the latter coincides with the x-axis in Figure 3.5. The acceleration of the fluid in region $2d$ makes θ negative, and $|\theta|$ grows monotonically across region $2d$. At the outer edge of region $2d$, it reaches the value given by equation (3.2.36), and then, according to (3.2.39), remains unchanged across the entire region $2c$. Equation (3.2.39) further shows that θ develops a singularity as $s \to 0+$. This behaviour is illustrated in Figure 3.6.

For future reference, we need to deduce an equivalent of equation (3.2.39) for the flow before the trailing edge. The boundary layer upstream of the trailing edge is described by the Blasius solution, and remains regular near the trailing edge. By substituting (3.1.10) into (3.1.7) and then into (3.2.34), we find that at the outer edge of the boundary layer

$$\tan\theta = Re^{-1/2}\frac{A_-}{2\sqrt{x}} + \cdots , \quad s < 0. \tag{3.2.40}$$

3.3 Perturbations in the Inviscid Flow

Following Prandtl's hierarchical strategy, we now consider the perturbations induced in the inviscid region 1 by the displacement effect of the boundary layer.[4] In the leading-order approximation the solution in region 1 is represented by the uniform flow (3.1.3).

---

[4]For a detailed discussion of the hierarchical strategy, see Section 1.2.

The displacement of the boundary layer produces $O(Re^{-1/2})$ perturbations in the flow, which means that in region 1 the velocity components (u, v) and the pressure p should be sought in the form of the asymptotic expansions

$$u(x, y; Re) = 1 + Re^{-1/2}u_1(x, y) + \cdots, \tag{3.3.1a}$$

$$v(x, y; Re) = Re^{-1/2}v_1(x, y) + \cdots, \tag{3.3.1b}$$

$$p(x, y; Re) = Re^{-1/2}p_1(x, y) + \cdots. \tag{3.3.1c}$$

By substituting (3.3.1) into the Navier–Stokes equations (3.1.2) and applying the limit

$$x = O(1), \quad y = O(1), \quad Re \to \infty,$$

we find that functions u_1, v_1, and p_1 satisfy the following equations:

$$\frac{\partial u_1}{\partial x} = -\frac{\partial p_1}{\partial x}, \quad \frac{\partial v_1}{\partial x} = -\frac{\partial p_1}{\partial y}, \quad \frac{\partial u_1}{\partial x} + \frac{\partial v_1}{\partial y} = 0. \tag{3.3.2}$$

These are known as the *equations of thin aerofoil theory*.[5] When dealing with the flow past a thin aerofoil, equations (3.3.2) are solved with the condition of attenuation of the perturbations in the far field

$$\left.\begin{array}{r} u_1 \to 0, \\ v_1 \to 0, \\ p_1 \to 0 \end{array}\right\} \quad \text{as} \quad x^2 + y^2 \to \infty, \tag{3.3.3}$$

and the impermeability condition of the aerofoil surface.[6] Notably, the attenuation condition (3.3.3) remains applicable to the flow considered here, but instead of the impermeability condition, we have to use the condition of matching with the solution in the boundary layer (region 2). It follows from (3.3.1) that in region 1 the streamline slope angle θ is given by

$$\tan\theta = \frac{v}{u} = Re^{-1/2}v_1(x, y) + \cdots. \tag{3.3.4}$$

By comparing (3.3.4) with (3.2.34) and using Prandtl's matching rule, we can conclude that along the plate surface and in the wake

$$v_1\Big|_{y=0} = \lim_{Y \to \infty} \frac{V_0}{U_0} \quad \text{if} \quad x > 0. \tag{3.3.5a}$$

Since the flow considered is symmetric, we can also claim that on the x-axis upstream of the plate

$$v_1\Big|_{y=0} = 0 \quad \text{if} \quad x < 0. \tag{3.3.5b}$$

The solution to the boundary-value problem (3.3.2), (3.3.3), and (3.3.5) may be constructed with the help of the procedure described in Section 2.1.3 in Part 2 of this

---

[5]A detailed discussion of the properties of inviscid flows past thin aerofoils may be found in Section 2.1 in Part 2 of this book series.

[6]See equations (2.1.11), (2.1.15), and (2.1.16) on pages 85 and 86 in Part 2 of this book series.

book series. In particular, the pressure in the boundary layer and in the wake may be calculated as[7]

$$p_1\Big|_{y=0} = \frac{1}{\pi} \int\limits_0^\infty \frac{G(\zeta)}{\zeta - x}\, d\zeta, \tag{3.3.6}$$

where

$$G(x) = \lim_{Y \to \infty} \frac{V_0}{U_0}.$$

To determine the behaviour of p_1 near the trailing edge, we have to find the asymptotic behaviour of the integral in (3.3.6) for small $|x - 1| \to 0$ (see Problem 5 in Exercises 7 in Part 2). However, there is another, much easier way, that we describe here.

It appears that due to the singularity in (3.2.39), the inviscid flow in a vicinity of the trailing edge may be studied independently of the rest of the flow in region 1. We start with the first equation in (3.3.2). Integration of this equation with the far-field conditions (3.3.3) yields the linearized Bernoulli equation

$$u_1 = -p_1. \tag{3.3.7}$$

Using (3.3.7) to eliminate u_1 from (3.3.2), we have

$$\frac{\partial p_1}{\partial x} = \frac{\partial v_1}{\partial y}, \qquad \frac{\partial p_1}{\partial y} = -\frac{\partial v_1}{\partial x}.$$

These are the Cauchy–Riemann equations representing the necessary and sufficient conditions for the function

$$f(z) = p_1(x, y) + iv_1(x, y)$$

to be an analytic function of the complex variable $z = x + iy$. It follows from (3.2.39) and (3.2.40) that the boundary condition (3.3.5a) may be written for function $f(z)$ as

$$\Im f\Big|_{y=0} = \begin{cases} \dfrac{A_-}{2} + \cdots & \text{as} \quad x \to 1 - 0, \\[2ex] -\dfrac{A_+}{3\lambda}(x - 1)^{-2/3} + \cdots & \text{as} \quad x \to 1 + 0. \end{cases} \tag{3.3.8}$$

This suggests that near the trailing edge, function $f(z)$ has to be sought in the form

$$f(z) = C\,(z - 1)^{-2/3} + \cdots \quad \text{as} \quad z \to 1. \tag{3.3.9}$$

Here, C is a complex constant: $C = C_r + iC_i$. To determine its real and imaginary parts, boundary condition (3.3.8) will be used.

---

[7]See Problem 2 in Exercises 7 in Part 2.

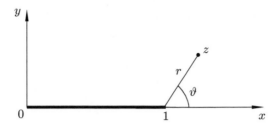

Fig. 3.7: Geometrical interpretation of equation (3.3.10).

We write (see Figure 3.7)

$$z - 1 = re^{i\vartheta}, \tag{3.3.10}$$

and then (3.3.9) assumes the form

$$f(z) = (C_r + iC_i) r^{-2/3} e^{-i2\vartheta/3} + \cdots . \tag{3.3.11}$$

First, consider a point z lying on the real axis downstream of the trailing edge (see Figure 3.7). For such point

$$\vartheta = 0, \qquad r = x - 1 = s,$$

which turns equation (3.3.11) into

$$f(z) = p_1(x, 0) + iv_1(x, 0) = (C_r + iC_i) s^{-2/3} + \cdots . \tag{3.3.12}$$

The imaginary part of (3.3.12) should satisfy the second condition in (3.3.8). Consequently,

$$C_i = -\frac{A_+}{3\lambda}. \tag{3.3.13}$$

To make use of the first condition in (3.3.8), we now assume that point z lies on the real axis upstream of the trailing edge, in which case

$$\vartheta = \pi, \qquad r = 1 - x. \tag{3.3.14}$$

Using (3.3.14) in (3.3.11), we have

$$f(z) = (C_r + iC_i)(1 - x)^{-2/3} e^{-i2\pi/3} =$$
$$= (C_r + iC_i)\left(\cos \frac{2}{3}\pi - i \sin \frac{2}{3}\pi \right)(1 - x)^{-2/3} + \cdots . \tag{3.3.15}$$

The separation of the real and imaginary parts in (3.3.15) yields

$$p_1(x, 0) = \left(C_r \cos \frac{2}{3}\pi + C_i \sin \frac{2}{3}\pi \right)(1 - x)^{-2/3} + \cdots , \tag{3.3.16}$$

$$v_1(x, 0) = \left(C_i \cos \frac{2}{3}\pi - C_r \sin \frac{2}{3}\pi \right)(1 - x)^{-2/3} + \cdots .$$

According to (3.3.8), the imaginary part of $f(z)$ should be finite upstream of the trailing edge, which is only possible if

$$C_i \cos \frac{2}{3}\pi - C_r \sin \frac{2}{3}\pi = 0.$$

Hence,

$$C_r = C_i \cot \frac{2}{3}\pi = \frac{A_+}{3\sqrt{3}\,\lambda}. \tag{3.3.17}$$

Now, we can substitute (3.3.13) and (3.3.17) into (3.3.16), and we find that on the plate surface upstream of the trailing edge

$$p_1(x,0) = -\frac{2A_+}{3\sqrt{3}\,\lambda}(1-x)^{-2/3} + \cdots \quad \text{as} \quad x \to 1-0. \tag{3.3.18}$$

The pressure in the wake downstream of the trailing edge is given by the real part of (3.3.12):

$$p_1(x,0) = \frac{A_+}{3\sqrt{3}\,\lambda}(x-1)^{-2/3} + \cdots \quad \text{as} \quad x \to 1+0. \tag{3.3.19}$$

3.4   Second-Order Perturbations in the Boundary Layer

Here, we perform the fourth step in Prandtl's hierarchical strategy, which consistes in the analysis of the perturbations produced in the boundary layer by the $O(Re^{-1/2})$ pressure perturbations in the external inviscid flow. In the leading-order approximation, the solution in the boundary layer (shown as region 2 in Figure 3.3) is represented in the form of asymptotic expansions (3.1.5). Now we write

$$u(x,y;Re) = U_0(x,Y) + Re^{-1/2}U_1(x,Y) + \cdots, \tag{3.4.1a}$$

$$v(x,y;Re) = Re^{-1/2}V_0(x,Y) + Re^{-1}V_1(x,Y) + \cdots, \tag{3.4.1b}$$

$$p(x,y,Re) = Re^{-1/2}P_1(x,Y) + \cdots. \tag{3.4.1c}$$

Remember that the asymptotic analysis of the Navier–Stokes equations (3.1.2) in region 2 is based on the limit

$$x = O(1), \quad Y = Re^{1/2}y = O(1), \quad Re \to \infty.$$

By substituting (3.4.1) into (3.1.2) and working with $O(Re^{-1/2})$ terms, we arrive at the following equations for the perturbation functions U_1, V_1, and P_1:

$$U_0\frac{\partial U_1}{\partial x} + U_1\frac{\partial U_0}{\partial x} + V_0\frac{\partial U_1}{\partial Y} + V_1\frac{\partial U_0}{\partial Y} = -\frac{\partial P_1}{\partial x} + \frac{\partial^2 U_1}{\partial Y^2}, \tag{3.4.2a}$$

$$\frac{\partial P_1}{\partial Y} = 0, \tag{3.4.2b}$$

$$\frac{\partial U_1}{\partial x} + \frac{\partial V_1}{\partial Y} = 0. \tag{3.4.2c}$$

These equations are applicable to the boundary layer on the plate surface as well as to the wake behind the plate. Let us consider the behaviour of the boundary layer as it approaches the trailing edge. Equation (3.4.2b) shows that the pressure (3.3.18) produced in the inviscid flow penetrates the boundary layer with no change. Hence, we can write

$$P_1(x) = k(-s)^{-2/3} + \cdots \quad \text{as} \quad s = x - 1 \to 0-, \tag{3.4.3}$$

where k is a constant given by

$$k = -\frac{2A_+}{3\sqrt{3}\,\lambda}.$$

The singularity in (3.4.3) causes the boundary layer upstream of the trailing edge to split into two parts: the locally inviscid main part of the boundary layer, and the viscous sublayer, shown in Figure 3.8 as regions 2b and 2a, respectively. We start our analysis with the viscous region 2a.

3.4.1 Viscous sublayer (region 2a)

For the flow in region 2a to be viscous, the first convective term on the left-hand side of the momentum equation (3.4.2a) should be of the same order of magnitude as the viscous term on the right-hand side:

$$U_0\frac{\partial U_1}{\partial x} \sim \frac{\partial^2 U_1}{\partial Y^2}. \tag{3.4.4}$$

Function U_0 on the left-hand side of (3.4.4) is given by the Blasius solution (3.1.7). The thickness of region 2a is expected to become progressively smaller as $s \to 0-$, which allows us to use equation (3.1.9). By substituting (3.1.9) into (3.1.7) and setting $x \to 1 - 0$, we find that in region 2a

$$U_0 = \lambda Y + \cdots. \tag{3.4.5}$$

With (3.4.5), the order-of-magnitude equation (3.4.4) takes the form

$$Y\frac{\partial U_1}{\partial x} \sim \frac{\partial^2 U_1}{\partial Y^2}.$$

Representing the derivatives in this equation by finite differences, we have

$$Y\frac{\Delta U_1}{\Delta x} \sim \frac{\Delta U_1}{Y^2}.$$

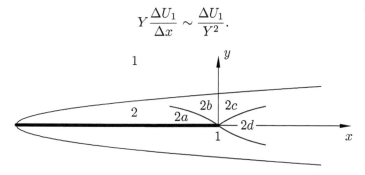

Fig. 3.8: Splitting of the boundary layer upstream of the trailing edge.

Solving this equation for Y, we can conclude that the thickness of region $2a$ is given by

$$Y \sim (-s)^{1/3}. \tag{3.4.6}$$

Here, it is taken into account that the distance Δx from the trailing edge may be calculated as $\Delta x = 1 - x = -s$.

To obtain an estimate for U_1 in region $2a$, we compare the convective term in the momentum equation (3.4.2a) with the pressure gradient:

$$U_0 \frac{\partial U_1}{\partial x} \sim \frac{dP_1}{dx}. \tag{3.4.7}$$

Differentiation of (3.4.3) yields

$$\frac{dP_1}{dx} = \frac{2}{3}(-s)^{-5/3} + \cdots . \tag{3.4.8}$$

It further follows from (3.4.5) and (3.4.6) that in region $2a$

$$U_0 \sim (-s)^{1/3}. \tag{3.4.9}$$

Substituting (3.4.8) and (3.4.9) into (3.4.7), we have

$$(-s)^{1/3} \frac{\partial U_1}{\partial x} \sim (-s)^{-5/3}. \tag{3.4.10}$$

It remains to approximate the derivative $\partial U_1/\partial s$ by finite differences

$$\frac{\partial U_1}{\partial x} \sim \frac{U_1}{\Delta s},$$

and we can cast equation (3.4.10) in the form

$$(-s)^{1/3} \frac{U_1}{\Delta s} \sim (-s)^{-5/3}.$$

We see that in region $2a$

$$U_1 \sim (-s)^{-1}. \tag{3.4.11}$$

The continuity equation (3.4.2c) allows us to introduce the stream function $\Psi_1(x, Y)$. It is related to the velocity components as

$$\frac{\partial \Psi_1}{\partial x} = -V_1, \qquad \frac{\partial \Psi_1}{\partial Y} = U_1. \tag{3.4.12}$$

An estimate for the stream function in region $2a$ may be obtained with the help of the second equation in (2.3.48). We have

$$\Psi_1 \sim U_1 Y. \tag{3.4.13}$$

The substitution of (3.4.11) and (3.4.6) into (3.4.13) shows that

$$\Psi_1 \sim (-s)^{-2/3}. \tag{3.4.14}$$

This concludes the order-of-magnitude analysis of the flow in region $2a$. Now we return to equations (3.4.2) and perform their analysis on a rigorous mathematical basis.

It follows from (3.4.14) and (3.4.6) that the solution in region $2a$ should be sought in the form

$$\Psi_1(x, Y) = (-s)^{-2/3} g_1(\xi) + \cdots \quad \text{as} \quad s \to 0-, \tag{3.4.15a}$$

where

$$\xi = \frac{Y}{(-s)^{1/3}}. \tag{3.4.15b}$$

The substitution of (3.4.15) into (3.4.12) yields

$$U_1 = (-s)^{-1} g_1'(\xi) + \cdots, \qquad V_1 = -(-s)^{-5/3} \frac{1}{3} \left(2g_1 + \xi g_1'\right) + \cdots. \tag{3.4.16}$$

Before performing the substitution of (3.4.16) into the momentum equation (3.4.2a), we also need to express the leading-order velocity components U_0, V_0 in terms of the variable ξ that is an order-one quantity in region $2a$. We have already established that U_0 is represented in region $2a$ by equation (3.4.5). The corresponding equation for V_0 may be deduced by substituting (3.1.9) into the second equation in (3.1.7), and setting $x \to 1 - 0$. We find that

$$V_0 = \frac{1}{4} \lambda Y^2 + \cdots. \tag{3.4.17}$$

It remains to solve (3.4.15b) for Y and substitute the result into (3.4.5) and (3.4.17). This yields

$$U_0 = (-s)^{1/3} \lambda \xi + \cdots, \qquad V_0 = (-s)^{2/3} \frac{1}{4} \lambda \xi^2 + \cdots. \tag{3.4.18}$$

The substitution of (3.4.16) together with (3.4.18) into the momentum equation (3.4.2a) results in the following ordinary differential equation for $g_1(\xi)$:

$$g_1''' - \frac{1}{3} \lambda \xi^2 g_1'' - \frac{2}{3} \lambda \xi g_1' + \frac{2}{3} \lambda g_1 = \frac{2}{3} k. \tag{3.4.19}$$

Now, we need to formulate the boundary conditions for this equation. Since region $2a$ is adjacent to the plate surface, the solution in this region should satisfy the no-slip conditions

$$U_1 = V_1 = 0 \quad \text{at} \quad Y = 0. \tag{3.4.20}$$

The substitution of (3.4.16) into (3.4.20) yields

$$g_1(0) = g_1'(0) = 0. \tag{3.4.21}$$

As a third boundary condition for (3.4.19) we shall use the requirement that function $g_1(\xi)$ does not grow exponentially as $\xi \to \infty$. This is because an exponential growth of $g_1(\xi)$ would preclude the matching of solutions in regions $2a$ and $2b$.[8] The solution of equation (3.4.19) satisfying this condition may be expressed in the form[9]

$$g_1'' = \frac{\Gamma(4/3)}{\Gamma(-1/3)} \frac{2k}{\sqrt[3]{3\lambda}} U\left(\frac{4}{3}, \frac{2}{3}; \frac{\lambda}{9} \xi^3\right), \tag{3.4.22}$$

where $U(a, b; z)$ is Tricomi's function. If one wants to find function $g_1(\xi)$, then equation (3.4.22) should be integrated twice with the initial conditions (3.4.21).

---

[8]See Section 1.4.2 in Part 2 of this book series.
[9]See Problem 1 in Exercises 11.

3.5   Triple-Deck Model

Thus far, the analysis of the flow past a flat plate was conducted in the framework of Prandtl's hierarchical strategy. We first considered the outer inviscid region 1 where the fluid motion was described by the Euler equations. At this point, the existence of the boundary layer was ignored, and the Euler equations were solved with the impermeability condition on the body surface. Once the inviscid flow solution was found, we turned to the analysis of the boundary layer, which constitutes the second step in the hierarchical procedure. The Prandtl equations, which describe the flow in the boundary layer (region 2), require that we know the pressure and the longitudinal velocity at the outer edge of the boundary layer. These are provided by the solution in the external inviscid region 1, and in the leading-order approximation, are independent of the solution in the boundary layer. After the leading-order solution in the boundary layer was found, we revisited the inviscid region 1 with the purpose of calculating the corrections provoked there due to the presence of the boundary layer. This is the third step in the hierarchical procedure where we aim to calculate the $O(Re^{-1/2})$ pressure perturbations produced in the inviscid flow by the displacement effect of the boundary layer. Finally, in the fourth step, the reaction of the boundary layer to the pressure perturbations was studied. Remember that to perform this task we represented the velocity components in the boundary layer in the form of asymptotic expansions (3.4.1a) and (3.4.1b). Let us consider the first of these:

$$u(x, y; Re) = U_0(x, Y) + Re^{-1/2}U_1(x, Y) + \cdots . \tag{3.5.1}$$

The analysis in Section 3.4, starting with the derivation of equations (3.4.2), was based on the assumption that the second term $Re^{-1/2}U_1$ in (3.5.1) was small compared with the leading-order term U_0. Thanks to the fact that factor $Re^{-1/2}$ in the second term tends to zero as $Re \to \infty$, this assumption holds for $s = x - 1 = O(1)$. However, it is clear that close to the trailing edge a proper ordering of the terms in (3.5.1) is violated. Indeed, let us examine the solution in the viscous region 2a. We can see from (3.4.16) that U_1 is growing as $s \to 0-$:

$$U_1 = (-s)^{-1}g_1'(\xi) + \cdots . \tag{3.5.2}$$

At the same time, according to (3.4.18), the leading-order term U_0 becomes progressively smaller

$$U_0 = (-s)^{1/3}\lambda\xi + \cdots . \tag{3.5.3}$$

Therefore, there exists a vicinity of the trailing edge where

$$U_0 \sim Re^{-1/2}U_1. \tag{3.5.4}$$

Substituting (3.5.2) and (3.5.3) into (3.5.4), we have

$$(-s)^{1/3} \sim Re^{-1/2}(-s)^{-1},$$

which is easily solved for $(-s)$ to give

$$|s| \sim Re^{-3/8}. \tag{3.5.5}$$

Remember that the second term $Re^{-1/2}U_1$ in (3.5.1) represents the boundary-layer response to the pressure perturbations in the inviscid part of the flow. The latter

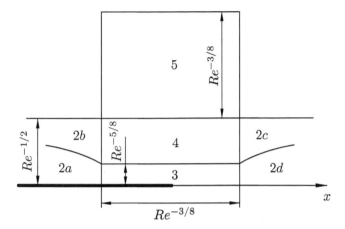

Fig. 3.9: Triple-deck region.

are produced by the displacement effect of the boundary layer and depend on the leading-order solution U_0 of the boundary-layer equations. In the trailing-edge region, defined by (3.5.5), the two terms in expansion (3.5.1) become indistinguishable from one another which suggests that in this region not only the pressure in the inviscid flow depends on the flow behaviour in the boundary layer but also the leading-order solution in the boundary layer depends on the induced pressure. Hence, this region is termed the region of *viscous-inviscid interaction.*

It has the three-tiered structure shown in Figure 3.9, being composed of three layers. Two of them, the *viscous sublayer* (region 3) and the *main part of the boundary layer* (region 4), are situated inside the boundary layer. They form as a results of splitting of the boundary layer upstream of the trailing edge and of the wake downstream of it onto viscous and locally inviscid parts. The lower deck (region 3) represents a 'continuation' of the viscous sublayers 2a and 2d into the region of viscous-inviscid interaction, while regions 2b and 2c form the middle deck (region 4) of the triple-deck structure. These should be supplemented by the upper deck (region 5), which serves to 'convert' the displacement effect of the boundary layer into the pressure perturbations.

3.5.1   Viscous sublayer (region 3)

We start the analysis of region 3 by introducing appropriate scalings of coordinates (x, y). As already mentioned, region 3 represents a continuation of the viscous sublayers 2a and 2d into the interaction region. The characteristic thickness of region 2a is given by (3.4.6). The correspondent equation for region 2d is (3.2.10). We can express both equations in the form

$$Y \sim |s|^{1/3}.$$

Taking into account that in the boundary layer $y = Re^{1/2}Y$, we have

$$y = Re^{-1/2}Y \sim Re^{-1/2}|s|^{1/3}. \tag{3.5.6}$$

According to (3.5.5), in the interaction region $|s| \sim Re^{-3/8}$, which being substituted into (3.5.6), yields

$$y \sim Re^{-1/2}Re^{-1/8} \sim Re^{-5/8}.$$

Hence, in region 3 the coordinates (x, y) should be scaled as

$$x = 1 + Re^{-3/8}x_*, \qquad y = Re^{-5/8}Y_*. \tag{3.5.7}$$

The form of asymptotic expansions for the velocity components u, v, and the pressure p in region 3 may be predicted as follows. We start with the longitudinal velocity component u. In region 2a it is given by the equation

$$u(x, y; Re) = (-s)^{1/3}\lambda\xi + Re^{-1/2}(-s)^{-1}g_1'(\xi) + \cdots, \tag{3.5.8}$$

which is obtained by substituting (3.5.2) and (3.5.3) into (3.5.1). The similarity variable ξ is given by (3.4.15b):

$$\xi = \frac{Y}{(-s)^{1/3}}. \tag{3.5.9}$$

We now express equations (3.5.8), (3.5.9) in terms of variables (3.5.7) of region 3. Remember that in the boundary layer $y = Re^{-1/2}Y$, which allows us to write (3.5.7) as

$$s = Re^{-3/8}x_*, \qquad Y = Re^{-1/8}Y_*. \tag{3.5.10}$$

The substitution of (3.5.10) into (3.5.8) yields

$$u(x, y; Re) = Re^{-1/8}\left[(-x_*)^{1/3}\lambda\xi + (-x_*)^{-1}g_1'(\xi) + \cdots\right] + \cdots,$$

and the similarity variable (3.5.9) may be expressed as

$$\xi = \frac{Y}{(-s)^{1/3}} = \frac{Re^{-1/8}Y_*}{(-Re^{-3/8}x_*)^{1/3}} = \frac{Y_*}{(-x_*)^{1/3}}. \tag{3.5.11}$$

It remains an order one quantity in region 3.

We see that in region 3 the longitudinal velocity component should be represented in the form

$$u(x, y; Re) = Re^{-1/8}U^*(x_*, Y_*) + \cdots$$

with U^* satisfying the following condition of matching with the solution in region 2a:

$$U^* = (-x_*)^{1/3}\lambda\xi + (-x_*)^{-1}g_1'(\xi) + \cdots \quad \text{as} \quad x_* \to -\infty. \tag{3.5.12}$$

The form of the asymptotic expansion of the lateral velocity component v may be predicted in the same way, but it is easier to use the continuity equation (3.1.2c). Writing (3.1.2c) in the order-of-magnitude form

$$\frac{u}{\Delta x} \sim \frac{v}{y},$$

we find that

$$v \sim \frac{y}{\Delta x}u \sim \frac{Re^{-5/8}}{Re^{-3/8}}Re^{-1/8} \sim Re^{-3/8}. \tag{3.5.13}$$

This suggests that in region 3 the lateral velocity component v should be sought in the form

$$v(x, y; Re) = Re^{-3/8}V^*(x_*, Y_*) + \cdots .$$

It remains to determine the form of the asymptotic expansion for the pressure p. We know that in region 1, the pressure p is represented by (3.3.1c):

$$p(x, y; Re) = Re^{-1/2}p_1(x, y) + \cdots . \tag{3.5.14}$$

As a result of the flow analysis in region 1, we found that in the boundary layer upstream of the interaction region, p_1 is given by equation (3.3.18):

$$p_1 = -\frac{2A_+}{3\sqrt{3}\,\lambda}\,(1-x)^{-2/3} + \cdots , \tag{3.5.15}$$

while downstream of the interaction region, we need to use equation (3.3.19):

$$p_1 = \frac{A_+}{3\sqrt{3}\,\lambda}(x-1)^{-2/3} + \cdots . \tag{3.5.16}$$

Substituting either (3.5.15) or (3.5.16) into (3.5.14), and taking into account that in the interaction region $|x - 1| \sim Re^{-3/8}$, we have

$$p \sim Re^{-1/2}p_1 \sim Re^{-1/2}\frac{1}{(1-x)^{2/3}} \sim Re^{-1/2}\frac{1}{(Re^{-3/8})^{2/3}} \sim Re^{-1/4}. \tag{3.5.17}$$

Thus, in region 3 the asymptotic expansion of the pressure has to be written as

$$p(x, y; Re) = Re^{-1/4}P^*(x_*, Y_*) + \cdots .$$

To conclude, we have established that in region 3 the velocity components (u, v) and the pressure p should be sought in the form of asymptotic expansions

$$u(x, y; Re) = Re^{-1/8}U^*(x_*, Y_*) + \cdots , \tag{3.5.18a}$$

$$v(x, y; Re) = Re^{-3/8}V^*(x_*, Y_*) + \cdots , \tag{3.5.18b}$$

$$p(x, y; Re) = Re^{-1/4}P^*(x_*, Y_*) + \cdots . \tag{3.5.18c}$$

The substitution of (3.5.18) together with (3.5.10) into the Navier–Stokes equations (3.1.2) yields

$$U^*\frac{\partial U^*}{\partial x_*} + V^*\frac{\partial U^*}{\partial Y_*} = -\frac{\partial P^*}{\partial x_*} + \frac{\partial^2 U^*}{\partial Y_*^2}, \tag{3.5.19a}$$

$$\frac{\partial P^*}{\partial Y_*} = 0, \tag{3.5.19b}$$

$$\frac{\partial U^*}{\partial x_*} + \frac{\partial V^*}{\partial Y_*} = 0. \tag{3.5.19c}$$

Now, let us formulate the boundary conditions for equations (3.5.19). Since region 3 is adjacent to the plate surface, the no-slip conditions should be posed on the plate surface upstream of the trailing edge:

$$U^* = V^* = 0 \quad \text{at} \quad Y_* = 0, \quad x_* < 0. \tag{3.5.20}$$

Downstream of the trailing edge, we need to use the symmetry conditions

$$V^* = \frac{\partial U^*}{\partial Y_*} = 0 \quad \text{at} \quad Y_* = 0, \quad x_* > 0. \tag{3.5.21}$$

Equations (3.5.19) are parabolic, and, as such, they require an initial condition for U^*, which is provided by the condition of matching with the solution in region 2a, given by (3.5.12). It is sufficient to retain only the leading-order term in (3.5.12). By using (3.5.9), we can write it as

$$U^* = \lambda Y_* + \cdots \quad \text{as} \quad x_* \to -\infty. \tag{3.5.22}$$

We also need to formulate a boundary condition for U^* at the outer edge of region 3. We will return to this task after the solution in the middle tier (region 4) is constructed.

3.5.2   Middle tier (region 4)

In region 4 the asymptotic analysis of the Navier–Stokes equations (3.1.2) is based on the limit

$$x_* = Re^{3/8}(x - 1) = O(1), \quad Y = Re^{1/2}y = O(1), \quad Re \to \infty. \tag{3.5.23}$$

The form of the solution in this region may be predicted by matching with the solution in region 2b (see Figure 3.8). The velocity components in region 2b are given by[10]

$$\left. \begin{aligned} u &= U_{00}(Y) + Re^{-1/2}(-s)^{-1}\frac{B}{\lambda}U'_{00}(Y) + \cdots, \\ v &= Re^{-1/2}V_{00}(Y) - Re^{-1}(-s)^{-2}\frac{B}{\lambda}U_{00}(Y) + \cdots. \end{aligned} \right\} \tag{3.5.24}$$

Let us express (3.5.24) in terms of the variables (3.5.23) of region 4. To perform this task we simply need to set $s = x - 1 = Re^{-3/8}x_*$ in (3.5.24). We have

$$u = U_{00}(Y) + Re^{-1/8}(-x_*)^{-1}\frac{B}{\lambda}U'_{00}(Y) + \cdots, \tag{3.5.25}$$

$$v = -Re^{-1/4}(-x_*)^{-2}\frac{B}{\lambda}U_{00}(Y) + \cdots.$$

This suggests that in region 4 the velocity components should be sought in the form

$$u(x, y; Re) = U_{00}(Y) + Re^{-1/8}\widetilde{U}_1(x_*, Y) + \cdots, \tag{3.5.26a}$$

$$v(x, y, Re) = Re^{-1/4}\widetilde{V}_1(x_*, Y) + \cdots. \tag{3.5.26b}$$

---

[10]See equations (3.5.69) in Problem 2, Exercises 11.

Notice that the leading-order term $U_{00}(Y)$ in (3.5.26a) is the Blasius velocity profile at the trailing edge of the plate. Comparing (3.5.26a) with (3.5.25), we can further see that

$$\widetilde{U}_1(x_*, Y) = (-x_*)^{-1}\frac{B}{\lambda}U'_{00}(Y) + \cdots \quad \text{as} \quad x_* \to -\infty. \tag{3.5.27}$$

Since the pressure does not change across the boundary layer, the estimate (3.5.17) remains valid, which allows us to seek the pressure in region 4 in the form

$$p(x, y, Re) = Re^{-1/4}\widetilde{P}_1(x_*, Y) + \cdots . \tag{3.5.28}$$

The substitution of (3.5.26) and (3.5.28) into the Navier–Stokes equations (3.1.2) results in

$$U_{00}\frac{\partial \widetilde{U}_1}{\partial x_*} + \widetilde{V}_1\frac{dU_{00}}{dY} = 0, \tag{3.5.29a}$$

$$\frac{\partial \widetilde{P}_1}{\partial Y} = 0, \tag{3.5.29b}$$

$$\frac{\partial \widetilde{U}_1}{\partial x_*} + \frac{\partial \widetilde{V}_1}{\partial Y} = 0. \tag{3.5.29c}$$

Equations (3.5.29a), (3.5.29c) have been analysed before, first, in connection with the self-induced separation in a supersonic flow (see Section 2.2.4), and then when dealing with incompressible flow separation from a smooth body surface (see Section 2.3.4). Remember that these equations are solved as follows. We start by eliminating $\partial\widetilde{U}_1/\partial x_*$ from (3.5.29a), (3.5.29c). This results in

$$U_{00}\frac{\partial \widetilde{V}_1}{\partial Y} - \widetilde{V}_1\frac{dU_{00}}{dY} = 0. \tag{3.5.30}$$

Now, we divide both terms in (3.5.30) by U_{00}^2. We have

$$\frac{1}{U_{00}(Y)}\frac{\partial \widetilde{V}_1}{\partial Y} - \frac{\widetilde{V}_1}{U_{00}^2}\frac{dU_{00}}{dY} = 0,$$

which may be written as

$$\frac{\partial}{\partial Y}\left(\frac{\widetilde{V}_1}{U_{00}}\right) = 0.$$

We see that the ratio $\widetilde{V}_1/U_{00}$ is a function of x_* only, say $G(x_*)$, that is

$$\frac{\widetilde{V}_1}{U_{00}} = G(x_*).$$

This allows us to express $\widetilde{V}_1$ in the form

$$\widetilde{V}_1 = G(x_*)\, U_{00}(Y). \tag{3.5.31}$$

Substituting (3.5.31) into the continuity equation (3.5.29c), we have

$$\frac{\partial \tilde{U}_1}{\partial x_*} = -G(x_*) \, U'_{00}(Y). \tag{3.5.32}$$

The behaviour of $\tilde{U}_1$ upstream of the interaction region is given by (3.5.27). Keeping this in mind, we introduce function $A(x_*)$ such that

$$A'(x_*) = -\lambda G(x_*), \qquad A\Big|_{x_*=-\infty} = 0, \tag{3.5.33}$$

and then equation (3.5.32) may be integrated with respect to x_* to yield

$$\tilde{U}_1 = \frac{1}{\lambda} A(x_*) \, U'_{00}(Y). \tag{3.5.34}$$

Finally, we combine (3.5.31) with (3.5.33) to cast the solution for $\tilde{V}_1$ in the form

$$\tilde{V}_1 = -\frac{1}{\lambda} A'(x_*) \, U_{00}(Y). \tag{3.5.35}$$

Now, we can return to the boundary-value problem (3.5.19)–(3.5.22) for the viscous sublayer (region 3), and formulate the missing boundary condition at the outer edge of region 3. This is done by matching the longitudinal velocity component u in regions 3 and 4 (see Figure 3.9). In the middle tier, u is represented by asymptotic expansion (3.5.26a). Substituting (3.5.34) into (3.5.26a), we have

$$u = U_{00}(Y) + Re^{-1/8}\frac{1}{\lambda} A(x_*) \, U'_{00}(Y) + \cdots. \tag{3.5.36}$$

We know that near the bottom of the boundary layer, the Blasius velocity profile behaves as

$$U_{00}(Y) = \lambda Y + \cdots \quad \text{as} \quad Y \to 0,$$

which, when substituted into (3.5.36), yields

$$u = \lambda Y + Re^{-1/8} A(x_*) + \cdots \quad \text{as} \quad Y \to 0. \tag{3.5.37}$$

To express (3.5.37) in terms of the variables of region 3, we need to take into account that the longitudinal coordinate x_* is common for regions 3 and 4, while the normal coordinate transformation is given by

$$Y = Re^{-1/8} Y_*. \tag{3.5.38}$$

Substituting (3.5.38) into (3.5.37), we have

$$u = Re^{-1/8}\big[\lambda Y_* + A(x_*)\big] + \cdots. \tag{3.5.39}$$

It remains to compare (3.5.39) with the asymptotic expansion (3.5.18a) of u in region 3, and we can conclude that the sought matching condition is written as

$$U^* = \lambda Y_* + A(x_*) + \cdots \quad \text{as} \quad Y_* \to \infty. \tag{3.5.40}$$

Lastly, to complete the analysis of the viscous-inviscid interaction process, we need to study the flow in the upper tier of the triple-deck region.

3.5.3   Upper tier (region 5)

Region 5 lies in the inviscid potential flow outside the boundary layer (see Figure 3.9). The dimensions of region 5 may be predicted with the help of the same arguments that were used in Section 2.3.4, where the flow separating from a smooth body surface was considered. The longitudinal extent of region 5 is predetermined by the length of the viscous-inviscid interactive region as a whole: $|x - 1| = O(Re^{-3/8})$. To determine the lateral size of region 5, we can use the fact that any inviscid potential flow of an incompressible fluid obeys the Laplace equation

$$\frac{\partial^2 \varphi}{\partial x^2} + \frac{\partial^2 \varphi}{\partial y^2} = 0$$

for the velocity potential φ.[11] The principle of least degeneration, applied to this equation, suggests that $y \sim |x - 1|$. Consequently, the asymptotic analysis of the Navier–Stokes equations (3.1.2) in region 5 should be performed based on the limit

$$x_* = \frac{x - 1}{Re^{-3/8}} = O(1), \quad y_* = \frac{y}{Re^{-3/8}} = O(1), \quad Re \to \infty. \tag{3.5.41}$$

To predict the form of the asymptotic expansions of the velocity components u, v, and the pressure p in region 5, we note that if there were no influence of the boundary layer, then the flow in this region would remain uniform:

$$u = 1, \quad v = 0, \quad p = 0.$$

The perturbations to this flow are caused by the displacement effect of the boundary layer. The latter is characterized by the angle ϑ made by the streamlines with the x-axis. It follows from (3.5.26) and (3.5.35) that in the middle tier (region 4 in Figure 3.9)

$$\tan \vartheta = \frac{v}{u} = Re^{-1/4} \frac{\widetilde{V}_1}{U_{00}} + \cdots = Re^{-1/4} \left[-\frac{1}{\lambda} A'(x_*) \right] + \cdots . \tag{3.5.42}$$

Equation (3.5.42) shows that ϑ does not change across region 4, which allows us to use (3.5.42) at the bottom of region 5. Since ϑ is small, we can expect the thin aerofoil theory to describe the flow in region 5. Keeping this in mind, we seek the solution in region 5 in the form of the asymptotic expansions [12]

$$u(x, y; Re) = 1 + Re^{-1/4} u_*(x_*, y_*) + \cdots , \tag{3.5.43a}$$

$$v(x, y; Re) = Re^{-1/4} v_*(x_*, y_*) + \cdots , \tag{3.5.43b}$$

$$p(x, y, Re) = Re^{-1/4} p_*(x_*, y_*) + \cdots . \tag{3.5.43c}$$

The substitution of (3.5.43) into the Navier–Stokes equations (3.1.2) results in the equations of the thin aerofoil theory:

$$\frac{\partial u_*}{\partial x_*} = -\frac{\partial p_*}{\partial x_*}, \quad \frac{\partial v_*}{\partial x_*} = -\frac{\partial p_*}{\partial y_*}, \quad \frac{\partial u_*}{\partial x_*} + \frac{\partial v_*}{\partial y_*} = 0. \tag{3.5.44}$$

---

[11]See Section 3.2 in Part 1 of this book series.

[12]For a detailed description of the thin aerofoil theory, see Section 2.1 in Part 2 of this book series.

The 'far-field' conditions for these equations

$$
\left.
\begin{aligned}
u_* &= -\Re\{Cz_*^{-2/3}\} + \cdots, \\
v_* &= \Im\{Cz_*^{-2/3}\} + \cdots, \\
p_* &= \Re\{Cz_*^{-2/3}\} + \cdots
\end{aligned}
\right\} \quad \text{as} \quad z_* = x_* + iy_* \to \infty \qquad (3.5.45)
$$

are deduced by matching with the solution in region 1.[13] To complete the formulation of the boundary-value problem for region 5, we also need to perform the matching with solution in the middle tier (region 4). It follows from (3.5.43) that in region 5 the streamline slope angle ϑ may be calculated as

$$
\tan\vartheta = \frac{v}{u} = \frac{Re^{-1/4}v_*(x_*, y_*) + \cdots}{1 + Re^{-1/4}u_*(x_*, y_*) + \cdots} = Re^{-1/4}v_*(x_*, y_*) + \cdots . \qquad (3.5.46)
$$

Comparing (3.5.46) with (3.5.42) and applying Prandtl's matching rule, we can conclude that

$$
v_*\Big|_{y_*=0} = -\frac{1}{\lambda}A'(x_*). \qquad (3.5.47)
$$

To find the solution to equations (3.5.44) subject to boundary conditions (3.5.45) and (3.5.47), it is convenient to cast the problem in complex variables. Elimination of $\partial u_*/\partial x_*$ from (3.5.44) leads to the Cauchy–Riemann equations

$$
\frac{\partial p_*}{\partial x_*} = \frac{\partial v_*}{\partial y_*}, \qquad \frac{\partial p_*}{\partial y_*} = -\frac{\partial v_*}{\partial x_*},
$$

which are the necessary and sufficient conditions for the function

$$
f_*(z_*) = p_*(x_*, y_*) + iv_*(x_*, y_*)
$$

to be an analytical function of complex variable $z_* = x_* + iy_*$. It follows from (3.5.45) that

$$
f_*(z_*) = Cz_*^{-2/3} + \cdots \quad \text{as} \quad z_* \to \infty.
$$

In Section 2.3.4, we demonstrated that if

1. function of complex variable $F(z_*)$ is analytic in the upper half of the complex z_*-plane, and
2. $F(z_*) \to 0$ as $z_* \to \infty$,

then for $F(z_*)$, equation (2.3.128) is valid. Since function $f_*(z_*)$, considered here, satisfies the above conditions, we can write

$$
f_*(z_*) = \frac{1}{2\pi i} \int\limits_{-\infty}^{\infty} \frac{f_*(\zeta) - \overline{f_*(\zeta)}}{\zeta - z_*} \, d\zeta, \qquad (3.5.48)
$$

---

[13]See Problem 3 in Exercises 11.

where z_* is an arbitrary point lying in the upper half of the complex plane. The numerator of the integrand in (3.5.48) can be calculated using boundary condition (3.5.47). We have

$$f_*(\zeta) - \overline{f_*(\zeta)} = 2iv_*\Big|_{y_*=0} = -2i\frac{1}{\lambda}\frac{dA}{d\zeta}. \tag{3.5.49}$$

The substitution of (3.5.49) into (3.5.48) yields

$$f_*(z_*) = -\frac{1}{\pi\lambda}\int\limits_{-\infty}^{\infty}\frac{A'(\zeta)}{\zeta - z_*}\,d\zeta. \tag{3.5.50}$$

Our main interest is in the pressure distribution along the boundary layer. It is obtained by assuming that z_* tends to a point $(x_*, 0)$ on the real axis, and using the Sokhotsky–Plemelj formula (2.3.131). We find that

$$p_*(x_*, 0) = -\frac{1}{\pi\lambda}\int\limits_{-\infty}^{\infty}\frac{A'(\zeta)}{\zeta - x_*}\,d\zeta. \tag{3.5.51}$$

3.5.4 Interaction problem and numerical results

We have shown that the region of viscous-inviscid interaction, which forms at the trailing edge of a flat plate, has a three-tiered structure. It is composed of the nonlinear viscous sublayer (region 3), the main part of the boundary layer (region 4), and the inviscid potential flow (region 5) situated outside the boundary layer. The interaction takes place between the lower and upper tiers, with the middle tier playing a passive role in the interaction process. Indeed, equation (3.5.42) shows that the middle tier (region 4) does not make any contribution to the displacement effect of the boundary layer, but simply transmits the deformation of the streamlines produced in the viscous sublayer (region 3) to the external inviscid region 5. It further follows from (3.5.29b) that the pressure perturbations induced in region 5 are transmitted across the middle tier (region 4) to the viscous sublayer (region 3) without change.

The fluid motion in the viscous sublayer is described by the boundary-layer equations (3.5.19a), (3.5.19c):

$$U^*\frac{\partial U^*}{\partial x_*} + V^*\frac{\partial U^*}{\partial Y_*} = -\frac{dP^*}{dx_*} + \frac{\partial^2 U^*}{\partial Y_*^2}, \tag{3.5.52a}$$

$$\frac{\partial U^*}{\partial x_*} + \frac{\partial V^*}{\partial Y_*} = 0. \tag{3.5.52b}$$

These should be solved with boundary conditions (3.5.20)–(3.5.22) and (3.5.40):

$$U^* = V^* = 0 \qquad\qquad \text{at} \quad Y_* = 0, \ x_* < 0, \tag{3.5.52c}$$

$$V^* = \frac{\partial U^*}{\partial Y_*} = 0 \qquad\qquad \text{at} \quad Y_* = 0, \ x_* > 0, \tag{3.5.52d}$$

$$U^* = \lambda Y_* + \cdots \qquad\qquad \text{as} \quad x_* \to -\infty, \tag{3.5.52e}$$

$$U^* = \lambda Y_* + A(x_*) + \cdots \quad \text{as} \quad Y_* \to \infty. \tag{3.5.52f}$$

If the pressure distribution $P^*(x_*)$ were given, as is the case in the classical Prandtl boundary-layer theory, then equations (3.5.52a), (3.5.52b) could be solved for the velocity components U^*, V^*, and, as a part of the solution, the displacement function $A(x_*)$ could be found using (3.5.52f). However, in the interaction region neither the pressure nor the displacement function is known in advance. Instead, the flow analysis in region 5 leads to equation (3.5.51) that relates $p_*(x_*, 0)$ and $A(x_*)$. Keeping in mind that the pressure does not change across regions 4 and 3, we can write this equation as

$$P^*(x_*) = -\frac{1}{\pi\lambda} \int_{-\infty}^{\infty} \frac{A'(\zeta)}{\zeta - x_*}\, d\zeta. \tag{3.5.52g}$$

This completes the formulation of the viscous-inviscid interaction problem.

Parameter λ may be excluded from (3.5.52) by means of the affine transformations

$$x_* = \lambda^{-5/4}\bar{X}, \qquad Y_* = \lambda^{-3/4}\bar{Y}, \qquad U^* = \lambda^{1/4}\bar{U},$$

$$V^* = \lambda^{3/4}\bar{V}, \qquad A = \lambda^{1/4}\bar{A}, \qquad P^* = \lambda^{1/2}\bar{P}.$$

As a result, the interaction problem assumes the following canonical form:

$$\left.\begin{aligned}
\bar{U}\frac{\partial\bar{U}}{\partial\bar{X}} + \bar{V}\frac{\partial\bar{U}}{\partial\bar{Y}} &= -\frac{d\bar{P}}{d\bar{X}} + \frac{\partial^2\bar{U}}{\partial\bar{Y}^2}, \\
\frac{\partial\bar{U}}{\partial\bar{X}} + \frac{\partial\bar{V}}{\partial\bar{Y}} &= 0, \\
\bar{P} = -\frac{1}{\pi}\int_{-\infty}^{\infty} &\frac{\bar{A}'(s)}{s - \bar{X}}\, ds, \\
\bar{U} = \bar{V} = 0 \qquad &\text{at} \quad \bar{Y} = 0, \ \bar{X} < 0, \\
\bar{V} = \frac{\partial\bar{U}}{\partial\bar{Y}} = 0 \qquad &\text{at} \quad \bar{Y} = 0, \ \bar{X} > 0, \\
\bar{U} = \bar{Y} + \cdots \qquad &\text{as} \quad \bar{X} \to -\infty, \\
\bar{U} = \bar{Y} + \bar{A}(\bar{X}) + \cdots \qquad &\text{as} \quad \bar{Y} \to \infty.
\end{aligned}\right\} \tag{3.5.53}$$

As mentioned, the theory of viscous-inviscid interaction near the trailing edge of a flat plate was developed independently by Stewartson (1969) and Messiter (1970). A numerical solution of the interaction problem (3.5.53) was first given by Jobe and Burggraf (1974). The results of calculations are displayed in Figure 3.10. Figure 3.10(a) shows the dimensionless skin friction

$$\bar{\tau}_w = \left.\frac{\partial\bar{U}}{\partial\bar{Y}}\right|_{\bar{Y}=0}$$

on the plate surface ($\bar{X} < 0$), and the longitudinal velocity $\bar{U}(\bar{X}, 0)$ along the axis of symmetry of the flow downstream of the trailing edge ($\bar{X} > 0$). The pressure distribution $\bar{P}(\bar{X})$ is shown in Figure 3.10(b). As expected, the flow experiences an

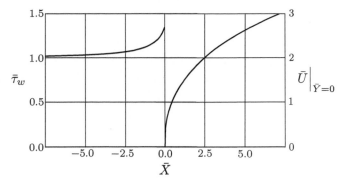

(a) Skin friction on the plate surface $\bar{X} < 0$, and longitudinal velocity distribution along the axis of symmetry downstream of the trailing edge ($\bar{X} > 0$).

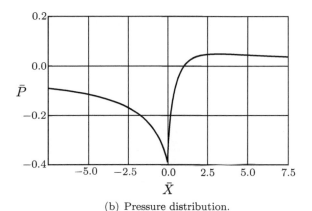

(b) Pressure distribution.

Fig. 3.10: Results of the numerical solution of the interaction problem (3.5.53).

acceleration when approaching the trailing edge. The pressure drops along the plate surface before it starts to recover in the wake, and the skin increases from $\bar{\tau}_w = 1$ in the Blasius boundary layer ($\bar{X} = -\infty$) to $\bar{\tau}(0) = 1.343$ at the trailing edge.

This results in an increase of the drag D of the flat plate. In fact, the drag coefficient C_D of the plate may be calculated as[14]

$$C_D = \frac{D}{\frac{1}{2}\rho V_\infty^2 L} = \frac{8\lambda}{Re^{1/2}} + \frac{d_0}{Re^{7/8}} + \cdots \quad \text{as} \quad Re \to \infty. \quad (3.5.54)$$

Here, the leading-order term is given by the Blasius solution (see Problem 3 in Exercises 1), while the second term represents an influence of the interaction region with the coefficient d_0 given by

$$d_0 = 4\lambda^{-1/4} \int\limits_{-\infty}^{0} \left(\bar{\tau}_w - 1\right) d\bar{X}.$$

[14]See Problem 4 in Exercises 11.

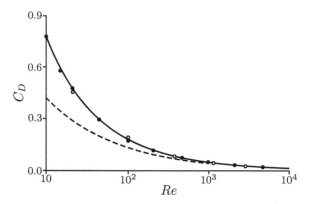

Fig. 3.11: Drag coefficient as a function of the Reynolds number; solid line: formula (3.5.54), dashed line: Blasius solution, open circles: experiments by Nishioka and Miyagi (1978), filled circles: Navier–Stokes calculations by Dennis and Dunwoody (1966), and Dennis and Chang (1969).

According to Jobe and Burggraf (1974), $d_0 = 5.388$.

In Figure (3.11), the theoretical predictions for the drag coefficient are compared with the experimental date by Nishioka and Miyagi (1978) and the numerical solution of the Navier–Stokes equations by Dennis and Dunwoody (1966), and Dennis and Chang (1969). We can see that even for the Reynolds number as small as $Re = 10$, the asymptotic theory which is based on $Re \to \infty$ gives a remarkably good agreement with both the experiment and Navier–Stokes calculations.

An important extension of the above theory concerns an influence of the angle of attack on the flow near the trailing edge of a thin aerofoil. The correspondent triple-deck problem was formulated by Brown and Stewartson (1970) and the numerical solution to this problem was provided by Chow and Melnik (1976), and Korolev (1991). The triple-deck theory not only confirms the validity of the Joukovskii–Kutta condition (see Section 3.6 in Part 1 of this book series), but also allows us to calculate the 'viscous corrections' to the circulation around the aerofoil.

Exercises 11

1. Consider ordinary differential equation (3.4.19), which describes the flow in the viscous region 2a (see Figure 3.8) before the region of viscous-inviscid interaction:

$$g_1''' - \frac{1}{3}\lambda\xi^2 g_1'' - \frac{2}{3}\lambda\xi g_1' + \frac{2}{3}\lambda g_1 = \frac{2}{3}k. \tag{3.5.55}$$

Your task is to solve this equation with the boundary conditions

$$g_1(0) = g_1'(0) = 0, \tag{3.5.56}$$

and the requirement that $g_1(\xi)$ does not grow exponentially as $\xi \to \infty$.

You may perform this task in the following steps:

(a) Differentiate equation (3.5.55) with respect to ξ, and show that function $h = g_1''(\xi)$ satisfies the equation

$$\frac{d^2 h}{d\xi^2} - \frac{1}{3}\lambda\xi^2 \frac{dh}{d\xi} - \frac{4}{3}\lambda\xi h = 0. \tag{3.5.57}$$

Now return to equation (3.5.55), and show that

$$\left.\frac{dh}{d\xi}\right|_{\xi=0} = \frac{2}{3}k. \tag{3.5.58}$$

(b) Introduce instead of ξ a new independent variable $z = \frac{1}{9}\lambda\xi^3$, and show that this turns (3.5.57) into the confluent hypergeometric equation

$$z\frac{d^2 h}{dz^2} + \left(\frac{2}{3} - z\right)\frac{dh}{dz} - \frac{4}{3}h = 0. \tag{3.5.59}$$

Show further that the boundary condition (3.5.58) is written in the new variables as

$$\frac{dh}{dz} = \frac{2k}{3\sqrt[3]{3\lambda}} z^{-2/3} + \cdots \quad \text{as} \quad z \to 0. \tag{3.5.60}$$

(c) Finally, show that the solution of equation (3.5.59), which does not grow exponentially as $z \to \infty$ and satisfies the boundary condition (3.5.60), has the form

$$h(\xi) = \frac{\Gamma(4/3)}{\Gamma(-1/3)} \frac{2k}{\sqrt[3]{3\lambda}} U\left(\frac{4}{3}, \frac{2}{3}; \frac{\lambda}{9}\xi^3\right), \tag{3.5.61}$$

where $U(a, b; z)$ is Tricomi's function.

Suggestion: In general, the confluent hypergeometric equation is written as

$$z\frac{d^2 w}{dz^2} + (b - z)\frac{dw}{dz} - aw = 0.$$

Two complementary solutions of this equation are Kummer's function $M(a, b, z)$ and Tricomi's function $U(a, b, z)$. You may use without proof the following properties of these function. Kummer's function is regular at $z = 0$ but grows exponentially as $z \to \infty$. As far as Tricomi's function is concerned, it displays an algebraic behaviour at large values of z:

$$U(a, b; z) = z^{-a}\left[1 + O(z^{-1})\right] \quad \text{as} \quad z \to \infty. \tag{3.5.62}$$

Near $z = 0$ it can be represented by

$$U(a, b; z) = \frac{\Gamma(1 - b)}{\Gamma(a - b + 1)} + \frac{\Gamma(b - 1)}{\Gamma(a)} z^{1-b} + O(z) \quad \text{as} \quad z \to 0,$$

where Γ stands for Euler's gamma function.

2. Now, turn your attention to the main part of the boundary layer (region 2b in Figure 3.8). Remember that in the boundary layer upstream of the viscous-inviscid interaction region, the velocity components are represented by the asymptotic expansions (3.4.1a), (3.4.1b):

$$u(x, y; Re) = U_0(x, Y) + Re^{-1/2}U_1(x, Y) + \cdots,$$
$$v(x, y; Re) = Re^{-1/2}V_0(x, Y) + Re^{-1}V_1(x, Y) + \cdots.$$

Here, the leading-order terms U_0, V_0 are given by the Blasius solution, while U_1 and V_1 should be found by solving equations (3.4.2):

$$U_0\frac{\partial U_1}{\partial x} + U_1\frac{\partial U_0}{\partial x} + V_0\frac{\partial U_1}{\partial Y} + V_1\frac{\partial U_0}{\partial Y} = -\frac{\partial P_1}{\partial x} + \frac{\partial^2 U_1}{\partial Y^2}, \qquad (3.5.63a)$$

$$\frac{\partial U_1}{\partial x} + \frac{\partial V_1}{\partial Y} = 0. \qquad (3.5.63b)$$

Your task is to find the solution of these equations in region 2b. You may perform this task in the following steps:

(a) Use (3.5.62) in equation (3.5.61) for $h = g_1''$ to show that the asymptotic behaviour of g_1'' at the outer edge of region 2a is given by

$$g_1''(\xi) = O(\xi^{-4}) \quad \text{as} \quad \xi \to \infty.$$

Keeping this in mind, deduce that

$$g_1(\xi) = B\xi + O(1) \quad \text{as} \quad \xi \to \infty, \qquad (3.5.64)$$

where

$$B = \frac{\Gamma(4/3)}{\Gamma(-1/3)}\frac{2k}{\sqrt[3]{3\lambda}}\int_0^\infty U\left(\frac{4}{3}, \frac{2}{3}; \frac{\lambda}{9}\xi^3\right)d\xi.$$

(b) Now, consider the velocity components U_1, V_1. You may use without proof the fact that in region 2a these are given by equations (3.4.16):

$$U_1 = (-s)^{-1}g_1'(\xi) + \cdots, \qquad V_1 = -(-s)^{-5/3}\frac{1}{3}(2g_1 + \xi g_1') + \cdots, \quad (3.5.65)$$

where $\xi = Y/(-s)^{1/3}$. Substitute (3.5.64) into (3.5.65), and deduce that in the overlap region that lies between regions 2a and 2b

$$U_1 = (-s)^{-1}B + \cdots, \qquad V_1 = -(-s)^{-2}BY + \cdots. \qquad (3.5.66)$$

(c) Guided by (3.5.66), represent the solution in region 2b in the form of asymptotic expansions

$$\left.\begin{array}{l} U_1(x, Y) = (-s)^{-1}U_{11}(Y) + \cdots, \\ V_1(x, Y) = (-s)^{-2}V_{11}(Y) + \cdots \end{array}\right\} \quad \text{as} \quad s \to 0-. \qquad (3.5.67)$$

What is the behaviour of functions $U_{11}(Y)$, $V_{11}(Y)$ as $Y \to 0$?

(d) Substitute (3.5.67) into (3.5.63), and deduce that

$$U_{00}U_{11} + V_{11}\frac{dU_{00}}{dY} = 0, \qquad U_{11} + \frac{dV_{11}}{dY} = 0. \qquad (3.5.68)$$

Hint: Remember that upstream of the interaction region, the pressure gradient is given by (3.4.8):

$$\frac{dP_1}{dx} = \frac{2}{3}(-s)^{-5/3} + \cdots \quad \text{as} \quad s \to 0-.$$

You also need to remember that in the leading-order approximation, the velocity components (U_0, V_0) are represented by the Blasius solution that is regular near the trailing edge of the plate, i.e.,

$$\left.\begin{array}{l} U_0(x, Y) = U_{00}(Y) + \cdots, \\ V_0(x, Y) = V_{00}(Y) + \cdots \end{array}\right\} \quad \text{as} \quad x \to 1 - 0.$$

(e) Finally, show that the solution to (3.5.68) is written as

$$U_{11} = \frac{B}{\lambda}U'_{00}(Y), \qquad V_{11} = -\frac{B}{\lambda}U_{00}(Y).$$

Hence, conclude that in region 2*b*

$$\left.\begin{array}{l} u = U_{00}(Y) + Re^{-1/2}(-s)^{-1}\dfrac{B}{\lambda}U'_{00}(Y) + \cdots, \\[2mm] v = Re^{-1/2}V_{00}(Y) - Re^{-1}(-s)^{-2}\dfrac{B}{\lambda}U_{00}(Y) + \cdots. \end{array}\right\} \qquad (3.5.69)$$

3. Consider the upper tier (region 5 in Figure 3.9) of the interaction region where the velocity components (u, v) and the pressure p are represented in the form of asymptotic expansions:

$$\left.\begin{array}{l} u(x, y; Re) = 1 + Re^{-1/4}u_*(x_*, y_*) + \cdots, \\ v(x, y; Re) = Re^{-1/4}v_*(x_*, y_*) + \cdots, \\ p(x, y, Re) = Re^{-1/4}p_*(x_*, y_*) + \cdots \end{array}\right\} \quad \text{as} \quad Re \to \infty.$$

Your task is to show that the far-field behaviour of functions u_*, v_*, and p_* is given by

$$\left.\begin{array}{l} u_* = -\Re\{Cz_*^{-2/3}\} + \cdots, \\ v_* = \Im\{Cz_*^{-2/3}\} + \cdots, \\ p_* = \Re\{Cz_*^{-2/3}\} + \cdots \end{array}\right\} \quad \text{as} \quad z_* = x_* + iy_* \to \infty.$$

To perform this task you may use without proof the following facts:

(a) In region 1 the velocity components (u, v) and the pressure p are written as

$$
\left.
\begin{array}{l}
u(x, y; Re) = 1 + Re^{-1/2}u_1(x, y) + \cdots, \\
v(x, y; Re) = Re^{-1/2}v_1(x, y) + \cdots, \\
p(x, y; Re) = Re^{-1/2}p_1(x, y) + \cdots
\end{array}
\right\}
\quad \text{as} \quad Re \to \infty.
$$

(b) It is known that $f(z) = p_1(x, y) + iv_1(x, y)$ is an analytic function of complex variable $z = x + iy$, which shows the following behaviour near the trailing edge of the plate:

$$
f(z) = C(z - 1)^{-2/3} + \cdots \quad \text{as} \quad z \to 1.
$$

(c) It is further known that the perturbations of the longitudinal velocity component and of the pressure are related to one another through the linearized Bernoulli equation

$$
u_1 = -p_1.
$$

(d) Finally, remember that the coordinates in region 5 are scaled as

$$
x = 1 + Re^{-3/8}x_*, \qquad y = Re^{-3/8}y_*.
$$

4. The viscous drag of a flat plate is calculated as

$$
D = 2 \int_0^L \hat{\tau}_w \, d\hat{x}, \tag{3.5.70}
$$

where $\hat{\tau}_w$ is the shear stress on the plate surface given by

$$
\hat{\tau}_w = \mu \frac{\partial \hat{u}}{\partial \hat{y}}\bigg|_{\hat{y}=0}.
$$

Keeping in mind the existence of the viscous-inviscid interaction region near the trailing edge of the plate, show that the drag coefficient of the plate is given by

$$
C_D = \frac{D}{\frac{1}{2}\rho V_\infty^2 L} = \frac{8\lambda}{Re^{1/2}} + \frac{d_0}{Re^{7/8}} + \cdots \quad \text{as} \quad Re \to \infty,
$$

with

$$
d_0 = 4\lambda^{-1/4} \int_{-\infty}^0 \left(\bar{\tau}_w - 1\right) d\bar{X}.
$$

Suggestion: Express the drag integral (3.5.70) in the form

$$
D = 2 \int_0^L \hat{\tau}_w^* \, d\hat{x} + 2 \int_0^L \left(\hat{\tau}_w - \hat{\tau}_w^*\right) d\hat{x}, \tag{3.5.71}
$$

where $\hat{\tau}_w^*$ is the skin friction in the Blasius boundary layer. Calculate the first integral in (3.5.71) as suggested in Problem 3 in Exercises 1. When calculating the second integral, remember that in the lower tier of the interaction region

$$
\hat{u} = V_\infty Re^{-1/8}\lambda^{1/4}\bar{U}, \qquad \hat{x} = LRe^{-3/8}\lambda^{-5/4}\bar{X}, \qquad \hat{y} = LRe^{-5/8}\lambda^{-3/4}\bar{Y}.
$$

5. Consider a transonic flow of a perfect gas past a flat plate. Assume that the plate is aligned with the free-stream velocity vector. Assume further that the Reynolds number Re is large. Using inspection analysis (see Section 2.2.3), demonstrate that the longitudinal extent of the triple-deck region, which forms at the trailing edge of the plate, may be estimated as

$$\Delta x \sim Re^{-3/10}.$$

   Hint: To 'adjust' the inspection analysis of Section 2.2.3 to transonic flow regime, we need to substitute the Ackeret formula (2.2.17) with its transonic analogue:[15]

$$\Delta p \sim \theta^{2/3}.$$

   For a detailed analysis of this problem, the interested reader is referred to Bodonyi and Kluwick (1998).

6. In Section 2.1.7 in Part 1 of this book series, Kármán flow of an incompressible fluid was analysed as one of the exact solutions of the Navier–Stokes equations. Remember that this is the flow around a plane disk placed in an infinite reservoir filled with a stagnant fluid. The disk is brought in rotation in its plane around the centre O; see Figure 3.12. Owing to the action of viscous forces, the fluid particles adjacent to the disk start to move following the disk surface, but their trajectories are not circular. Because of inertia, the fluid particles will also tend to move away from the axis of rotation. As a result, radial jets are created on both sides of the disk as shown in Figure 3.13.

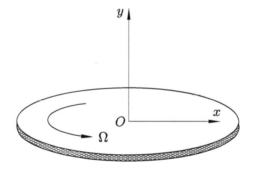

Fig. 3.12: Kármán flow.

   Suppose that the disk has a finite radius R, and define the Reynolds number as

$$Re = \frac{R V_{\max}}{\nu},$$

where $V_{\max}$ is the maximum velocity in the jet as it approaches the disk edge (see Figure 3.13). Your task is to study the flow in the interaction region that

---

[15]See the first equation in (4.3.20) on page 221 in Part 2 of this book series.

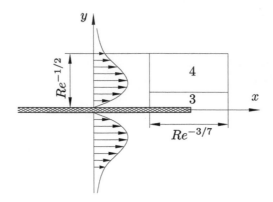

Fig. 3.13: Flow near the disk edge.

forms near the disk edge. Assume that the Reynolds number is large and, using the inspection analysis, show that the radial extent of the interaction region is

$$\Delta x \sim Re^{-3/7}.$$

Suggestion: The Ackeret formula (2.2.17) is not applicable to the flow considered here. Instead, the pressure perturbations should be found using the following observations. In the absence of radial motion of the fluid outside the jet, the pressure remains constant along the outer edge of the jet. Therefore, the pressure variations in the viscous sublayer (region 3 in Figure 3.13) can only be created by the centrifugal forces acting in the main part of the jet (region 4). This is described by balancing the following two terms in the lateral momentum equation:

$$u\frac{\partial v}{\partial x} \sim \frac{\partial p}{\partial y}. \tag{3.5.72}$$

In region 4, the lateral velocity component v can be expressed in the form

$$v = \theta u, \tag{3.5.73}$$

where θ is the streamline slope angle. Remember that in region 4, the longitudinal velocity u is independent of x in the leading-order approximation, and the streamline slope angle is independent of y. Hence, substitution of (3.5.73) into (3.5.72) leads to

$$\frac{\partial p}{\partial y} \sim u^2 \frac{d\theta}{dx}. \tag{3.5.74}$$

Represent the derivatives in (3.5.74) by finite differences

$$\frac{\Delta p}{Re^{-1/2}} \sim u^2 \frac{\theta}{\Delta x},$$

and take into account that in the main part of the jet $u = O(1)$ and θ is given by equation (2.2.16).

For a detailed analysis of the flow, the interested reader is referred to the original papers of Shidlovskii (1977), and Smith and Duck (1977).

4

Incipient Separation Near Corners

In Chapter 2, we were concerned with boundary-layer separation leading to formation of large-scale recirculation eddies. The present Chapter 4 discusses the incipience of separation, namely, the transition from fully attached flow to the flow with a small separation bubble. The latter may be caused by different type of roughness on the body surface, say, a hump or a dent. Here, we consider the formation of the separation bubble near a corner.

4.1 Problem Formulation

This chapter describes a theory that is applicable to a variety of physical situations. However, for presentational purposes it is convenient to deal with a particular flow layout. We consider two-dimensional flow of a perfect gas past a 'compression ramp' built of two flat plates, AB and BC; see Figure 4.1. We assume that the first of these, plate AB, is aligned with the oncoming flow, and the second, plate BC, makes an angle θ with AB.

To perform the flow analysis, it is convenient to use Cartesian coordinates $(\hat{x}, \hat{y})$ with $\hat{x}$ measured from the corner B parallel to plate AB. We denote the distance between the leading edge A and the corner point B by L. The gas velocity, density, dynamic viscosity coefficient, and the pressure in the unperturbed flow upstream of the plate AB are denoted as V_∞, ρ_∞, μ_∞, and p_∞, respectively. Using these quantities, the non-dimensional variables are introduces as follows:

$$\left.\begin{aligned} \hat{x} = Lx, \qquad \hat{y} = Ly, \qquad\qquad \hat{u} = V_\infty u, \qquad \hat{v} = V_\infty v, \\ \hat{\rho} = \rho_\infty \rho, \qquad \hat{p} = p_\infty + \rho_\infty V_\infty^2 p, \qquad \hat{h} = V_\infty^2 h, \qquad \hat{\mu} = \mu_\infty \mu. \end{aligned}\right\} \qquad (4.1.1)$$

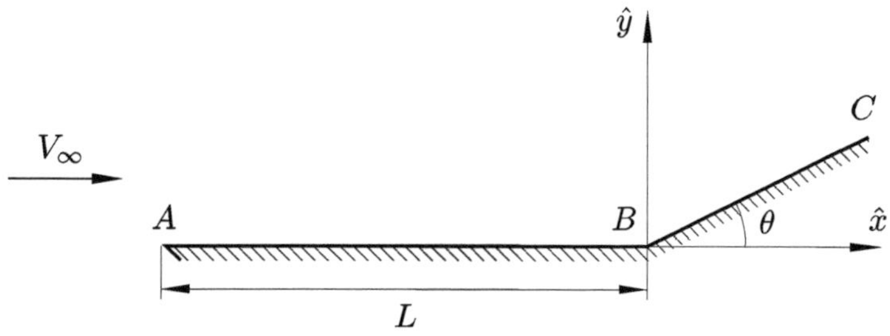

Fig. 4.1: Compression ramp flow.

Fluid Dynamics: Part 3: Boundary Layers. © Anatoly I. Ruban, 2018. Published 2018 by Oxford University Press. 10.1093/oso/9780199681754.001.0001

The substitution of (4.1.1) into the Navier–Stokes equations (1.10.1) renders these in the non-dimensional form:

$$\rho\left(u\frac{\partial u}{\partial x} + v\frac{\partial u}{\partial y}\right) = -\frac{\partial p}{\partial x} + \frac{1}{Re}\left\{\frac{\partial}{\partial x}\left[\mu\left(\frac{4}{3}\frac{\partial u}{\partial x} - \frac{2}{3}\frac{\partial v}{\partial y}\right)\right]\right.$$
$$\left. + \frac{\partial}{\partial y}\left[\mu\left(\frac{\partial u}{\partial y} + \frac{\partial v}{\partial x}\right)\right]\right\}, \quad (4.1.2a)$$

$$\rho\left(u\frac{\partial v}{\partial x} + v\frac{\partial v}{\partial y}\right) = -\frac{\partial p}{\partial y} + \frac{1}{Re}\left\{\frac{\partial}{\partial y}\left[\mu\left(\frac{4}{3}\frac{\partial v}{\partial y} - \frac{2}{3}\frac{\partial u}{\partial x}\right)\right]\right.$$
$$\left. + \frac{\partial}{\partial x}\left[\mu\left(\frac{\partial u}{\partial y} + \frac{\partial v}{\partial x}\right)\right]\right\}, \quad (4.1.2b)$$

$$\rho\left(u\frac{\partial h}{\partial x} + v\frac{\partial h}{\partial y}\right) = u\frac{\partial p}{\partial x} + v\frac{\partial p}{\partial y} + \frac{1}{Re}\left\{\frac{1}{Pr}\left[\frac{\partial}{\partial x}\left(\mu\frac{\partial h}{\partial x}\right) + \frac{\partial}{\partial y}\left(\mu\frac{\partial h}{\partial y}\right)\right]\right.$$
$$\left. + \mu\left(\frac{4}{3}\frac{\partial u}{\partial x} - \frac{2}{3}\frac{\partial v}{\partial y}\right)\frac{\partial u}{\partial x} + \mu\left(\frac{4}{3}\frac{\partial v}{\partial y} - \frac{2}{3}\frac{\partial u}{\partial x}\right)\frac{\partial v}{\partial y} + \mu\left(\frac{\partial u}{\partial y} + \frac{\partial v}{\partial x}\right)^2\right\}, \quad (4.1.2c)$$

$$\frac{\partial \rho u}{\partial x} + \frac{\partial \rho v}{\partial y} = 0, \quad (4.1.2d)$$

$$h = \frac{1}{(\gamma-1)M_\infty^2}\frac{1}{\rho} + \frac{\gamma}{\gamma-1}\frac{p}{\rho}. \quad (4.1.2e)$$

In what follows, we assume that the Reynolds number

$$Re = \frac{\rho_\infty V_\infty L}{\mu_\infty}$$

is large, while the free-stream Mach number

$$M_\infty = \frac{V_\infty}{\sqrt{\gamma p_\infty/\rho_\infty}}$$

remains finite. Both subsonic ($M_\infty < 1$) and supersonic ($M_\infty > 1$) flow regimes will be studied. We begin with the subsonic flow regime.

4.2   Subsonic Flow

The incipient separation is expected to take place when the angle θ reaches a certain value. We assume, subject to subsequent confirmation, that this happens when θ is large compared to $Re^{-1/2}$, but small compared to unity. Correspondingly, we write

$$\theta = \varepsilon(Re)\,\theta_0, \quad (4.2.1)$$

where θ_0 is an order one quantity, positive or negative, and $\varepsilon(Re)$ satisfies the conditions

$$\varepsilon(Re) \gg Re^{-1/2}, \quad (4.2.2a)$$
$$\varepsilon(Re) \ll 1. \quad (4.2.2b)$$

4.2.1   Inviscid flow region

As usual, we first consider the inviscid region 1 (see Figure 4.2) where the asymptotic analysis of the Navier–Stokes equations (4.1.2) is based on the limit

$$x = O(1), \quad y = O(1), \quad Re \to \infty.$$

Under condition (4.2.2a) the influence of the boundary layer on the flow in region 1 can be disregarded, that is, in the leading-order approximation, the perturbations are due to the body shape, not the displacement effect of the boundary layer. In view of (4.2.2b), these perturbations are weak, which allows us to use the thin aerofoil theory.

Taking into account that in the unperturbed free-stream flow

$$u = 1, \quad v = 0, \quad \rho = 1, \quad p = 0, \quad h = \frac{1}{(\gamma - 1)M_\infty^2},$$

we represent the solution in region 1 in the form

$$\left.\begin{aligned}
u(x, y; Re) &= 1 + \varepsilon u_1(x, y) + \cdots, \\
v(x, y; Re) &= \varepsilon v_1(x, y) + \cdots, \\
\rho(x, y; Re) &= 1 + \varepsilon \rho_1(x, y) + \cdots, \\
p(x, y; Re) &= \varepsilon p_1(x, y) + \cdots, \\
h(x, y; Re) &= \frac{1}{(\gamma - 1)M_\infty^2} + \varepsilon h_1(x, y) + \cdots.
\end{aligned}\right\} \tag{4.2.3}$$

The substitution of (4.2.3) into the Navier–Stokes equations (4.1.2) results in

$$\frac{\partial u_1}{\partial x} = -\frac{\partial p_1}{\partial x} \qquad\qquad \text{(x-momentum equation)}, \tag{4.2.4a}$$

$$\frac{\partial v_1}{\partial x} = -\frac{\partial p_1}{\partial y} \qquad\qquad \text{(y-momentum equation)}, \tag{4.2.4b}$$

$$\frac{\partial h_1}{\partial x} = \frac{\partial p_1}{\partial x} \qquad\qquad \text{(energy equation)}, \tag{4.2.4c}$$

$$\frac{\partial u_1}{\partial x} + \frac{\partial \rho_1}{\partial x} + \frac{\partial v_1}{\partial y} = 0 \qquad \text{(continuity equation)}, \tag{4.2.4d}$$

$$h_1 = \frac{\gamma}{\gamma - 1} p_1 - \frac{1}{(\gamma - 1)M_\infty^2} \rho_1 \qquad \text{(state equation)}. \tag{4.2.4e}$$

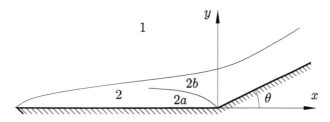

Fig. 4.2: Subsonic flow past a compression ramp.

The solution to these equations can be constructed as follows. Integrating the x-momentum equation (4.2.4a) and the energy equation (4.2.4d) with the free-stream conditions

$$
\left.\begin{array}{l}
u_1 \to 0, \\
p_1 \to 0, \\
h_1 \to 0
\end{array}\right\} \quad \text{as} \quad x^2 + y^2 \to \infty,
$$

we have

$$
u_1 = -p_1, \qquad h_1 = p_1. \tag{4.2.5}
$$

The second equation in (4.2.5) may be used to eliminate h_1 in the state equation (4.2.4e), which yields

$$
\rho_1 = M_\infty^2 p_1. \tag{4.2.6}
$$

Now, we can use the first equation in (4.2.5) together with (4.2.6) to eliminate u_1 and ρ_1 from the continuity equation (4.2.4d). We find that

$$
(1 - M_\infty^2) \frac{\partial p_1}{\partial x} - \frac{\partial v_1}{\partial y} = 0. \tag{4.2.7}
$$

The second equation relating p_1 and v_1 is the y-momentum equation (4.2.4b). The Prandtl–Glauert transformation[1]

$$
p_1 = \frac{1}{\sqrt{1 - M_\infty^2}} \tilde{p}_1, \qquad y = \frac{1}{\sqrt{1 - M_\infty^2}} \tilde{y} \tag{4.2.8}
$$

allows us to turn (4.2.4b) and (4.2.7) into the Cauchy–Riemann equations

$$
\frac{\partial \tilde{p}_1}{\partial x} = \frac{\partial v_1}{\partial \tilde{y}}, \qquad \frac{\partial \tilde{p}_1}{\partial \tilde{y}} = -\frac{\partial v_1}{\partial x},
$$

which represent the necessary and sufficient conditions for the function

$$
f(z) = \tilde{p}_1(x, \tilde{y}) + i v_1(x, \tilde{y})
$$

to be an analytic function of the complex variable $z = x + i\tilde{y}$.

When dealing with flow regimes where the boundary layer is either fully attached or the separation region is localized within a small neighbourhood of the corner, we can impose the impermeability condition on the body surface:

$$
\left. \frac{v}{u} \right|_{y=y_w(x)} = \frac{dy_w}{dx}. \tag{4.2.9}
$$

Here, the equation of the body surface is written as

$$
y_w(x) = \begin{cases} 0 & \text{if} \quad x < 0, \\ \varepsilon \theta_0 x & \text{if} \quad x > 0. \end{cases} \tag{4.2.10}
$$

---

[1]See Section 2.2.2 in Part 2 of this book series.

The velocity components u and v are given by (4.2.3). By substituting these together with (4.2.10) into (4.2.9) and working with the $O(\varepsilon)$ terms, we find that

$$
v_1\Big|_{y=0} = \begin{cases} 0 & \text{if} \quad x < 0, \\ \theta_0 & \text{if} \quad x > 0. \end{cases} \tag{4.2.11}
$$

Condition (4.2.11) determines uniquely the leading-order term of the asymptotic expansion of function $f(z)$ in the vicinity of the corner:[2]

$$
f(z) = -\frac{\theta_0}{\pi} \ln z + \cdots \quad \text{as} \quad z \to 0. \tag{4.2.12}
$$

Separating the real and imaginary parts of (4.2.12) and returning to the original variables (4.2.8), we can conclude that near the corner

$$
p_1 = -\frac{\theta_0}{\pi\beta} \ln \sqrt{x^2 + \beta^2 y^2} + \cdots, \qquad v_1 = -\frac{\theta_0}{\pi} \arctan \frac{\beta y}{x} + \cdots, \tag{4.2.13}
$$

where $\beta = \sqrt{1 - M_\infty^2}$.

Now, let us see how the pressure perturbations influence the flow in the boundary layer (region 2 in Figure 4.2).

4.2.2   Boundary layer before the corner

The asymptotic analysis of the Navier–Stokes equations (4.1.2) in region 2 is conducted based on the limit

$$
x = O(1), \quad Y = Re^{1/2}y = O(1), \quad Re \to \infty.
$$

Keeping in mind that the pressure acting on the boundary layer is an $O(\varepsilon)$ quantity, we represent the fluid-dynamic functions in region 2 in the form of asymptotic expansions

$$
\left.\begin{aligned}
u(x, y; Re) &= U_0(x, Y) + \varepsilon U_1(x, Y) \cdots, \\
v(x, y; Re) &= Re^{-1/2}V_0(x, Y) + \varepsilon Re^{-1/2}V_1(x, Y) + \cdots, \\
\rho(x, y; Re) &= \rho_0(x, Y) + \varepsilon \rho_1(x, Y) + \cdots, \\
p(x, y; Re) &= \varepsilon P_1(x, Y) + \cdots, \\
h(x, y; Re) &= h_0(x, Y) + \varepsilon h_1(x, Y) + \cdots, \\
\mu(x, y; Re) &= \mu_0(x, Y) + \varepsilon \mu_1(x, Y) + \cdots.
\end{aligned}\right\} \tag{4.2.14}
$$

Here, the leading-order terms U_0, V_0, ρ_0, and h_0 satisfy the classical boundary-layer equations (1.10.5) with zero pressure gradient. Remember that these equations admit a self-similar solution (1.10.12), provided that the body surface is thermally isolated or the surface temperature is constant. However, we do not need to restrict ourselves to these particular flow conditions. To proceed further we only need to know that

[2]See Problem 7 in Exercises 7 in Part 2 of this book series.

the leading-order solution remains smooth as the boundary layer approaches the corner, and the functions U_0, V_0, ρ_0, and μ_0 may be represented in the form of Taylor expansions:

$$\left.\begin{aligned}
U_0(x,Y) &= U_{00}(Y) + xU_{01}(Y) + \cdots, \\
V_0(x,Y) &= V_{00}(Y) + xV_{01}(Y) + \cdots, \\
\rho_0(x,Y) &= \rho_{00}(Y) + x\rho_{01}(Y) + \cdots, \\
h_0(x,Y) &= h_{00}(Y) + xh_{01}(Y) + \cdots, \\
\mu_0(x,Y) &= \mu_{00}(Y) + x\mu_{01}(Y) + \cdots
\end{aligned}\right\} \quad \text{as} \quad x \to 0. \qquad (4.2.15)$$

Near the body surface, the leading-order terms in (4.2.15) exhibit the following behaviour:

$$\left.\begin{aligned}
U_{00}(Y) &= \lambda Y + \cdots, \\
V_{00}(Y) &= O(Y^2) + \cdots, \\
\rho_{00}(Y) &= \rho_w + \cdots, \\
\mu_{00}(Y) &= \mu_w + \cdots
\end{aligned}\right\} \quad \text{as} \quad Y \to 0, \qquad (4.2.16)$$

where λ, ρ_w, and μ_w are positive constants representing the dimensionless skin friction, the density, and the viscosity coefficient on the body surface immediately before the corner.

Now, let us consider the perturbation terms in (4.2.14). To deduce equations for U_1, V_1, P_1, ρ_1, and h_1, we substitute (4.2.14) into the Navier–Stokes equations (4.1.2) and work with the $O(\varepsilon)$ terms. We have

$$\rho_0 U_0 \frac{\partial U_1}{\partial x} + (\rho_0 U_1 + \rho_1 U_0)\frac{\partial U_0}{\partial x} + \rho_0 V_0 \frac{\partial U_1}{\partial Y} + (\rho_0 V_1 + \rho_1 V_0)\frac{\partial U_0}{\partial Y}$$
$$= -\frac{\partial P_1}{\partial x} + \frac{\partial}{\partial Y}\left(\mu_0 \frac{\partial U_1}{\partial Y} + \mu_1 \frac{\partial U_0}{\partial Y}\right), \qquad (4.2.17a)$$

$$\frac{\partial P_1}{\partial Y} = 0, \qquad (4.2.17b)$$

$$\rho_0 U_0 \frac{\partial h_1}{\partial x} + (\rho_0 U_1 + \rho_1 U_0)\frac{\partial h_0}{\partial x} + \rho_0 V_0 \frac{\partial h_1}{\partial Y} + (\rho_0 V_1 + \rho_1 V_0)\frac{\partial h_0}{\partial Y}$$
$$= U_0 \frac{\partial P_1}{\partial x} + \frac{1}{Pr}\frac{\partial}{\partial Y}\left(\mu_0 \frac{\partial h_1}{\partial Y} + \mu_1 \frac{\partial h_0}{\partial Y}\right)$$
$$+ 2\mu_0 \frac{\partial U_0}{\partial Y}\frac{\partial U_1}{\partial Y} + \mu_1 \left(\frac{\partial U_0}{\partial Y}\right)^2, \qquad (4.2.17c)$$

$$\frac{\partial}{\partial x}(\rho_0 U_1 + \rho_1 U_0) + \frac{\partial}{\partial Y}(\rho_0 V_1 + \rho_1 V_0) = 0, \qquad (4.2.17d)$$

$$h_1 = \frac{\gamma}{\gamma-1}\frac{P_1}{\rho_0} - \frac{1}{(\gamma-1)M_\infty^2}\frac{\rho_1}{\rho_0^2}. \qquad (4.2.17e)$$

As expected, the y-momentum equation (4.2.17b) shows that the pressure P_1 does not change across the boundary layer. It therefore can be found by setting $y = 0$ in the first equation in (4.2.13):

$$P_1 = -\frac{\theta_0}{\pi\sqrt{1 - M_\infty^2}} \ln(-x) + \cdots \quad \text{as} \quad x \to 0-. \tag{4.2.18}$$

The singularity in the pressure gradient

$$\frac{dP_1}{dx} = \frac{\theta_0}{\pi\sqrt{1 - M_\infty^2}} \frac{1}{(-x)} + \cdots \tag{4.2.19}$$

causes the boundary layer to split into two regions, viscous near-wall sublayer (shown as region $2a$ in Figure 4.2) and main part of the boundary layer (region $2b$).

Viscous sublayer (region $2a$)

In region $2a$ the following three terms in the x-momentum equation (4.2.17a) should be in balance:

$$\rho_0 U_0 \frac{\partial U_1}{\partial x} \sim \frac{dP_1}{dx} \sim \frac{\partial}{\partial Y}\left(\mu_0 \frac{\partial U_1}{\partial Y}\right). \tag{4.2.20}$$

Since region $2a$ lies close to the body surface, we can use for ρ_0, U_0, and μ_0 their asymptotic representations (4.2.16). We can also use equation (4.2.19) for dP_1/dx, which renders (4.2.20) in the form

$$Y \frac{\partial U_1}{\partial x} \sim \frac{1}{(-x)} \sim \frac{\partial^2 U_1}{\partial Y^2}. \tag{4.2.21}$$

Equations (4.2.21) are dealt with in the same way as the corresponding equations of the trailing-edge flow theory.[3] We find that the characteristic thickness of region $2a$ is given by

$$Y \sim (-x)^{1/3}, \tag{4.2.22}$$

and the longitudinal velocity perturbation function U_1 is estimated as

$$U_1 \sim (-x)^{-1/3}. \tag{4.2.23}$$

Similarly, it may be deduced from the energy equation (4.2.17c) that[4]

$$h_1 = O\big[(-x)^{-1/3}\big]. \tag{4.2.24}$$

Then, using (4.2.16), (4.2.18), and (4.2.24) in the state equation (4.2.17e), we can see that in region $2a$

$$\rho_1 = O\big[(-x)^{-1/3}\big]. \tag{4.2.25}$$

Now, let us consider the continuity equation (4.2.17d). It allows us to introduce the stream function $\Psi_1(x, Y)$ such that

$$\frac{\partial \Psi_1}{\partial x} = -\rho_0 V_1 - \rho_1 V_0, \qquad \frac{\partial \Psi_1}{\partial Y} = \rho_0 U_1 + \rho_1 U_0. \tag{4.2.26}$$

[3]See equations (3.4.4) and (3.4.7).
[4]See Problem 1 in Exercises 12.

An estimate for Ψ_1 in region $2a$ may be easily obtained with the help of the second equation in (4.2.26). We have

$$\Psi_1 \sim \rho_0 U_1 Y.$$

Here $\rho_0 = \rho_w$, and estimates of U_1 and Y are given by (4.2.23) and (4.2.22), respectively. We see that

$$\Psi_1 = O(1). \tag{4.2.27}$$

We are ready now to turn from the order-of-magnitude arguments to a rigorous mathematical analysis of the flow in region $2a$. Guided by (4.2.27) and (4.2.22), we seek the stream function Ψ_1 in the form

$$\Psi_1(x, Y) = g_1(\xi) + \cdots \quad \text{as} \quad x \to 0-, \tag{4.2.28a}$$

where

$$\xi = \frac{Y}{(-x)^{1/3}}. \tag{4.2.28b}$$

To find U_1 and V_1, we need to substitute (4.2.28) into (4.2.26). When performing the substitution, we use the fact that the leading-order solution (4.2.15), (4.2.16) is written in region $2a$ as

$$U_0 = (-x)^{1/3}\lambda\xi + \cdots, \quad V_0 = O\big[(-x)^{2/3}\big], \quad \rho_0 = \rho_w, \quad \mu_0 = \mu_w. \tag{4.2.29}$$

We find that

$$U_1 = (-x)^{-1/3}\frac{1}{\rho_w}g_1'(\xi) + \cdots, \quad V_1 = -(-x)^{-1}\frac{1}{3\rho_w}\xi g_1'(\xi) + \cdots. \tag{4.2.30}$$

The equation

$$\frac{\mu_w}{\rho_w}g_1''' - \frac{1}{3}\lambda\xi^2 g_1'' = \frac{\theta_0}{\pi\sqrt{1 - M_\infty^2}} \tag{4.2.31}$$

for function $g_1(\xi)$ is deduced by substituting (4.2.31), (4.2.19), (4.2.25), and (4.2.29) into the x-momentum equation (4.2.17a). Equation (4.2.31) may be treated as a first order differential equation for g_1''. Its general solution is found to be

$$g_1''(\xi) = \frac{\rho_w\theta_0}{\pi\mu_w\sqrt{1 - M_\infty^2}}\left(Ce^{\varkappa\xi^3} - e^{\varkappa\xi^3}\int_\xi^\infty e^{-\varkappa s^3}\,ds\right), \tag{4.2.32}$$

where $\varkappa = \lambda\rho_w/9\mu_w$ and C is an arbitrary constant. Clearly, C has to be set to zero, because if $C \neq 0$, then the solution (4.2.32) exhibits an exponential growth at large values of ξ, which makes the matching with the main part of the boundary layer (shown as region $2b$ in Figure 4.2) impossible.[5] Thus, we can conclude that

$$g_1''(\xi) = -\frac{\rho_w\theta_0}{\pi\mu_w\sqrt{1 - M_\infty^2}}e^{\varkappa\xi^3}\int_\xi^\infty e^{-\varkappa s^3}\,ds. \tag{4.2.33}$$

---

[5]See Section 1.4.2 in Part 2 of this book series.

The perturbation velocity components U_1 and V_1 can now be determined by integrating (4.2.33) with the no-slip condition on the body surface,

$$g_1'(0) = 0, \qquad (4.2.34)$$

and using the result in equations (4.2.30).

Main part of the boundary layer (region 2*b*)

The main part of the boundary layer is shown as region 2*b* in Figure 4.2. In this region the asymptotic analysis of equations (4.2.17) has to be performed using the limit

$$Y = O(1), \quad x \to 0-.$$

The form of asymptotic expansions of U_1 and V_1 in region 2*b* may be predicted by means of matching with the solution in region 2*a*. It follows from (4.2.33) that[6]

$$g_1'(\xi) = A_1 + \frac{3\theta_0}{\pi\lambda\sqrt{1 - M_\infty^2}}\frac{1}{\xi} + \cdots \quad \text{as} \quad \xi \to \infty, \qquad (4.2.35)$$

where A_1 is a constant. By substituting (4.2.35) into (4.2.30), we find that near the outer edge of region 2*a*

$$U_1 = (-x)^{-1/3}\frac{A_1}{\rho_w} + \cdots, \qquad V_1 = -(-x)^{-1}\frac{A_1}{3\rho_w}\xi + \cdots. \qquad (4.2.36)$$

Equations (4.2.36) are valid in the overlap region that lies between regions 2*a* and 2*b*. By using (4.2.28b), we can express (4.2.36) as

$$U_1 = (-x)^{-1/3}\frac{A_1}{\rho_w} + \cdots, \qquad V_1 = -(-x)^{-4/3}\frac{A_1}{3\rho_w}Y + \cdots. \qquad (4.2.37)$$

This suggests that the solution in region 2*b* should be sought in the form

$$U_1(x, Y) = (-x)^{-1/3}U_{11}(Y) + \cdots, \qquad (4.2.38a)$$

$$V_1(x, Y) = (-x)^{-4/3}V_{11}(Y) + \cdots. \qquad (4.2.38b)$$

It also follows from (4.2.37) that

$$\left.\begin{array}{l} U_{11}(Y) = \dfrac{A_1}{\rho_w} + \cdots, \\[2ex] V_{11}(Y) = -\dfrac{A_1}{3\rho_w}Y + \cdots \end{array}\right\} \quad \text{as} \quad Y \to 0. \qquad (4.2.39)$$

To predict the form of asymptotic expansion of the enthalpy perturbation function h_1 we compare the following two terms in the energy equation (4.2.17c):

$$\rho_0 U_0 \frac{\partial h_1}{\partial x} \sim \rho_0 V_1 \frac{\partial h_0}{\partial Y}.$$

Here, V_1 is given by (4.2.38b), and U_0 and $\partial h_0/\partial Y$ are order one quantities. Hence, we can conclude that in region 2*b*

$$h_1 \sim (-x)^{-1/3}. \qquad (4.2.40)$$

---

[6]See Problem 2 in Exercises 12.

Using (4.2.40) and (4.2.18) in the state equation (4.2.17e), we can further see that

$$\rho_1 \sim (-x)^{-1/3}. \qquad (4.2.41)$$

It follows from (4.2.40) and (4.2.41) that the asymptotic expansions of h_1 and ρ_1 in region $2b$ are written as

$$h_1(x, Y) = (-x)^{-1/3} h_{11}(Y) + \cdots, \qquad (4.2.42a)$$
$$\rho_1(x, Y) = (-x)^{-1/3} \rho_{11}(Y) + \cdots. \qquad (4.2.42b)$$

The substitution of (4.2.15), (4.2.38), (4.2.42), and (4.2.19) into equations (4.2.17) results in

$$\frac{1}{3} U_{00} U_{11} + U_{00}' V_{11} = 0 \qquad \text{(x-momentum equation)}, \qquad (4.2.43a)$$

$$\frac{1}{3} U_{00} h_{11} + h_{00}' V_{11} = 0 \qquad \text{(energy equation)}, \qquad (4.2.43b)$$

$$\frac{1}{3} \rho_{00} U_{11} + \frac{1}{3} U_{00} \rho_{11} + \rho_{00}' V_{11} + \rho_{00} V_{11}' = 0 \quad \text{(continuity equation)}, \qquad (4.2.43c)$$

$$h_{11} = -\frac{1}{(\gamma - 1) M_\infty^2} \frac{\rho_{11}}{\rho_{00}^2} \qquad \text{(state equation)}. \qquad (4.2.43d)$$

When solving this set of equations we can use the fact that functions $h_0(Y)$ and $\rho_0(Y)$, given by (4.2.15), are related to one another by means of the state equation (1.10.5d). Setting $x = 1$, we have

$$h_{00} = \frac{1}{(\gamma - 1) M_\infty^2} \frac{1}{\rho_{00}}. \qquad (4.2.44)$$

We use the following elimination procedure. We first substitute (4.2.43d) and (4.2.44) into the energy equation (4.2.43b). This leads to

$$\frac{1}{3} U_{00} \rho_{11} + \rho_{00}' V_{11} = 0,$$

showing that the continuity equation (4.2.43c) may be written as

$$\frac{1}{3} U_{11} + V_{11}' = 0. \qquad (4.2.45)$$

Now, using (4.2.45), we can eliminate U_{11} from the x-momentum equation (4.2.43a). This results in

$$U_{00} V_{11}' - U_{00}' V_{11} = 0. \qquad (4.2.46)$$

Then, dividing both terms in (4.2.46) by U_{00}^2, we have

$$\frac{V_{11}'}{U_{00}} - \frac{U_{00}'}{U_{00}^2} V_{11} = 0,$$

or, equivalently,

$$\frac{d}{dY} \left(\frac{V_{11}}{U_{00}} \right) = 0.$$

This means that the ratio V_{11}/U_{00} remains constant across region 2b:

$$\frac{V_{11}}{U_{00}} = const. \tag{4.2.47}$$

The value of the constant on the right-hand side of equation (4.2.47) is found by taking into account that, according to (4.2.16) and (4.2.39),

$$U_{00} = \lambda Y + \cdots, \quad V_{11} = -\frac{A_1}{3\rho_w}Y + \cdots \quad \text{as} \quad Y \to 0.$$

We have

$$const = -\frac{A_1}{3\lambda\rho_w},$$

and we can conclude that in region 2b

$$V_{11} = -\frac{A_1}{3\lambda\rho_w}U_{00}(Y). \tag{4.2.48}$$

The solution for U_{11} is found by substituting (4.2.48) into (4.2.45):

$$U_{11} = \frac{A_1}{\lambda\rho_w}U'_{00}(Y). \tag{4.2.49}$$

Displacement effect of the boundary layer

The displacement effect of the boundary layer is characterized by the streamline slope angle

$$\vartheta = \arctan\frac{v}{u}.$$

In the boundary layer the velocity components (u, v) are given by (4.2.14). Hence,

$$\vartheta = \frac{Re^{-1/2}V_0 + \varepsilon Re^{-1/2}V_1 + \cdots}{U_0 + \varepsilon U_1 + \cdots}$$

$$= Re^{-1/2}\left[\frac{V_0}{U_0} + \varepsilon\left(\frac{V_1}{U_0} - \frac{V_0}{U_0^2}U_1\right)\right] + \cdots. \tag{4.2.50}$$

The first term in (4.2.50) is regular and cannot provoke the viscous-inviscid interaction. Consequently, in what follows, we concentrate on the second term

$$\vartheta = \varepsilon Re^{-1/2}\left(\frac{V_1}{U_0} - \frac{V_0}{U_0^2}U_1\right) + \cdots \tag{4.2.51}$$

that is provoked by the singular pressure gradient (4.2.19).

In the viscous sublayer (region 2a), equations (4.2.29) and (4.2.30) are valid. Using these in (4.2.51) and setting $x \to 0-$, we have

$$\vartheta = -\varepsilon Re^{-1/2}(-x)^{-4/3}\frac{1}{3\lambda\rho_w}g'_1(\xi) + \cdots. \tag{4.2.52}$$

On the body surface, function $g_1(\xi)$ satisfies condition (4.2.34), and we see that ϑ is zero. At the outer edge of region 2a, where g_1' is represented by (4.2.35), the streamline slope angle reaches the value

$$\vartheta = -\varepsilon Re^{-1/2}(-x)^{-4/3}\frac{A_1}{3\lambda\rho_w} + \cdots . \tag{4.2.53}$$

After that, ϑ remains constant across the main part of the boundary layer (region 2b). Indeed, in region 2b, the leading-order velocity components (U_0, V_0) and their perturbations (U_1, V_1) are given by (4.2.15) and (4.2.38), respectively. Substituting these into (4.2.51), we have

$$\vartheta = \varepsilon Re^{-1/2}(-x)^{-4/3}\frac{V_{11}}{U_{00}} + \cdots ,$$

which, owing to (4.2.48), proves to coincide with (4.2.53).

4.2.3   Viscous-inviscid interaction region

In Section 4.2.2, when constructing the solution for the boundary layer on the surface of plate AB, we assumed that the pressure gradient is given by equation (4.2.19). However, close to the corner B, the displacement effect of the boundary layer leads to a strong distortion of the original pressure gradient. Indeed, with the streamline slope angle at the 'bottom' of the inviscid region 1, given by (4.2.53), the thin aerofoil theory predicts the pressure perturbation to be[7]

$$p \sim \varepsilon Re^{-1/2}(-x)^{-4/3}.$$

Correspondingly, the pressure gradient induced by the displacement effect of the boundary layer may be estimated as

$$\frac{dp}{dx} \sim \varepsilon Re^{-1/2}(-x)^{-7/3}.$$

It becomes comparable with the original pressure gradient (4.2.19) in a vicinity of the corner point, where

$$|x| = O\big(Re^{-3/8}\big). \tag{4.2.54}$$

Thus, the characteristic length of the interaction region is the same as in the case of self-induced boundary-layer separation from a smooth surface (see Chapter 2) and in the flow near the trailing edge of a flat (Chapter 3).

   To construct a theory that describes the transition from an attached flow to a flow with separation at the corner, it is necessary to choose the parameter $\varepsilon(Re)$ appropriately. Remember that this parameter has been introduced by equation (4.2.1), which defines the ramp angle θ, and thus determines the level of perturbations in the boundary layer. Upstream of the interaction region, the solution for the boundary

---

[7]See Section 2.2 in Part 2 of this book series.

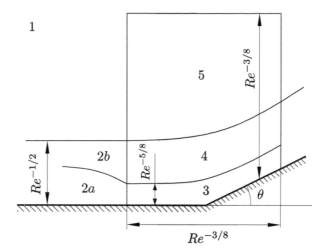

Fig. 4.3: Three-tiered structure of the interaction region.

layer is expressed by asymptotic expansions (4.2.14). In particular, the longitudinal velocity component is written as

$$u(x, y; Re) = U_0(x, Y) + \varepsilon U_1(x, Y) \cdots. \tag{4.2.55}$$

Here, the leading-order term U_0 stands for the unperturbed attached flow. At any finite distance from the corner, it is much larger than the second term, so that the perturbations introduced into the boundary layer are insufficient to cause the appearance of flow reversal characteristic of the separation process. However, as $|x|$ is reduced, the relationship between the two terms in (4.2.55) changes. An especially rapid growth of the perturbations is observed in the viscous sublayer (region $2a$), where U_0 and U_1 are given by (4.2.29) and (4.2.30), respectively:

$$U_0 = (-x)^{1/3} \lambda \xi + \cdots, \qquad U_1 = (-x)^{-1/3} \frac{1}{\rho_w} g_1'(\xi) + \cdots. \tag{4.2.56}$$

We require the two terms in (4.2.55) to become same order quantities in the interaction region. Hence, we write

$$(-x)^{1/3} \sim \varepsilon(-x)^{-1/3}. \tag{4.2.57}$$

Here, $|x|$ is given by (4.2.54). We see that

$$\varepsilon = Re^{-1/4}. \tag{4.2.58}$$

The interaction region has a conventional three-tiered structure composed of the viscous near-wall sublayer (shown as region 3 in Figure 4.3), the main part of the boundary layer (region 4), and an inviscid potential flow (region 5) situated outside the boundary layer. We shall first consider region 4, which represents a continuation of region $2b$ into the viscous-inviscid interaction region.

Main part of the boundary layer (region 4)

The asymptotic analysis of the Navier–Stokes equations in region 4 is based on the limit

$$x_* = \frac{x}{Re^{-3/8}} = O(1), \quad Y = \frac{y}{Re^{-1/2}} = O(1), \quad Re \to \infty. \tag{4.2.59}$$

To predict the form of the asymptotic expansions of the fluid-dynamic functions in region 4, we rely, as before, on the matching with the solution in region 2b. Substituting (4.2.15), (4.2.18), (4.2.38), (4.2.42), and (4.2.58) into (4.2.14), we have the solution in region 2b in the form

$$\left.\begin{aligned}
u &= U_{00}(Y) + Re^{-1/4}(-x)^{-1/3}U_{11}(Y) + \cdots, \\
v &= Re^{-3/4}(-x)^{-4/3}V_{11}(Y) + \cdots, \\
\rho &= \rho_{00}(Y) + Re^{-1/4}(-x)^{-1/3}\rho_{11}(Y) + \cdots, \\
h &= h_{00}(Y) + Re^{-1/4}(-x)^{-1/3}h_{11}(Y) + \cdots, \\
p &= -Re^{-1/4}\frac{\theta_0}{\pi\sqrt{1-M_\infty^2}}\ln(-x) + \cdots.
\end{aligned}\right\} \tag{4.2.60}$$

Now, we express (4.2.60) in terms of the variables (4.2.59) of region 4. This is done by making the substitution $x = Re^{-3/8}x_*$. We have

$$\left.\begin{aligned}
u &= U_{00}(Y) + Re^{-1/8}(-x_*)^{-1/3}U_{11}(Y) + \cdots, \\
v &= Re^{-1/4}(-x_*)^{-4/3}V_{11}(Y) + \cdots, \\
\rho &= \rho_{00}(Y) + Re^{-1/8}(-x_*)^{-1/3}\rho_{11}(Y) + \cdots, \\
h &= h_{00}(Y) + Re^{-1/8}(-x_*)^{-1/3}h_{11}(Y) + \cdots, \\
p &= Re^{-1/4}\ln Re\frac{3\theta_0}{8\pi\sqrt{1-M_\infty^2}} - Re^{-1/4}\frac{\theta_0}{\pi\sqrt{1-M_\infty^2}}\ln(-x_*) + \cdots.
\end{aligned}\right\} \tag{4.2.61}$$

It follows from (4.2.61) that the solution in region 4 should be sought in the form

$$\left.\begin{aligned}
u &= U_{00}(Y) + Re^{-1/8}\tilde{U}_1(x_*,Y) + \cdots, \\
v &= Re^{-1/4}\tilde{V}_1(x_*,Y) + \cdots, \\
h &= h_{00}(Y) + Re^{-1/8}\tilde{h}_1(x_*,Y) + \cdots, \\
\rho &= \rho_{00}(Y) + Re^{-1/8}\tilde{\rho}_1(x_*,Y) + \cdots, \\
p &= Re^{-1/4}\ln Re\frac{3\theta_0}{8\pi\sqrt{1-M_\infty^2}} + Re^{-1/4}\tilde{P}_1(x_*,Y) + \cdots.
\end{aligned}\right\} \tag{4.2.62}$$

We can also use (4.2.61) to formulate the 'upstream' boundary conditions for functions $\tilde{U}_1$, $\tilde{V}_1$, $\tilde{h}_1$, $\tilde{\rho}_1$, and $\tilde{P}_1$. In particular, we see that

$$\tilde{U}_1 = (-x_*)^{-1/3}U_{11}(Y) + \cdots \quad \text{as} \quad x_* \to -\infty. \tag{4.2.63}$$

The substitution of (4.2.62) into the Navier–Stokes equations (4.1.2) leads to the same set of equations (2.2.65) found to describe the flow in the middle tier in the case of self-induced separation of the boundary layer in supersonic flow. The procedure for solving these equations was discussed in Section 2.2.4. In particular, we found that functions $\tilde{U}_1$ and $\tilde{V}_1$ were given by (2.2.182) and (2.2.70), respectively:

$$\tilde{U}_1 = \frac{1}{\lambda}A(x_*)U'_{00}(Y), \qquad \tilde{V}_1 = -\frac{1}{\lambda}\frac{dA}{dx_*}U_{00}(Y). \tag{4.2.64}$$

Function $A(x_*)$ remains arbitrary at this stage of the analysis, except that it follows from (4.2.63), (4.2.49), and the first equation in (4.2.64) that

$$A(x_*) = \frac{A_1}{\rho_w}(-x_*)^{-1/3} + \cdots \quad \text{as} \quad x_* \to -\infty.$$

Viscous sublayer (region 3)

In the viscous sublayer, the fluid-dynamic functions are represented in the form of asymptotic expansions

$$\left.\begin{array}{c} u = Re^{-1/8}U^*(x_*, Y_*) + \cdots, \qquad v = Re^{-3/8}V^*(x_*, Y_*) + \cdots, \\[2mm] p = Re^{-1/4}\ln Re\frac{3\theta_0}{8\pi\sqrt{1-M_\infty^2}} + Re^{-1/4}P^*(x_*, Y_*) + \cdots, \\[2mm] \rho = \rho_w + O(Re^{-1/8}), \qquad \mu = \mu_w + O(Re^{-1/8}). \end{array}\right\} \tag{4.2.65}$$

Here, the independent variables x_*, Y_* are introduced through the scalings

$$x_* = \frac{x}{Re^{-3/8}}, \qquad Y_* = \frac{y}{Re^{-5/8}}. \tag{4.2.66}$$

The substitution of (4.2.65), (4.2.66) into the Navier–Stokes equations (4.1.2) yields

$$\rho_w\left(U^*\frac{\partial U^*}{\partial x_*} + V^*\frac{\partial U^*}{\partial Y_*}\right) = -\frac{\partial P^*}{\partial x_*} + \mu_w\frac{\partial^2 U^*}{\partial Y_*^2}, \tag{4.2.67a}$$

$$\frac{\partial P^*}{\partial Y_*} = 0, \tag{4.2.67b}$$

$$\frac{\partial U^*}{\partial x_*} + \frac{\partial V^*}{\partial Y_*} = 0. \tag{4.2.67c}$$

Equations (4.2.67) require an initial condition for U^* that may be formulated by matching with the solution in region 2a (see Figure 4.3). Substituting (4.2.56) and (4.2.58) into (4.2.55), we have the longitudinal velocity component in region 2a in the form

$$u = (-x)^{1/3}\lambda\xi + Re^{-1/4}(-x)^{-1/3}\frac{1}{\rho_w}g'_1(\xi) + \cdots. \tag{4.2.68}$$

Now, we need to cast (4.2.68) in terms of the variables (4.2.66) of region 3:

$$u = Re^{-1/8}\left[(-x_*)^{1/3}\lambda\xi + (-x_*)^{-1/3}\frac{1}{\rho_w}g'_1(\xi) + \cdots\right] + \cdots. \tag{4.2.69}$$

It remains to compare (4.2.69) with the asymptotic expansion of u in (4.2.65), and we can conclude that the sought initial condition is written as

$$U^* = (-x_*)^{1/3}\lambda\xi + (-x_*)^{-1/3}\frac{1}{\rho_w}g_1'(\xi) + \cdots \quad \text{as} \quad x_* \to -\infty, \tag{4.2.70a}$$

where

$$\xi = \frac{Y}{(-x)^{1/3}} = \frac{Y_*}{(-x_*)^{1/3}}. \tag{4.2.70b}$$

Since the flow in region 3 is viscous, we need to pose the no-slip conditions on the ramp surface:

$$U^* = V^* = 0 \quad \text{at} \quad Y_* = H(x_*). \tag{4.2.71}$$

Here, function $H(x_*)$ gives the body surface shape (4.2.10) in terms of the variables (4.2.66) of region 3:

$$H(x_*) = \begin{cases} 0 & \text{if} \quad x_* < 0, \\ \theta_0 x_* & \text{if} \quad x_* > 0. \end{cases} \tag{4.2.72}$$

Finally, we need to formulate a boundary condition for U^* at the outer edge of region 3, which is done by matching with the solution in region 4 (see Figure 4.3). To perform the matching we substitute the first equation in (4.2.64) into the asymptotic expansion for u in (4.2.62). This leads to the following representation of u in region 4:

$$u = U_{00}(Y) + Re^{-1/8}\frac{1}{\lambda}A(x_*)U_{00}'(Y) + \cdots . \tag{4.2.73}$$

We are interested in the behaviour of u at the 'bottom' of region 4. Remember that

$$U_{00}(Y) = \lambda Y + \cdots \quad \text{as} \quad Y \to 0. \tag{4.2.74}$$

By using (4.2.74) in (4.2.73) and taking into account that $Y = Re^{-1/8}Y_*$, we find that at the outer edge of region 3

$$u = Re^{-1/8}\big[\lambda Y_* + A(x_*) + \cdots\big] + \cdots . \tag{4.2.75}$$

It remains to compare (4.2.75) with the asymptotic expansion for u in (4.2.65), and we can conclude that

$$U^* = \lambda Y_* + A(x_*) + \cdots \quad \text{as} \quad Y_* \to \infty. \tag{4.2.76}$$

Upper tier (region 5)

The flow analysis in region 5 is aimed at formulation of the *interaction law* that relates the pressure acting on the boundary layer to the displacement function $A(x_*)$. Region 5 lies outside the boundary layer, and is defined by the limit

$$x_* = \frac{x}{Re^{-3/8}} = O(1), \quad y_* = \frac{y}{Re^{-3/8}} = O(1), \quad Re \to \infty. \tag{4.2.77}$$

To predict the form of the asymptotic expansions of the fluid-dynamic functions in region 5, we use the matching with the solution in region 1 (see Figure 4.3). By substituting (4.2.13), (4.2.5), and (4.2.6) into (4.2.3) and expressing the resulting equations in terms of variables (4.2.77) of region 5, we find that in the overlap region that lies between regions 1 and 5

$$
\left.
\begin{aligned}
u &= 1 - Re^{-1/4}\ln Re\,\frac{3\theta_0}{8\pi\beta} + Re^{-1/4}\frac{\theta_0}{\pi\beta}\ln\sqrt{x_*^2 + \beta^2 y_*^2} + \cdots, \\[4pt]
v &= -Re^{-1/4}\frac{\theta_0}{\pi}\arctan\frac{\beta y_*}{x_*} + \cdots, \\[4pt]
p &= Re^{-1/4}\ln Re\,\frac{3\theta_0}{8\pi\beta} - Re^{-1/4}\frac{\theta_0}{\pi\beta}\ln\sqrt{x_*^2 + \beta^2 y_*^2} + \cdots, \\[4pt]
\rho &= 1 + Re^{-1/4}\ln Re\,M_\infty^2\frac{3\theta_0}{8\pi\beta} - Re^{-1/4}M_\infty^2\frac{\theta_0}{\pi\beta}\ln\sqrt{x_*^2 + \beta^2 y_*^2} + \cdots, \\[4pt]
h &= \frac{1}{(\gamma-1)M_\infty^2} + Re^{-1/4}\ln Re\,\frac{3\theta_0}{8\pi\beta} - Re^{-1/4}\frac{\theta_0}{\pi\beta}\ln\sqrt{x_*^2 + \beta^2 y_*^2} + \cdots,
\end{aligned}
\right\} \tag{4.2.78}
$$

where $\beta = \sqrt{1 - M_\infty^2}$. This suggests that the solution in region 5 should be sought in the form

$$
\left.
\begin{aligned}
u &= 1 - Re^{-1/4}\ln Re\,\frac{3\theta_0}{8\pi\beta} + Re^{-1/4}u_*(x_*, y_*) + \cdots, \\[4pt]
v &= Re^{-1/4}v_*(x_*, y_*) + \cdots, \\[4pt]
p &= Re^{-1/4}\ln Re\,\frac{3\theta_0}{8\pi\beta} + Re^{-1/4}p_*(x_*, y_*) + \cdots, \\[4pt]
\rho &= 1 + Re^{-1/4}\ln Re\,M_\infty^2\frac{3\theta_0}{8\pi\beta} + Re^{-1/4}\rho_*(x_*, y_*) + \cdots, \\[4pt]
h &= \frac{1}{(\gamma-1)M_\infty^2} + Re^{-1/4}\ln Re\,\frac{3\theta_0}{8\pi\beta} + Re^{-1/4}h_*(x_*, y_*) + \cdots.
\end{aligned}
\right\} \tag{4.2.79}
$$

It further follows from (4.2.78) that the far-field behaviour of functions p_* and v_* is given by

$$
p_* = -\frac{\theta_0}{\pi\beta}\ln\sqrt{x_*^2 + \beta^2 y_*^2} + \cdots, \qquad v_* = -\frac{\theta_0}{\pi}\arctan\frac{\beta y_*}{x_*} + \cdots. \tag{4.2.80}
$$

The substitution of (4.2.79) into the Navier–Stokes equations (4.1.2) leads to the linearized Euler equations:

$$
\frac{\partial u_*}{\partial x_*} = -\frac{\partial p_*}{\partial x_*}, \tag{4.2.81a}
$$

$$
\frac{\partial v_*}{\partial x_*} = -\frac{\partial p_*}{\partial y_*}, \tag{4.2.81b}
$$

$$
\frac{\partial h_*}{\partial x_*} = \frac{\partial p_*}{\partial x_*}, \tag{4.2.81c}
$$

$$
\frac{\partial u_*}{\partial x_*} + \frac{\partial \rho_*}{\partial x_*} + \frac{\partial v_*}{\partial y_*} = 0, \tag{4.2.81d}
$$

$$
h_* = \frac{\gamma}{\gamma-1}p_* - \frac{1}{(\gamma-1)M_\infty^2}\rho_*, \tag{4.2.81e}
$$

with which we have been dealing on a number of occasions. These equations are solved as follows. We first substitute the state equation (4.2.81e) into the energy equation (4.2.81c). We find that

$$\frac{\partial \rho_*}{\partial x_*} = M_\infty^2 \frac{\partial p_*}{\partial x_*}. \tag{4.2.82}$$

Now, we use (4.2.82) and (4.2.81a) to eliminate $\partial u_*/\partial x_*$ and $\partial \rho_*/\partial x_*$ from the continuity equation (4.2.81d), which results in

$$\beta^2 \frac{\partial p_*}{\partial x_*} - \frac{\partial v_*}{\partial y_*} = 0. \tag{4.2.83}$$

The second equation relating p_* and v_* is the y-momentum equation (4.2.81b). The Prandtl–Glauert transformation[8]

$$p_* = \frac{1}{\beta} \tilde{p}_*, \qquad y_* = \frac{1}{\beta} \tilde{y}_* \tag{4.2.84}$$

allows us to turn (4.2.81b) and (4.2.83) into the Cauchy–Riemann equations

$$\frac{\partial \tilde{p}_1}{\partial x_*} = \frac{\partial v_*}{\partial \tilde{y}_*}, \qquad \frac{\partial \tilde{p}_*}{\partial \tilde{y}_*} = -\frac{\partial v_*}{\partial x_*},$$

which represent the necessary and sufficient conditions for the function

$$f(z_*) = \tilde{p}_*(x_*, \tilde{y}_*) + i v_*(x_*, \tilde{y}_*)$$

to be an analytic function of the complex variable $z_* = x_* + i\tilde{y}_*$.

It follows from (4.2.80) that the far-field boundary condition for function $f(z_*)$ is

$$f(z_*) = -\frac{\theta_0}{\pi} \ln z_* + \cdots \quad \text{as} \quad z_* \to \infty. \tag{4.2.85}$$

We also need to formulate the condition of matching with the solution in the middle tier (region 4). Substituting the second equation in (4.2.64) into the asymptotic expansion for v in (4.2.62), we have the solution for the lateral velocity in region 4 in the form

$$v = -Re^{-1/4} \frac{1}{\lambda} \frac{dA}{dx_*} U_{00}(Y) + \cdots . \tag{4.2.86}$$

Taking into account that $U_{00}(\infty) = 1$ and comparing (4.2.86) with the asymptotic expansion (4.2.79) for v in region 5, we can conclude that

$$v_*\Big|_{y_*=0} = -\frac{1}{\lambda} \frac{dA}{dx_*}. \tag{4.2.87}$$

Since function $f(z_*)$ does not tend to zero as $z_* \to \infty$, we shall consider instead its derivative

$$F(z_*) = \frac{df}{dz_*} = \frac{\partial \tilde{p}_*}{\partial x_*} + i \frac{\partial v_*}{\partial x_*}.$$

---

[8]See Section 2.2.2 in Part 2 of this book series.

Now, function $F(z_*)$ satisfies condition (2.3.121), which allows us to use equation (2.3.128):

$$F(z_*) = \frac{1}{2\pi i} \int\limits_{-\infty}^{\infty} \frac{F(\zeta) - \overline{F(\zeta)}}{\zeta - z_*} \, d\zeta. \qquad (4.2.88)$$

It follows from (4.2.87) that the numerator of the integrand in (4.2.88) is given by

$$F(\zeta) - \overline{F(\zeta)} = 2i \, \Im\{F(\zeta)\} = 2i\left(-\frac{1}{\lambda}\frac{d^2 A}{d\zeta^2} \right),$$

which, being substituted into (4.2.88), yields

$$F(z_*) = \frac{1}{2\pi i} \int\limits_{-\infty}^{\infty} 2i\left(-\frac{1}{\lambda}\frac{d^2 A}{d\zeta^2} \right) \frac{d\zeta}{\zeta - z_*}. \qquad (4.2.89)$$

Equation (4.2.89) is applicable to any point z_* that lies in the upper half of the complex z_*-plane. To find the pressure gradient acting on the boundary layer, we assume that z_* tends to a point $(x_*, 0)$ lying on the real axis. The corresponding limiting value of (4.2.89) is obtained with the help of the Sokhotsky–Plemelj formula (2.3.131), which yields

$$\left.\frac{\partial \tilde{p}_*}{\partial x_*}\right|_{y_*=0} = -\frac{1}{\pi\lambda} \int\limits_{-\infty}^{\infty} \frac{A''(\zeta)}{\zeta - x_*} \, d\zeta.$$

It remains to return to the original variables (4.2.84), and we can conclude that

$$\left.\frac{\partial p_*}{\partial x_*}\right|_{y_*=0} = -\frac{1}{\pi\lambda\beta} \int\limits_{-\infty}^{\infty} \frac{A''(\zeta)}{\zeta - x_*} \, d\zeta. \qquad (4.2.90)$$

4.2.4 Interaction problem

We are now in a position to formulate the viscous-inviscid interaction problem. We found that in the viscous sublayer the flow is described by the boundary-layer equations (4.2.67):

$$\rho_w\left(U^* \frac{\partial U^*}{\partial x_*} + V^* \frac{\partial U^*}{\partial Y_*} \right) = -\frac{dP^*}{dx_*} + \mu_w \frac{\partial^2 U^*}{\partial Y_*^2}, \qquad (4.2.91a)$$

$$\frac{\partial U^*}{\partial x_*} + \frac{\partial V^*}{\partial Y_*} = 0. \qquad (4.2.91b)$$

The pressure gradient dP^*/dx_* in (4.2.91a) is not known in advance, and has to be found as a part of the solution using the interaction law

$$\frac{dP^*}{\partial x_*} = -\frac{1}{\pi\lambda\beta} \int\limits_{-\infty}^{\infty} \frac{A''(\zeta)}{\zeta - x_*} \, d\zeta. \qquad (4.2.92)$$

Here, it has been taken into account that the pressure does not change across the boundary layer, which allows us to apply equation (4.2.90) directly to the pressure gradient in the viscous sublayer (region 3).

The boundary conditions for equations (4.2.91) are no-slip conditions (4.2.71) on the ramp surface,

$$U^* = V^* = 0 \quad \text{at} \quad Y_* = H(x_*), \tag{4.2.93}$$

the condition (4.2.70) of matching with the solution in region 2a,

$$U^* = \lambda Y_* + \cdots \quad \text{as} \quad x_* \to -\infty, \tag{4.2.94}$$

and the condition (4.2.76) of matching with the solution in region 4,

$$U^* = \lambda Y_* + A(x_*) + \cdots \quad \text{as} \quad Y_* \to \infty. \tag{4.2.95}$$

In addition, we require that downstream of the interaction region the flow returns to the unperturbed state:

$$U^* = \lambda Y_* + \cdots \quad \text{as} \quad x_* \to \infty. \tag{4.2.96}$$

This precludes those solutions with large-scale separation we dealt with in Chapter 2.

The viscous-inviscid interaction problem (4.2.91)–(4.2.96) involves four parameters: the dimensionless skin friction immediately before the interaction region λ, the fluid density ρ_w and viscosity coefficient μ_w on the body surface, and the 'compressibility' parameter $\beta = \sqrt{1 - M_\infty^2}$. We perform the substitution of variables

$$
\left.
\begin{aligned}
x_* &= \frac{\mu_w^{-1/4} \rho_w^{-1/2}}{\lambda^{5/4} \beta^{3/4}} \bar{X}, & Y_* &= \frac{\mu_w^{1/4} \rho_w^{-1/2}}{\lambda^{3/4} \beta^{1/4}} \bar{Y} + H(x_*), \\
U^* &= \frac{\mu_w^{1/4} \rho_w^{-1/2}}{\lambda^{-1/4} \beta^{1/4}} \bar{U}, & V^* &= \frac{\mu_w^{3/4} \rho_w^{-1/2}}{\lambda^{-3/4} \beta^{-1/4}} \bar{V} + U^* \frac{dH}{dx_*}, \\
P^* &= \frac{\mu_w^{1/2}}{\lambda^{-1/2} \beta^{1/2}} \bar{P}, & A &= \frac{\mu_w^{1/4} \rho_w^{-1/2}}{\lambda^{-1/4} \beta^{1/4}} \bar{A} - H(x_*), \\
H &= \frac{\mu_w^{1/4} \rho_w^{-1/2}}{\lambda^{3/4} \beta^{1/4}} \bar{H}, & \zeta &= \frac{\mu_w^{-1/4} \rho_w^{-1/2}}{\lambda^{5/4} \beta^{3/4}} s,
\end{aligned}
\right\} \tag{4.2.97}
$$

that combines standard affine transformations of the triple-deck theory (2.2.171) with Prandtl's transposition (1.4.60). The latter introduces the body-fitted coordinates $(\bar{X}, \bar{Y})$ with $\bar{X}$ measured along the body contour and $\bar{Y}$ in the normal direction.

As a result, the interaction problem (4.2.91)–(4.2.96) assumes the form

$$
\left.
\begin{aligned}
\bar{U} \frac{\partial \bar{U}}{\partial \bar{X}} + \bar{V} \frac{\partial \bar{U}}{\partial \bar{Y}} &= -\frac{d\bar{P}}{d\bar{X}} + \frac{\partial^2 \bar{U}}{\partial \bar{Y}^2}, \\
\frac{\partial \bar{U}}{\partial \bar{X}} + \frac{\partial \bar{V}}{\partial \bar{Y}} &= 0, \\
\frac{d\bar{P}}{d\bar{X}} &= -\frac{1}{\pi} \int\limits_{-\infty}^{\infty} \frac{\bar{A}''(s) - \bar{H}''(s)}{s - \bar{X}} \, ds, \\
\bar{U} = \bar{V} = 0 & \quad \text{at} \quad \bar{Y} = 0, \\
\bar{U} = \bar{Y} + \cdots & \quad \text{as} \quad \bar{X} \to -\infty, \\
\bar{U} = \bar{Y} + \bar{A}(\bar{X}) + \cdots & \quad \text{as} \quad \bar{Y} \to \infty, \\
\bar{U} = \bar{Y} + \cdots & \quad \text{as} \quad \bar{X} \to \infty.
\end{aligned}
\right\} \tag{4.2.98}
$$

Here, function

$$\bar{H}(\bar{X}) = \begin{cases} 0 & \text{if } \bar{X} < 0, \\ \alpha\bar{X} & \text{if } \bar{X} > 0 \end{cases} \tag{4.2.99}$$

represents the body shape in the new variables. The scaled ramp angle

$$\alpha = \frac{\mu_w^{-1/2}}{\lambda^{1/2}\beta^{1/2}}\,\theta_0$$

is a single controlling parameter of the flow. We shall call it the *asymptotic similarity parameter*, since unlike conventional similarity parameters, the Reynolds number, the Mach number, etc., parameter α has emerged as a result of an asymptotic analysis of the flow, which, by its nature, involves a degree of approximation.

The asymptotic theory of viscous-inviscid interaction was first applied to the corner flows by Stewartson (1970).[9] In addition to formulating the problem (4.2.98), he was also able to solve analytically its linearized version that was applicable to small values of α when the Blasius boundary layer was only slightly perturbed.[10] The linear theory is, however, insufficient to describe the process of formation of the separation region at the corner. A numerical solution of the nonlinear problem was obtained by Ruban (1976), Smith and Merkin (1982), and Kravtsova *et al.* (2005).

Numerical results for concave corner

The computations show that, in the case of concave corner ($\alpha > 0$), the flow experiences deceleration before the corner. This leads to a decrease of the skin friction

$$\bar{\tau}_w = \left.\frac{\partial \bar{U}}{\partial \bar{Y}}\right|_{\bar{Y}=0},$$

which happens more rapidly for larger values of α; see Figure 4.4(a). The minimum skin friction is reached at the point $\bar{X} = 0$, and downstream of this point $\bar{\tau}_w$ first increases up to a value exceeding unity, and then returns to its unperturbed value before the interaction region ($\bar{\tau}_w = 1$). As far as the pressure is concerned, it first rises with $\hat{X}$ and reaches a maximum at a point behind the corner ($\bar{X} = 0$), and then starts to fall. An increase of α leads to a reduction of the minimal skin friction. When $\bar{\tau}_w$ becomes negative, a small separation bubble forms near the corner. Interestingly enough, increasing α leads to a displacement of the pressure maximum further downstream of the corner point (see Figure 4.4b) such that when the separation bubble forms, the pressure gradient remains positive in the entire separation region. The transition between attached and separated flows is observed at

$$\alpha_1^* = 2.315.$$

Further increase of α leads to an increase of the size of the separation bubble, with the separation and reattachment points moving from the corner in the upstream and downstream directions, respectively.

[9]See also Stewartson's note published in 1971.
[10]See Section 4.3.1 and Problem 5 in Exercises 12.

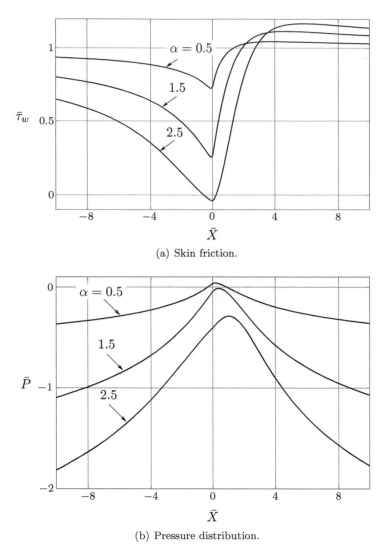

(a) Skin friction.

(b) Pressure distribution.

Fig. 4.4: Results of the numerical solution of the interaction problem (4.2.98) for concave corner.

Numerical results for convex corner

The flow past a convex corner ($\alpha < 0$) displays the opposite trends. In this case, instead of decelerating before the corner, the flow experiences an acceleration, which results in an increase of the skin friction (see Figure 4.5a) and a decrease of the pressure (Figure 4.5b). Behind the corner the skin friction displays a non-monotonic behaviour. It first decreases below unity, but then recovers towards the unperturbed value $\bar{\tau} = 1$, despite a monotonic growth of the pressure. For all $\alpha < 0$, the minimal value of the skin friction, $\bar{\tau}_{\min}$, is found to be less than unity. It decreases monotonically with increasing

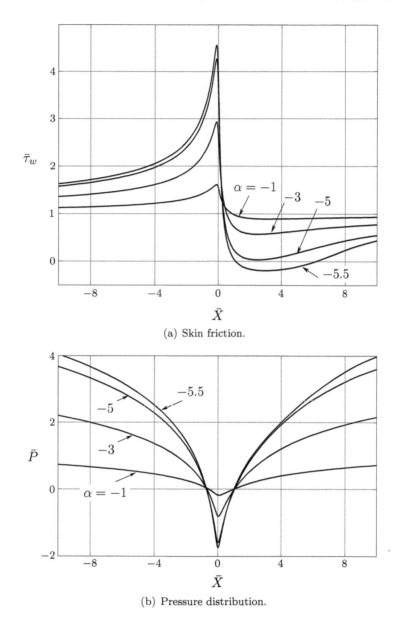

(a) Skin friction.

(b) Pressure distribution.

Fig. 4.5: Results of the numerical solution of the interaction problem (4.2.98) for convex corner.

$|\alpha|$ and becomes zero at

$$\alpha_2^* = -5.088.$$

For $\alpha < \alpha_2^*$ a separation bubble forms in the boundary layer. It is of interest to note that, in this case, the entire region of recirculating flow is located behind the corner.

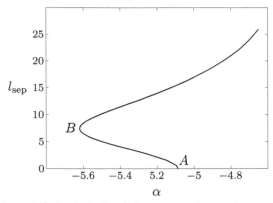

Fig. 4.6: Dependence of the length of the separation region on the ramp angle α.

Another interesting feature of the flow past a convex corner is that the solution of the interaction problem (4.2.98) appears to be non-unique. This is illustrated by Figure 4.6, where the length l_{sep} of the separation region, defined as the distance between the separation and reattachment points, is plotted as a function of α. Point A on the graph corresponds to the incipient separation. First, the solution shows the

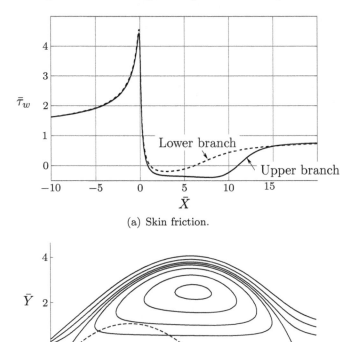

(a) Skin friction.

(b) Streamline pattern.

Fig. 4.7: Comparison of the upper and lower branch solutions for $\alpha = -5.5$.

'expected' behaviour, where the length of the separation region grows as $|\alpha|$ increases. This continues until point B, where α assumes the value

$$\alpha_3^* = -5.6199.$$

To extend the solution beyond point B, we now have to decrease α. As a result, the second solution branch is formed. Figure 4.7 compares the two solutions for $\alpha = -5.5$. It should be noted that before the corner, where the flow experiences a sharp acceleration, the two solutions are almost indistinguishable from one another (see Figure 4.7a). The main difference is in the separation region that is identified by negative skin friction. The solution on the upper branch has a relatively large separation region. Figure 4.7(b) displays the streamlines in this region. For comparison, the dashed line shows the shape of the separation region given by the solution on the lower branch.

4.3   Supersonic Compression Ramp Flow

In this section, we return to the problem depicted in Figure 4.1 but here we assume that the oncoming flow is supersonic, that is, $M_\infty > 1$. We expect again the region of viscous-inviscid interaction to form near the corner point B. However, unlike in subsonic case, the flow outside the interaction region behaves differently. The perturbations produced by a deflection of surface BC cannot propagate upstream either through the supersonic inviscid part of the flow or through the boundary layer. In fact, the description of the boundary layer upstream of the interaction region, given in Section 2.2.2 for the self-induced separation, remains applicable for the compression ramp flow.

  To cause the boundary-layer separation, a certain level of pressure perturbations Δp should be introduced in the flow by the deflection of surface BC. According to (2.2.28), it is estimated as

$$\Delta p \sim Re^{-1/4}.$$

Consequently, using the Ackeret formula,[11] we can claim that this is achieved if the ramp angle

$$\theta = Re^{-1/4}\theta_0,$$

where θ_0 is an order one parameter.

  The viscous-inviscid interaction region has a conventional three-tiered structure, shown in Figure 4.3. As far as the mathematical analysis of the flow in the three tiers is concerned, it is conducted in the same way as done in the case of self-induced separation in the supersonic flow (see Section 2.2.4), and may be summarized as follows. To describe the interaction process, we have to solve the boundary layer equations (2.2.87), (2.2.88) for the lower tier (region 3 in Figure 4.3):

$$\rho_w U^* \frac{\partial U^*}{\partial x_*} + \rho_w V^* \frac{\partial U^*}{\partial Y_*} = -\frac{dP^*}{dx_*} + \mu_w \frac{\partial^2 U^*}{\partial Y_*^2}, \qquad (4.3.1a)$$

$$\frac{\partial U^*}{\partial x_*} + \frac{\partial V^*}{\partial Y_*} = 0. \qquad (4.3.1b)$$

---

[11]See equation (3.2.35) on page 174 in Part 2 of this book series

The pressure P^* in (4.3.1a) is not known in advance, but has to be found as a part of the solution by using the Ackeret formula (2.2.91)

$$P^* = -\frac{1}{\lambda\sqrt{M_\infty^2 - 1}}\frac{dA}{dx_*}, \qquad (4.3.1c)$$

which is deduced by means of flow analysis in the upper tier (region 5). Remember that function $A(x_*)$ is known as the displacement function. It may be found from equation (2.2.92), which represents the asymptotic behaviour of U^* at the outer edge of region 3:

$$U^*(x_*, Y_*) = \lambda Y_* + A(x_*) + \cdots \quad \text{as} \quad Y_* \to \infty. \qquad (4.3.1d)$$

When solving the viscous-inviscid interaction problem, we also need to pose the matching condition (2.2.90) with solution in the boundary layer before the interaction region:

$$U^* = \lambda Y_* \quad \text{at} \quad x_* = -\infty, \qquad (4.3.1e)$$

and the no-slip conditions on the ramp surface:

$$U^* = V^* = 0 \quad \text{at} \quad Y_* = H(x_*). \qquad (4.3.1f)$$

Here, the ramp-shape function $H(x_*)$ is written as

$$H(x_*) = \begin{cases} 0 & \text{if} \quad x_* < 0, \\ \theta_0 x_* & \text{if} \quad x_* > 0. \end{cases}$$

Finally, to preclude the solutions with large-scale separation, we require that the flow returns to the unperturbed state downstream of the interaction region:

$$U^* = \lambda Y_* + \cdots \quad \text{as} \quad x_* \to \infty. \qquad (4.3.1g)$$

To express the viscous-inviscid interaction problem (4.3.1) in canonical form, we apply again transformations (4.2.97), now with $\beta = \sqrt{M_\infty^2 - 1}$. This turns (4.3.1) into

$$\left.\begin{aligned} &\bar{U}\frac{\partial\bar{U}}{\partial\bar{X}} + \bar{V}\frac{\partial\bar{U}}{\partial\bar{Y}} = -\frac{d\bar{P}}{d\bar{X}} + \frac{\partial^2\bar{U}}{\partial\bar{Y}^2}, \\ &\frac{\partial\bar{U}}{\partial\bar{X}} + \frac{\partial\bar{V}}{\partial\bar{Y}} = 0, \\ &\qquad \bar{P} = \frac{d\bar{H}}{d\bar{X}} - \frac{d\bar{A}}{d\bar{X}}, \\ &\bar{U} = \bar{V} = 0 \qquad\qquad \text{at} \quad \bar{Y} = 0, \\ &\bar{U} = \bar{Y} + \cdots \qquad\quad \text{as} \quad \bar{X} \to -\infty, \\ &\bar{U} = \bar{Y} + \bar{A}(\bar{X}) + \cdots \quad \text{as} \quad \bar{Y} \to \infty, \\ &\bar{U} = \bar{Y} + \cdots \qquad\quad \text{as} \quad \bar{X} \to \infty. \end{aligned}\right\} \qquad (4.3.2)$$

Here, function

$$\bar{H}(\bar{X}) = \begin{cases} 0 & \text{if } \bar{X} < 0, \\ \alpha\bar{X} & \text{if } \bar{X} > 0 \end{cases}$$

expresses the body shape in the new variables. The scaled ramp angle

$$\alpha = \frac{\mu_w^{-1/2}}{\lambda^{1/2}\beta^{1/2}}\theta_0$$

is a single controlling parameter of the flow.

4.3.1 Linear problem

For small values of parameter α, the interaction problem (4.3.2) can be solved analytically. Let us start by setting $\alpha = 0$. In this case, the ramp degenerates into a flat surface. The latter cannot produce any perturbations in the Blasius boundary layer, and the solution of (4.3.2) may be written as

$$\bar{U} = \bar{Y}, \quad \bar{V} = \bar{P} = \bar{A} = 0. \tag{4.3.3}$$

The validity of (4.3.3) is verified by direct substitution of (4.3.3) into (4.3.2). If α is small, but non-zero, then we have to perturb (4.3.3). We seek the solution in the form

$$\left. \begin{aligned} \bar{U} &= \bar{Y} + \alpha\bar{U}'(\bar{X}, \bar{Y}) + \cdots, & \bar{V} &= \alpha\bar{V}'(\bar{X}, \bar{Y}) + \cdots, \\ \bar{P} &= \alpha\bar{P}'(\bar{X}) + \cdots, & \bar{A} &= \alpha\bar{A}'(\bar{X}) + \cdots. \end{aligned} \right\} \tag{4.3.4}$$

The substitution of (4.3.4) into (4.3.2) reduces the interaction problem to

$$\bar{Y}\frac{\partial\bar{U}'}{\partial\bar{X}} + \bar{V}' = -\frac{d\bar{P}'}{d\bar{X}} + \frac{\partial^2\bar{U}'}{\partial\bar{Y}^2}, \tag{4.3.5a}$$

$$\frac{\partial\bar{U}'}{\partial\bar{X}} + \frac{\partial\bar{V}'}{\partial\bar{Y}} = 0, \tag{4.3.5b}$$

$$\bar{P}' = \frac{d\bar{H}'}{d\bar{X}} - \frac{d\bar{A}'}{d\bar{X}}, \tag{4.3.5c}$$

$$\bar{U}' = \bar{V}' = 0 \quad \text{at} \quad \bar{Y} = 0, \tag{4.3.5d}$$

$$\bar{U}' \to 0 \qquad \text{as} \quad \bar{X} \to -\infty, \tag{4.3.5e}$$

$$\bar{U}' = \bar{A}' \qquad \text{at} \quad \bar{Y} = \infty, \tag{4.3.5f}$$

$$\bar{U}' \to 0 \qquad \text{as} \quad \bar{X} \to \infty, \tag{4.3.5g}$$

where

$$\bar{H}'(\bar{X}) = \begin{cases} 0 & \text{if } \bar{X} < 0, \\ \bar{X} & \text{if } \bar{X} > 0. \end{cases}$$

It should be noted that now all the equations and boundary conditions are linear, and their coefficients are independent of $\bar{X}$. This allows us to use the method of Fourier transforms. We define the Fourier transform $\breve{U}(k, \bar{Y})$ of function $\bar{U}'(\bar{X}, \bar{Y})$ as

$$\breve{U}(k, \bar{Y}) = \int\limits_{-\infty}^{\infty} \bar{U}'(\bar{X}, \bar{Y}) e^{-ik\bar{X}} \, d\bar{X},$$

and similarly for other functions.

Of course, the Fourier transforms are applicable to functions that decay to zero as $|\bar{X}| \to \infty$. Unfortunately, the pressure $\bar{P}$ does not satisfy this condition. Indeed, it follows from (4.3.5e), (4.3.5f), and (4.3.5g) that the displacement function $\bar{A}'$ tends to zero when $\bar{X} \to -\infty$ and when $\bar{X} \to \infty$. At the same time, $d\bar{H}'/d\bar{X}$ is the Heaviside step function

$$\frac{d\bar{H}'}{d\bar{X}} = \begin{cases} 0 & \text{if } \bar{X} < 0, \\ 1 & \text{if } \bar{X} > 0. \end{cases}$$

Consequently, using the interaction law (4.3.5c), we see that the pressure $\bar{P}'$ grows from $\bar{P}' = 0$ at $\bar{X} = -\infty$ to $\bar{P}' = 1$ at $\bar{X} = \infty$. Keeping this in mind, we shall be dealing instead with the pressure gradient

$$R(\bar{X}) = \frac{d\bar{P}'}{d\bar{X}}.$$

Applying the Fourier transform to the momentum equation (4.3.5a) we have

$$ik\bar{Y}\breve{U} + \breve{V} = -\breve{R} + \frac{d^2\breve{U}}{d\bar{Y}^2}, \tag{4.3.6}$$

where $\breve{R}(k)$ is the Fourier transform of the pressure gradient $R(\bar{X})$. Similarly, the continuity equation (4.3.5b) turns into

$$ik\breve{U} + \frac{d\breve{V}}{d\bar{Y}} = 0. \tag{4.3.7}$$

Now, we differentiate the interaction law (4.3.5c) with respect to $\bar{X}$, which results in

$$R = \delta(\bar{X}) - \frac{d^2\bar{A}'}{d\bar{X}^2},$$

where $\delta(\bar{X})$ is Dirac's delta function. Written in terms of the Fourier transforms, the above equation assumes the form

$$\breve{R} = 1 - (ik)^2 \breve{A}. \tag{4.3.8}$$

It remains to apply Fourier transforms to boundary conditions (4.3.5d) and (4.3.5f). We have

$$\breve{U} = \breve{V} = 0 \quad \text{at} \quad \bar{Y} = 0, \tag{4.3.9}$$

$$\breve{U} = \breve{A} \qquad \text{at} \quad \bar{Y} = \infty. \tag{4.3.10}$$

The boundary-value problem (4.3.6)–(4.3.10) may be solved in the following way. We start by eliminating $\bar{V}$ from (4.3.6) and (4.3.7). For this purpose we differentiate the momentum equation (4.3.6) with respect to $\bar{Y}$. Since the pressure gradient is independent of $\bar{Y}$, this results in

$$ik\bar{Y}\frac{d\check{U}}{d\bar{Y}} + ik\check{U} + \frac{d\check{V}}{d\bar{Y}} = \frac{d^3\check{U}}{d\bar{Y}^3}. \tag{4.3.11}$$

Now, we use the continuity equation (4.3.7) in (4.3.11). We see that the second and third term on the left-hand side cancel each other, and we can write

$$ik\bar{Y}\frac{d\check{U}}{d\bar{Y}} = \frac{d^3\check{U}}{d\bar{Y}^3}. \tag{4.3.12}$$

Since (4.3.12) is a third order differential equation, it requires an additional boundary condition. The latter may be obtained by setting $\bar{Y} = 0$ in (4.3.6) and using condition (4.3.9) for $\check{V}$. We have

$$\left.\frac{d^2\check{U}}{d\bar{Y}^2}\right|_{\bar{Y}=0} = \check{R}. \tag{4.3.13}$$

With the help of the interaction law (4.3.8), the boundary condition (4.3.13) can be expressed in the form

$$\left.\frac{d^2\check{U}}{d\bar{Y}^2}\right|_{\bar{Y}=0} = 1 - (ik)^2\check{A}. \tag{4.3.14}$$

The substitution of the independent variable

$$z = (ik)^{1/3}\bar{Y} \tag{4.3.15}$$

allows us to turn equation (4.3.12) into the Airy equation

$$\frac{d^2w}{dz^2} - zw = 0 \tag{4.3.16}$$

for the function

$$w = \frac{d\check{U}}{dz}. \tag{4.3.17}$$

The boundary conditions for equation (4.3.16) are

$$\left.\frac{dw}{dz}\right|_{z=0} = (ik)^{-2/3} - (ik)^{4/3}\check{A}, \tag{4.3.18}$$

$$\int_0^\infty w\,dz = \check{A}. \tag{4.3.19}$$

The first of these follows directly from condition (4.3.14). To deduce condition (4.3.19), we first integrate (4.3.17) with initial condition (4.3.9) for $\check{U}$, and then use condition (4.3.10).

The general solution of equation (4.3.16) may be written in the form

$$w = C_1 Ai(z) + C_2 Bi(z),$$ (4.3.20)

where $Ai(z)$ and $Bi(z)$ are two complementary solutions of the Airy equation. The properties of these functions are well known (see, for example, Abramowitz and Stegun, 1965). In particular, it is known that at large values of z

$$\left.\begin{aligned} Ai(z) &= \frac{z^{-1/4}}{2\sqrt{\pi}} e^{-\zeta} + \cdots , \\ Bi(z) &= \frac{z^{-1/4}}{\sqrt{\pi}} e^{\zeta} + \cdots \end{aligned}\right\} \quad \text{as} \quad z \to \infty,$$ (4.3.21)

where $\zeta = \frac{2}{3} z^{3/2}$.

To find constants C_1 and C_2 in (4.3.20) we notice, first of all, that for the integral in (4.3.19) to converge, the function $w(z)$ should tend to zero as $z \to \infty$. It follows from (4.3.21) that the behaviour of $Ai(z)$ and $Bi(z)$ at large values of z depends on the direction along which z tends to infinity in the complex plane. The latter is defined by a choice of analytical branch of $(ik)^{1/3}$ in (4.3.15). For our purposes, it is convenient to make the branch cut in the complex k-plane along the positive imaginary semi-axis (see Figure 4.8a). Then, with k written in the form $k = |k|e^{i\vartheta}$, we define $(ik)^{1/3}$ as

$$(ik)^{1/3} = \left(e^{i\pi/2}|k|e^{i\vartheta}\right)^{1/3} = |k|^{1/3} e^{i(\pi/6+\vartheta/3)}.$$ (4.3.22)

Here, ϑ belongs to the interval

$$\vartheta \in \left(-\frac{3}{2}\pi, \frac{1}{2}\pi\right).$$

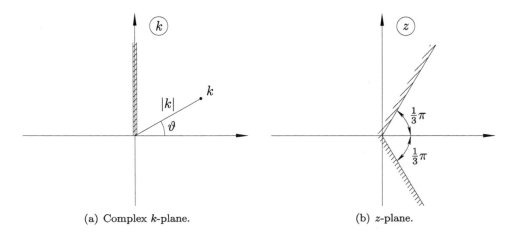

(a) Complex k-plane.   (b) z-plane.

Fig. 4.8: How to choose an analytical branch of $(ik)^{1/3}$.

Hence, it follows from (4.3.22) and (4.3.15) that

$$\arg z \in \left(\frac{1}{3}\pi , \; -\frac{1}{3}\pi \right), \tag{4.3.23}$$

as shown in Figure 4.8(b). Thus, for all z in sector (4.3.23) the real part of ζ is positive, and the Airy function $Ai(z)$ decays as $\bar{Y} \to \infty$ but $Bi(z)$ is exponentially growing. Consequently, we have to set $C_2 = 0$ in (4.3.20), and we will have

$$w = C_1 Ai(z). \tag{4.3.24}$$

The substitution of (4.3.24) into (4.3.18) and (4.3.19) results in the following set of linear algebraic equations for C_1 and $\breve{A}$:[12]

$$\frac{1}{3^{1/3}\Gamma(1/3)} \, C_1 = (ik)^{4/3}\breve{A} - (ik)^{-2/3}, \qquad \frac{1}{3}C_1 = \breve{A}.$$

These equations are easily solved to show that the Fourier transform of the displacement function is given by

$$\breve{A} = \frac{(ik)^{-2/3}}{(ik)^{4/3} - 3^{2/3}/\Gamma(1/3)}. \tag{4.3.25}$$

To find the displacement function itself, we need to take the inverse Fourier integral

$$\bar{A}'(\bar{X}) = \frac{1}{2\pi} \int\limits_{-\infty}^{\infty} \breve{A}(k)e^{ik\bar{X}} \, dk = \frac{1}{2\pi} \int\limits_{-\infty}^{\infty} \frac{(ik)^{-2/3}}{(ik)^{4/3} - 3^{2/3}/\Gamma(1/3)}e^{ik\bar{X}} \, dk. \tag{4.3.26}$$

Here, the integration is performed along the real axis in the complex k-plane. For $\bar{X} < 0$, the integration path can be closed by a semicircle C_R that lies in the lower half-plane (see Figure 4.9). Indeed, according to Jordan's lemma, for all negative $\bar{X}$, the integral

$$\int\limits_{C_R} \frac{(ik)^{-2/3}}{(ik)^{4/3} - 3^{2/3}/\Gamma(1/3)}e^{ik\bar{X}} \, dk$$

tends to zero as the radius R of the semicircle C_R tends to infinity. Thus, we can write

$$\bar{A}'(\bar{X}) = -\frac{1}{2\pi} \int\limits_{C} \frac{(ik)^{-2/3}}{(ik)^{4/3} - 3^{2/3}/\Gamma(1/3)}e^{ik\bar{X}} \, dk, \tag{4.3.27}$$

where C is the closed contour composed of the interval $[-R, R]$ of the real axis and the semicircle C_R, now traced in the anti-clockwise direction, that is, opposite to the direction shown in Figure 4.9. The integral in (4.3.27) can be calculated with the help

---

[12]Here it is taken into account that

$$Ai'(0) = -\frac{1}{3^{1/3}\Gamma(1/3)}, \qquad \int\limits_{0}^{\infty} Ai(z) \, dz = \frac{1}{3},$$

where Γ denotes the gamma function.

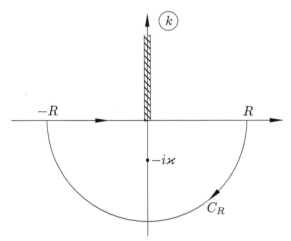

Fig. 4.9: Integration contour for integral in (4.3.27).

of the residue theorem. We can see that the integrand has a single pole in the lower half-plane that lies on the negative imaginary semiaxis at point $k = -i\varkappa$, where $\varkappa$ is given by

$$\varkappa = \frac{\sqrt{3}}{\left[\Gamma(1/3)\right]^{3/4}}.$$

We find that

$$\bar{A}' = -\frac{3}{4\varkappa}e^{\varkappa\bar{X}} \quad \text{for all } \bar{X} < 0. \tag{4.3.28}$$

Now, we can substitute (4.3.28) into the interaction law (4.3.5c). We see that before the corner the pressure grows exponentially:[13]

$$\bar{P}' = \frac{3}{4}e^{\varkappa\bar{X}}.$$

It reaches the value of $\bar{P}' = \frac{3}{4}$ at $\bar{X} = 0$, and has to increase further by $\frac{1}{4}$ behind the corner to become $\bar{P}' = 1$ at $\bar{X} = \infty$. To find the distribution of $\bar{P}'$ on the interval $\bar{X} \in (0, \infty)$, one has to calculate the integral (4.3.26) numerically. We leave this task with the reader to perform.

4.3.2   Numerical solution of the nonlinear problem

Figure 4.10 shows the results of a numerical solution of the viscous-inviscid interaction problem (4.3.2). We see that the pressure shows a monotonic growth through the interaction region. It starts from zero at $\bar{X} = -\infty$, and increases to $\bar{P} = \alpha$ as $\bar{X} \to \infty$. The behaviour of the skin friction depends on the ramp angle α. For smaller values of α, it first decreases towards the corner, where $\bar{\tau}_w$ reaches a minimum. Then, behind

---

[13]Compare this behaviour with that (2.2.106) predicted by Lighthill (1953).

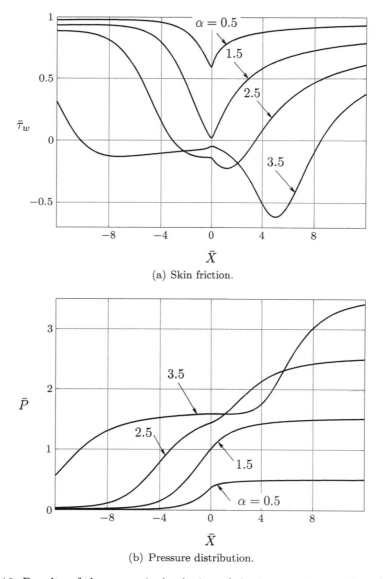

(a) Skin friction.

(b) Pressure distribution.

Fig. 4.10: Results of the numerical solution of the interaction problem (4.3.2).

the corner, it displays a monotonic growth, returning to the unperturbed value $\bar{\tau}_w = 1$ as $\bar{X} \to \infty$. The minimum value of $\bar{\tau}_w$ decreases with α, and becomes zero at

$$\alpha^* = 1.56,$$

which signifies the transition from an attached flow to the flow with a separation bubble. Inside the bubble the skin friction is negative, which means that the fluid near the wall is moving in the direction opposite to the rest of the flow. The recirculation of the fluid in the separation bubble is rather slow. For this reason, the pressure

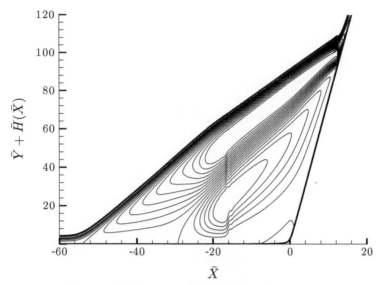

Fig. 4.11: The streamline pattern for $\alpha = 7.5$.

develops a 'plateau' that is clearly seen in the solution for $\alpha = 3.5$ (see Figure 4.10b). By this stage the behaviour of the skin friction (Figure 4.10a) changes drastically. Near the separation point it reproduces the solution of the self-induced separation problem (see Figure 2.18), with a minimum observed behind the separation point, and pressure 'plateau' level given by (2.2.174). The skin friction recovers partially before the corner, but then develops a second minimum some distance upstream from the reattachment point. It becomes progressively sharper as α increases (see Korolev *et al.*, 2002), and is accompanied by a sharp rise of the pressure. For a detailed discussion of the reattachment process, the interested reader is referred to Neiland (1970a).

It interesting to note that as α becomes larger, the separation region does not remain a single eddy. Instead, a fragmentation of the separation bubble is observed. It divides into counter-rotating eddies; the number of these increases with α. This is demonstrated by Figure 4.11, where the streamline pasttern is shown for $\alpha = 7.5$.

Exercises 12

1. Consider the subsonic flow past a corner (see Figure 4.2). Remember that in the boundary layer (region 2), the fluid-dynamic functions are represented by the asymptotic expansions (4.2.14), with the perturbation functions U_1, V_1, P_1, ρ_1, and h_1 obeying equations (4.2.17). Your task is to show that in the viscous sublayer (region 2a), the following estimate holds for perturbation of the enthalpy:

$$h_1 = O\big[(-x)^{-1/3}\big].$$

To perform this task, first consider the continuity equation (4.2.17d):

$$\frac{\partial}{\partial x}\big(\rho_0 U_1 + \rho_1 U_0\big) + \frac{\partial}{\partial Y}\big(\rho_0 V_1 + \rho_1 V_0\big) = 0.$$

Assume, subject to subsequent confirmation, that two dominant terms in this equation are

$$\frac{\partial}{\partial x}(\rho_0 U_1) \sim \frac{\partial}{\partial Y}(\rho_0 V_1),$$

and deduce that

$$V_1 \sim \frac{U_1}{(-x)}Y \sim (-x)^{-1}. \tag{4.3.29}$$

You may use without proof the fact that estimates for Y and U_1 are given by (4.2.22) and (4.2.23), respectively:

$$Y \sim (-x)^{1/3}, \qquad U_1 \sim (-x)^{-1/3}.$$

Now, turn to the energy equation (4.2.17c):

$$\rho_0 U_0 \frac{\partial h_1}{\partial x} + (\rho_0 U_1 + \rho_1 U_0)\frac{\partial h_0}{\partial x} + \rho_0 V_0 \frac{\partial h_1}{\partial Y} + (\rho_0 V_1 + \rho_1 V_0)\frac{\partial h_0}{\partial Y}$$
$$= U_0 \frac{\partial P_1}{\partial x} + \frac{1}{Pr}\frac{\partial}{\partial Y}\left(\mu_0 \frac{\partial h_1}{\partial Y} + \mu_1 \frac{\partial h_0}{\partial Y}\right)$$
$$+ 2\mu_0 \frac{\partial U_0}{\partial Y}\frac{\partial U_1}{\partial Y} + \mu_1\left(\frac{\partial U_0}{\partial Y}\right)^2,$$

where the following two terms

$$\rho_0 U_0 \frac{\partial h_1}{\partial x} \sim \rho_0 V_1 \frac{\partial h_0}{\partial Y}$$

should be compared. The second of these describes the variation of the enthalpy due to displacement of the streamlines from the body surface. Keeping in mind that at the 'bottom' of the boundary layer

$$U_0 = \lambda Y + \cdots, \qquad h_0 = h_{00}(0) + h'_{00}(0)Y + \cdots,$$

show that in region 2a

$$h_1 = O\big[(-x)^{-1/3}\big].$$

2. Consider equation (4.2.33):

$$g''_1(\xi) = -\frac{\rho_w \theta_0}{\pi \mu_w \sqrt{1 - M_\infty^2}} e^{\varkappa \xi^3}\int_\xi^\infty e^{-\varkappa s^3}\, ds.$$

By applying the method of integration by parts to the integral on the right-hand side of this equation,[14] show that

$$g'_1(\xi) = A_1 + \frac{3\theta_0}{\pi \lambda \sqrt{1 - M_\infty^2}}\frac{1}{\xi} + \cdots \quad \text{as} \quad \xi \to \infty,$$

where A_1 is a constant.

[14]See Section 1.1.2 in Part 2 of this book series.

Confirm this result by balancing the second term on the left-hand side of equation (4.2.31):

$$\frac{\mu_w}{\rho_w} g_1''' - \frac{1}{3}\lambda\xi^2 g_1'' = \frac{\theta_0}{\pi\sqrt{1-M_\infty^2}}$$

with the term on the right-hand side.

3. Adjust the formulation of the viscous-inviscid interaction problem (3.5.52) for symmetric flow past an aerofoil with wedge-shaped trailing edge (see Figure 4.12).

Fig. 4.12: The flow past a thin symmetric aerofoil.

4. Return to Problem 6 in Exercises 12 in Part 2 of this book series, where equation (3.2.38) represents a generalization of the Ackeret formula for the case of a supersonic flow past a thin body with an impinging shock wave. Use this equation to establish an analogy between the following two flows. The first is the supersonic flow past a compression ramp that is described by equations (4.3.2). The second is the flow in the viscous-inviscid interaction region that forms as a result of a weak shock wave impinging on the boundary layer on a flat plate (see Figures 2.5 and 2.6). What should the strength of the impinging shock be to cause the incipient separation of the boundary layer on a flat plate?

5. In the subsonic version of the linearized viscous-inviscid interaction problem (4.3.5), the interaction law is written as

$$\frac{d\bar{P}'}{d\bar{X}} = -\frac{1}{\pi}\int\limits_{-\infty}^{\infty} \frac{d^2\bar{A}'/ds^2 - d^2\bar{H}'/ds^2}{s - \bar{X}}\, ds, \tag{4.3.30}$$

while the rest of the equations in (4.3.5) remain unchanged. Here, your task is to show that the Fourier transform of (4.3.30) is written as

$$\check{R} = ik\frac{k^2\check{A}+1}{|k|},$$

where $\check{R}$ is the Fourier transform of the pressure gradient $d\bar{P}'/d\bar{X}$ and $\check{A}$ is the Fourier transform of the displacement function $\bar{A}'$.

You may perform this task in the following steps:

(a) Return to equations (4.2.81b), (4.2.83), which describe the flow in the upper tier (region 5). Eliminate v_* from these equations by cross-differentiation, and show that the pressure p_* satisfies the equation

$$\beta^2 \frac{\partial^2 p_*}{\partial x_*^2} + \frac{\partial^2 p_*}{\partial y_*^2} = 0. \tag{4.3.31}$$

Now, combine equation (4.2.81b) with boundary condition (4.2.87) to show that

$$\left.\frac{\partial p_*}{\partial y_*}\right|_{y_*=0} = \frac{1}{\lambda} \frac{d^2 A}{dx_*^2}. \tag{4.3.32}$$

(b) Corresponding to (4.2.97), perform the following affine transformations in the upper tier

$$p_* = \frac{\mu_w^{1/2}}{\lambda^{-1/2} \beta^{1/2}} \, \bar{p}, \qquad A = \frac{\mu_w^{1/4} \rho_w^{-1/2}}{\lambda^{-1/4} \beta^{1/4}} (\bar{A} - \bar{H}),$$

$$x_* = \frac{\mu_w^{-1/4} \rho_w^{-1/2}}{\lambda^{5/4} \beta^{3/4}} \, \bar{X}, \qquad y_* = \frac{\mu_w^{-1/4} \rho_w^{-1/2}}{\lambda^{5/4} \beta^{7/4}} \, \bar{y},$$

and show that this turns (4.3.31) and (4.3.32) into

$$\frac{\partial^2 \bar{p}}{\partial \bar{X}^2} + \frac{\partial^2 \bar{p}}{\partial \bar{y}^2} = 0,$$

$$\left.\frac{\partial \bar{p}}{\partial \bar{y}}\right|_{\bar{y}=0} = \frac{d^2 \bar{A}}{d\bar{X}^2} - \frac{d^2 \bar{H}}{d\bar{X}^2}.$$

(c) Assume that the ramp angle α is small, and seek the solution for $\bar{p}$ in the form

$$\bar{p} = \alpha \bar{p}' + \cdots .$$

Argue that function $\bar{p}'$ should be found by solving the following boundary-value problem:

$$\left.\begin{aligned} \frac{\partial^2 \bar{p}'}{\partial \bar{X}^2} + \frac{\partial^2 \bar{p}'}{\partial \bar{y}^2} &= 0, \\ \left.\frac{\partial \bar{p}'}{\partial \bar{y}}\right|_{\bar{y}=0} &= \frac{d^2 \bar{A}'}{d\bar{X}^2} - \frac{d^2 \bar{H}'}{d\bar{X}^2}, \\ \bar{p}' &\to 0 \quad \text{as} \quad \bar{X}^2 + \bar{y}^2 \to \infty. \end{aligned}\right\} \tag{4.3.33}$$

After solving this problem, the pressure in the viscous sublayer (region 3) may be found by setting

$$\bar{P}' = \bar{p}'\big|_{\bar{y}=0}.$$

(d) Show that the boundary-value problem (4.3.33) is written in terms of Fourier transforms as

$$-k^2 \breve{p} + \frac{d^2 \breve{p}}{d\bar{y}^2} = 0, \tag{4.3.34a}$$

$$\left.\frac{d\breve{p}}{d\bar{y}}\right|_{\bar{y}=0} = -k^2 \breve{A} - 1, \tag{4.3.34b}$$

$$\breve{p} \to 0 \quad \text{as} \quad \bar{y} \to \infty. \tag{4.3.34c}$$

Argue that the solution of equation (4.3.34a) that satisfies condition (4.3.34c) is written as

$$\check{p} = Ce^{-|k|\bar{y}}. \tag{4.3.35}$$

(e) Find constant C by substituting (4.3.35) into boundary condition (4.3.34b), and deduce that the Fourier transform of the pressure gradient $d\bar{P}'/d\bar{X}$ in region 3 is given by

$$\check{R} = ik\frac{k^2\check{A}+1}{|k|}. \tag{4.3.36}$$

(f) Use (4.3.36) instead of (4.3.8) in (4.3.13) to show that, in subsonic flow, equation (4.3.25) for the Fourier transform of the displacement function $\bar{A}'$ should be substituted by

$$\check{A} = -\frac{(ik)^{1/3}/|k|}{(ik)^{1/3}|k| + 3^{2/3}/\Gamma(1/3)}.$$

5

Marginal Separation Theory

Marginal separation theory was developed independently by Ruban (1981, 1982) and Stewartson *et al.* (1982) in their studies of the boundary-layer separation at the leading edge of a thin aerofoil. Later, it became clear that the theory describes a variety of flows where small separation bubbles form on a smooth part of the body surface. These include supersonic flows on a surface with large curvature (Fomina, 1983), separation of three-dimensional boundary layer on the surface of a paraboloid at an angle of attack (Brown, 1985), incipient separation in a near-wall jet (Zametaev, 1986) and in the boundary layer on the surface of a fast rotating cylinder (Negoda and Sychev, 1986), etc. For a discussion of various aspects of marginal separation theory, the interested reader is referred to Braun and Kluwick (2004) and Braun and Scheichl (2014).

This chapter follows the original papers by Ruban (1981, 1982) and Stewartson *et al.* (1982).

5.1   Experimental Observations

The leading edge separation was first observed by Jones (1934) in wind tunnel and real flight experiments. Since then, many researchers have been involved in experimental study of the flow round the leading edge of an aerofoil. A comprehensive account of these efforts may be found, for example, in Tani's (1964) review. The experiments show that for thick aerofoils (with thickness to chord ratio larger than 15%), the boundary layer first separates near the trailing edge. For thin aerofoils (with thickness to chord ratio smaller than 12%), this is the leading edge separation that results in a deterioration of the aerodynamic characteristics of the aerofoil.

For small values of the angle of attack α, the flow over a thin aerofoil remains fully attached, and the pressure has its maximum at the front stagnation point O; see Fig. 5.1(a). As we move from point O around the aerofoil nose, the pressure first drops rapidly to reach a minimum at point M on the upper side of the aerofoil, and then starts to recover, so that downstream of point M the boundary layer finds itself under an adverse pressure gradient. Its magnitude increases with angle of attack, and the boundary layer separates at $\alpha = \alpha_s$.

When it happens, we observe the appearance of a closed region of recirculating flow on the upper surface of the aerofoil (Fig. 5.1a). This region is referred to as a *short bubble*. Its length does not exceed 1% of the aerofoil chord, and, therefore, it has an extremely weak influence on the flow field and the values of the aerodynamic forces acting on the aerofoil. However, the short separation bubble only exists within an interval $\alpha \in (\alpha_s, \alpha_c)$, and when the angle of attack reaches α_c, the bubble suddenly

Fluid Dynamics: Part 3: Boundary Layers. © Anatoly I. Ruban, 2018. Published 2018 by Oxford University Press. 10.1093/oso/9780199681754.001.0001

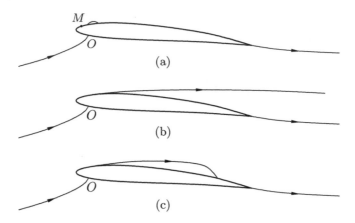

Fig. 5.1: The flow past a thin aerofoil: (a) with a short separation bubble; (b) with an extended separation region that forms after the bubble bursts; (c) with a long separation bubble.

bursts. As a result, a transition to a new flow regime takes place with an extended separation region that covers the entire upper surface of the aerofoil (Fig. 5.1b) or flow with the so-called *long bubble* (Fig. 5.1c). In either case the flow transformation is accompanied by an abrupt decrease in the lift produced by the aerofoil and a significant increase in the drag. Hence the name, the *leading-edge stall*. If encountered in real flight, the consequences are most likely to be catastrophic, which explains why the problem attracted significant attention of aeronautic engineers.

We now proceed to the theoretical analysis of the described phenomenon. To start, we use Prandtl's hierarchical strategy.

5.2   Inviscid Flow Region

Let an aerofoil, placed in a two-dimensional flow of incompressible fluid, be of a relative thickness ε, so that its equation may be written as

$$y' = \begin{cases} \varepsilon Y_+(x') & \text{for the upper surface,} \\ \varepsilon Y_-(x') & \text{for the lower surface.} \end{cases} \tag{5.2.1}$$

Here, we use the Cartesian coordinate system $O'x'y'$ with the origin O' placed at the leading edge and the x'-axis directed tangentially to the aerofoil's middle line, shown in Figure 5.2 as the dashed line. The coordinates x' and y' are made dimensionless by referring them to the aerofoil chord L.

We assume that the nose of the aerofoil is parabolic, in which case parameter ε in (5.2.1) may be chosen such that

$$Y_\pm(x') = \pm\sqrt{2x'} + \cdots \quad \text{as} \quad x' \to 0.$$

With this choice, the radius of curvature of the leading edge of the aerofoil is calculated as $r = L\varepsilon^2$.

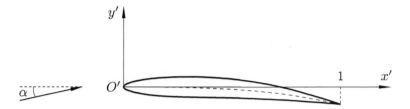

Fig. 5.2: Incompressible flow past a thin aerofoil.

Since our interest is in the leading-edge flow region, we shall define the Reynolds number as

$$Re = \frac{V_\infty r}{\nu}.$$

In what follows, we shall assume that the Reynolds number is large, and the thickness of the aerofoil is small:

$$Re \to \infty, \qquad \varepsilon \to 0.$$

We introduce the dimensionless fluid-dynamic functions through the usual scalings

$$\hat{u} = V_\infty u, \quad \hat{v} = V_\infty v, \quad \hat{p} = p_\infty + \rho V_\infty^2 p,$$

and start the flow analysis with the main inviscid region where $x' = O(1)$, $y' = O(1)$. If the angle of attack α is an $O(\varepsilon)$ quantity, that is

$$\alpha = \varepsilon \alpha_*,$$

where $\alpha_* = O(1)$, then the flow in this region is described by the thin aerofoil theory.[1] According to this theory, the velocity components (u, v) and the pressure p are represented by the asymptotic expansions

$$u = 1 + \varepsilon u_1(x', y') + \cdots, \tag{5.2.2a}$$
$$v = \varepsilon v_1(x', y') + \cdots, \tag{5.2.2b}$$
$$p = \varepsilon p_1(x', y') + \cdots. \tag{5.2.2c}$$

The solution for function $f(z) = p_1 + iv_1$ is given by equation (2.1.52) in Part 2. Combining (2.1.52) with the linearized Bernoulli equation, $p_1 = -u_1$, we can express this solution in the form

$$u_1 - iv_1 = -\frac{1}{2\pi} \int_0^1 \frac{Y'_+(\zeta) - Y'_-(\zeta)}{\zeta - z} \, d\zeta$$

$$- \sqrt{\frac{z-1}{z}} \left[i\alpha_* + \frac{1}{2\pi i} \int_0^1 \sqrt{\frac{\zeta}{1-\zeta}} \, \frac{Y'_+(\zeta) + Y'_-(\zeta)}{\zeta - z} \, d\zeta \right], \tag{5.2.3}$$

where $z = x' + iy'$.

---

[1]See Section 2.1 in Part 2 of this book series.

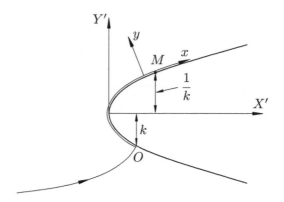

Fig. 5.3: The flow near the leading edge of a thin aerofoil.

Of course, the pressure perturbations (5.2.2c) are too small to cause the boundary-layer separation. However, the perturbations grow as the leading edge is approached. Indeed, we can see that (5.2.3) develops a singularity at the leading edge of the aerofoil:[2]

$$u_1 - iv_1 = \frac{k - i}{\sqrt{2z}} + \cdots \quad \text{as} \quad z \to 0 \tag{5.2.4}$$

with k given by

$$k = \sqrt{2} \left[\alpha_* - \frac{1}{2\pi} \int_0^1 \frac{Y'_+(\zeta) + Y'_-(\zeta)}{\sqrt{\zeta(1 - \zeta)}} \, d\zeta \right]. \tag{5.2.5}$$

This singularity makes the thin aerofoil theory inapplicable to a vicinity of the leading edge, where $|z| = O(\varepsilon^2)$. In this vicinity, the second term in (5.2.2a) becomes comparable with the first one.

To study the flow near the leading edge we introduce the scaled variables

$$x' = \varepsilon^2 X', \qquad y' = \varepsilon^2 Y',$$

and consider the flow region where X' and Y' are order one quantities. In this region (we shall call it region 1), the aerofoil contour is represented by the infinite parabola $Y' = \pm\sqrt{2X'}$; see Figure 5.3. The solution of the Euler equations for the inviscid flow past the parabola was discussed in Problems 5 and 6 in Exercises 11 in Part 1 of this book series. It was found that the tangential velocity on the surface of the parabola is given by

$$U_e = \frac{Y' + k}{\sqrt{Y'^2 + 1}}. \tag{5.2.6}$$

Here, Y' has to be thought of as the distance from the point on the surface of the parabola, where U_e is to be found, to the axis of symmetry of the parabola. Parameter k

---

[2]The asymptotic behaviour of the first integral in (5.2.3) may be obtained as suggested in Problem 2(b) in Exercises 7 in Part 2 of this book series. When dealing with the second integral in (5.2.3) we can simply set $z = 0$ in the integrand and take into account that $\sqrt{(z-1)/z} \to i/\sqrt{z}$ as $z \to 0$.

represents the degree of non-symmetry of the flow. It can be obtained by matching with the solution (5.2.2), (5.2.4) in the outer inviscid region, and proves to coincide with k in (5.2.4).

Note that the stagnation point O, where $U_e = 0$, is the point with $Y' = -k$. By differentiating (5.2.6) with respect to Y' and setting the derivative to zero, we can find that the maximum of U_e is achieved at point M, where $Y' = 1/k$; see Figure 5.3. At this point,

$$U_e\Big|_B = \sqrt{1 + k^2}.$$

Downstream of point M, the tangential velocity $U_e(x)$ shows a monotonic decay, and tends to unity as $x \to \infty$.

5.3 Boundary Layer

We now consider the boundary layer that forms on the aerofoil surface. In what follows we call it region 2. When analysing the fluid motion in this region, it is convenient to use the body-fitted coordinates. Remember that these are curvilinear orthogonal coordinates (x, y) with x measured along the aerofoil surface from the front stagnation point O, as shown in Figure 5.3, and y in the normal direction. We denote the velocity components in these coordinates as V_τ and V_n, respectively, and the stream function as ψ. All the variables are assumed dimensionless. We take the radius $r = \varepsilon^2 L$ of the leading edge of the aerofoil as the unit of length; the velocity components are referred to V_∞, the stream function to rV_∞, and the pressure increment $\hat{p} - p_\infty$ with respect to its value in the free-stream is referred to ρV_∞^2. The Navier–Stokes equations are written in these variables as[3]

$$\frac{V_\tau}{H_1}\frac{\partial V_\tau}{\partial x} + V_n\frac{\partial V_\tau}{\partial y} + \frac{\kappa V_\tau V_n}{H_1} = -\frac{1}{H_1}\frac{\partial p}{\partial x} + \frac{1}{Re}\left[\frac{1}{H_1}\frac{\partial}{\partial x}\left(\frac{1}{H_1}\frac{\partial V_\tau}{\partial x}\right)\right.$$
$$\left. + \frac{\partial^2 V_\tau}{\partial y^2} + \kappa\frac{\partial}{\partial y}\left(\frac{V_\tau}{H_1}\right) + \frac{\kappa}{H_1^2}\frac{\partial V_n}{\partial x} + \frac{1}{H_1}\frac{\partial}{\partial x}\left(\frac{\kappa V_n}{H_1}\right)\right], \tag{5.3.1a}$$

$$\frac{V_\tau}{H_1}\frac{\partial V_n}{\partial x} + V_n\frac{\partial V_n}{\partial y} - \frac{\kappa V_\tau^2}{H_1} = -\frac{\partial p}{\partial y} + \frac{1}{Re}\left[\frac{1}{H_1}\frac{\partial}{\partial x}\left(\frac{1}{H_1}\frac{\partial V_n}{\partial x}\right)\right.$$
$$\left. + \frac{\partial^2 V_n}{\partial y^2} + \kappa\frac{\partial}{\partial y}\left(\frac{V_n}{H_1}\right) - \frac{\kappa}{H_1^2}\frac{\partial V_\tau}{\partial x} - \frac{1}{H_1}\frac{\partial}{\partial x}\left(\frac{\kappa V_\tau}{H_1}\right)\right], \tag{5.3.1b}$$

$$\frac{1}{H_1}\frac{\partial V_\tau}{\partial x} + \frac{\partial V_n}{\partial y} + \frac{\kappa V_n}{H_1} = 0. \tag{5.3.1c}$$

Here, κ is the local curvature of the body contour, and H_1 is the Lamé coefficient:

$$H_1 = 1 + \kappa(x)y.$$

The velocity components are related to the stream function through the equations

---

[3]For a derivation of these equations, the interested reader is referred to pages 86 and 87 in Part 1 of this book series.

$$\frac{\partial \psi}{\partial x} = -H_1 V_n, \qquad \frac{\partial \psi}{\partial y} = V_\tau. \tag{5.3.2}$$

In region 2 the asymptotic expansion of the stream function has the form

$$\psi = Re^{-1/2}\Psi(x, Y) + \cdots, \quad \text{with} \quad y = Re^{-1/2}Y. \tag{5.3.3}$$

The substitution of (5.3.3) into (5.3.2) and then into the Navier–Stokes equations (5.3.1), results in the classical boundary-layer equation

$$\frac{\partial \Psi}{\partial Y}\frac{\partial^2 \Psi}{\partial x \partial Y} - \frac{\partial \Psi}{\partial x}\frac{\partial^2 \Psi}{\partial Y^2} = -\frac{dp_e}{dx} + \frac{\partial^3 \Psi}{\partial Y^3}. \tag{5.3.4a}$$

It should be solved with the initial condition at the stagnation point O,

$$\Psi = 0 \quad \text{at} \quad x = 0, \tag{5.3.4b}$$

the no-slip conditions on the aerofoil surface,

$$\Psi = \frac{\partial \Psi}{\partial Y} = 0 \quad \text{at} \quad Y = 0, \tag{5.3.4c}$$

and the condition of matching with the solution in the inviscid flow region,

$$\frac{\partial \Psi}{\partial Y} = U_e(x) \quad \text{at} \quad Y = \infty. \tag{5.3.4d}$$

The pressure gradient dp_e/dx on the right-hand side of equation (5.3.4a) does not depend on Y, and may be calculated with the help of the Bernoulli equation,

$$p_e + \frac{1}{2}U_e^2 = \frac{1}{2}. \tag{5.3.5}$$

Differentiating (5.3.5), we have

$$\frac{dp_e}{dx} = -U_e\frac{dU_e}{dx}.$$

The results of the numerical solution of problem (5.3.4) are shown in Figures 5.4 in the form of the skin friction distribution along the aerofoil surface. The skin friction is calculated as

$$\tau_w = \frac{\partial^2 \Psi}{\partial Y^2}\bigg|_{Y=0}. \tag{5.3.6}$$

It appears that there exists a critical value of parameter $k = k_0 = 1.1575$ that divides the solutions into two families. If $k > k_0$, then the solution terminates at a finite position $x = x_s$ where the skin friction turns zero. At this position, the solution develops Goldstein's (1948) singularity, which does not allow the solution to be continued downstream of x_s. The larger the parameter k, the earlier on the aerofoil surface the singularity is encountered.

If, on the other hand, $k < k_0$, then the solution exists for all values of x with the skin friction remaining positive. Interestingly enough, in this case, the skin friction displays a minimum, which tends to zero as $k \to k_0 - 0$ (see Figure 5.4b). We denote the coordinate of the point where τ_w first becomes zero by x_0. The calculations show that for the boundary layer on the parabola surface $x_0 = 8.265$.

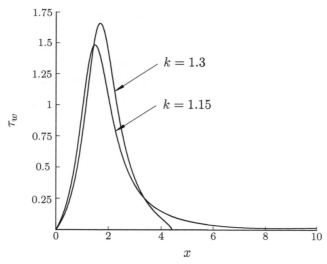

(a) Skin friction distribution on the aerofoil nose for $k = 1.15$ and $k = 1.3$.

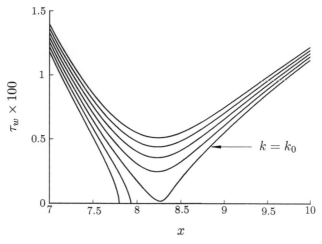

(b) Local behaviour of the solution near point $x = x_0$; the graphs of function $\tau(x)$ are plotted here with interval $\Delta k = 0.0003$.

Fig. 5.4: Results of the numerical solution of problem (5.3.4).

5.3.1 Theoretical analysis of the boundary layer

According to the classical viewpoint, which dates back to Prandtl's original 1904 work, the point of zero skin friction gives the position of the flow separation: $x = x_s$. It is therefore of considerable interest to study the behaviour of the solution of the problem (5.3.4) in the vicinity of this point.

Figure 5.4 shows that for $k > k_0$, the point of zero skin friction x_s lies upstream of x_0, but tends to x_0 as $k \to k_0 + 0$. We first consider the region lying before the line $x = x_s$, and our task is to find an asymptotic expansion of the stream function Ψ as $x \to x_s - 0$. We note that at any point (x, Y) in the region upstream of $x = x_s$, the longitudinal velocity component $u = \partial\Psi/\partial Y$ is positive. Consequently, the boundary-layer equation (5.3.4a) possesses the usual properties of equations of parabolic type. Its solution $\Psi(x, Y)$ near $x = x_s$ depends upon the boundary conditions upstream of this point, namely, on the velocity distribution $U_e(x)$ at the outer edge of the boundary layer on the entire interval $x \in [0, x_s]$. Meanwhile, the asymptotic procedure utilized below is restricted to analysis of a small vicinity of line $x = x_s$, and, therefore, it does not take into account all the boundary conditions affecting the solution for $\Psi(x, Y)$. This means that the sought asymptotic expansion of $\Psi(x, Y)$ must include the 'eigensolutions of the local problem'. The coefficients multiplying the eigenfunctions are expected to remain arbitrary from the viewpoint of the local analysis.

We start by noticing that the pressure gradient dp_e/dx is a smooth function. Near point $x = x_s$ it may be represented in the form of Taylor expansion

$$\frac{dp_e}{dx} = \lambda_0 + \lambda_1 s + \cdots \quad \text{as} \quad s \to 0. \tag{5.3.7}$$

Here, $s = x - x_s$ is the distance from the point $x = x_s$. The leading-order term, λ_0 in (5.3.7), coincides with the pressure gradient at point $x = x_s$. For all $k \geq k_0$, when the zero skin point exists, $\lambda_0 > 0$.

Setting $Y = 0$ in (5.3.4a) and using the no-slip conditions (5.3.4c), we find that at any point on the aerofoil surface

$$\frac{\partial^3 \Psi}{\partial Y^3} = \frac{dp_e}{dx}. \tag{5.3.8}$$

At the point of zero skin friction, in addition to the no-slip conditions (5.3.4c), we also know that

$$\tau_w(x_s) = \left.\frac{\partial^2 \Psi}{\partial Y^2}\right|_{Y=0} = 0. \tag{5.3.9}$$

Using (5.3.7) in (5.3.8), and integrating the resulting equation with (5.3.4c) and (5.3.9), we see that the leading-order term of the asymptotic representation of $\Psi(x, Y)$ near the point $x = x_s$ may be written as

$$\Psi = \frac{1}{6}\lambda_0 Y^3 + \cdots . \tag{5.3.10}$$

We can now determine the characteristic thickness of the viscous flow region that forms inside the boundary layer upstream of the point $x = x_s$. We call it region $2a$. In this region, the viscous term on the right-hand side of equation (5.3.4a) should be comparable with either term on the left-hand side. Using, for example, the first of these, we can write

$$\frac{\partial\Psi}{\partial Y}\frac{\partial^2\Psi}{\partial x\partial Y} \sim \frac{\partial^3\Psi}{\partial Y^3}. \tag{5.3.11}$$

It follows from (5.3.10) that the factor $\partial\Psi/\partial Y$ in the convective term on the left-hand side of (5.3.11) may be estimated as an order $O(Y^2)$ quantity. Hence, we can write

$$Y^2 \frac{\partial^2 \Psi}{\partial x \partial Y} \sim \frac{\partial^3 \Psi}{\partial Y^3}. \tag{5.3.12}$$

By approximating the derivatives in (5.3.12) by finite differences, we arrive at the following simple algebraic equation:

$$Y^2 \frac{\Psi}{(x-x_s)Y} \sim \frac{\Psi}{Y^3},$$

which, being solved for Y, shows that the thickness of the viscous region $2a$ decreases, as the point $x = x_s$ is approached, according to

$$Y = O\big[(-s)^{1/4}\big]. \tag{5.3.13}$$

Guided by (5.3.10) and (5.3.13), we seek the asymptotic expansion of $\Psi(x,Y)$ in region $2a$ in the form

$$\Psi(x,Y) = (-s)^{3/4}\frac{1}{6}\lambda_0\eta^3 + (-s)^\alpha f_1(\eta)$$
$$+ (-s)^{2\alpha-3/4}f_2(\eta) + \cdots \quad \text{as} \quad s \to 0-. \tag{5.3.14}$$

The first term in (5.3.14) is obtained by expressing (5.3.10) in terms of a new independent variable

$$\eta = \frac{Y}{(-s)^{1/4}}, \tag{5.3.15}$$

which, according to (5.3.13), remains an order one quantity in the viscous region. The second term represents an eigenfunction with an unknown eigenvalue α. The third term is a forced term, which arises in the expansion (5.3.14) due to the nonlinearity of the boundary-layer equation (5.3.4a).

By substituting (5.3.14) together with (5.3.15) into (5.3.4a), and setting $s \to 0-$, we find that function $f_1(\eta)$ satisfies the following ordinary differential equation:

$$f_1''' - \frac{1}{8}\lambda_0\eta^3 f_1'' + \frac{1}{2}\lambda_0\left(\alpha + \frac{1}{4}\right)\eta^2 f_1' - \lambda_0\alpha\eta f_1 = 0. \tag{5.3.16}$$

On the aerofoil surface it satisfies the conditions

$$f_1(0) = f_1'(0) = 0, \tag{5.3.17}$$

which are deduced by substituting (5.3.14), (5.3.15) into the no-slip conditions (5.3.4c).

Equation (5.3.16) is linear and homogeneous. Three complementary solutions to this equation, $f_{11}(\eta)$, $f_{12}(\eta)$, and $f_{13}(\eta)$, can be chosen such that

$$\begin{aligned}
f_{11}(0) &= 1, & f_{11}'(0) &= 0, & f_{11}''(0) &= 0, \\
f_{12}(0) &= 0, & f_{12}'(0) &= 1, & f_{12}''(0) &= 0, \\
f_{13}(0) &= 0, & f_{13}'(0) &= 0, & f_{13}''(0) &= 1.
\end{aligned} \tag{5.3.18}$$

The first two solutions do not satisfy the boundary conditions (5.3.17) and must be rejected. As far as the third solution is concerned, it appears to be $f_{13} = \frac{1}{2}\eta^2$, which is easily verified by direct substitution into (5.3.16). Thus, we can conclude that a non-trivial solution of equation (5.3.16) with boundary conditions (5.3.17) exists for all α, and may be written in the form

$$f_1(\eta) = \frac{1}{2}a_0\eta^2, \tag{5.3.19}$$

where a_0 is an arbitrary constant.

To determine the eigenvalue α, we have to consider the third term of the expansion (5.3.14). It satisfies the equation

$$f_2''' - \frac{1}{8}\lambda_0\eta^3 f_2'' + \lambda_0\left(\alpha - \frac{1}{4}\right)\eta^2 f_2' - \lambda_0\left(2\alpha - \frac{3}{4}\right)\eta f_2 = \frac{1}{4}(1 - 2\alpha)a_0^2\eta^2, \tag{5.3.20}$$

which also has to be solved with the no-slip conditions

$$f_2(0) = f_2'(0) = 0. \tag{5.3.21}$$

Equation (5.3.20) is not homogeneous; in addition to three complementary solutions $f_{21}(\eta)$, $f_{22}(\eta)$, and $f_{23}(\eta)$ of the homogeneous part of the equation, it requires a particular integral $f_{2p}(\eta)$. We can choose, for example,

$$f_{2p}(\eta) = \frac{a_0^2}{2\lambda_0}\eta,$$

and then the general solution to (5.3.20) is written as

$$f_2(\eta) = C_1 f_{21}(\eta) + C_2 f_{22}(\eta) + C_3 f_{23}(\eta) + \frac{a_0^2}{2\lambda_0}\eta. \tag{5.3.22}$$

Here C_1, C_2 and C_3 are arbitrary constants. Choosing again the complementary solutions such that

$$\begin{aligned} f_{21}(0) &= 1, & f_{21}'(0) &= 0, & f_{21}''(0) &= 0, \\ f_{22}(0) &= 0, & f_{22}'(0) &= 1, & f_{22}''(0) &= 0, \\ f_{23}(0) &= 0, & f_{23}'(0) &= 0, & f_{23}''(0) &= 1, \end{aligned}$$

and applying the no-slip conditions (5.3.21), we have

$$C_1 = 0, \qquad C_2 = -\frac{a_0^2}{2\lambda_0}.$$

Taking further into account that $f_{23}(\eta) = \frac{1}{2}\eta^2$, we can express (5.3.22) in the form

$$f_2(\eta) = \frac{a_0^2}{2\lambda_0}(\eta - f_{22}) + \frac{1}{2}b_0\eta^2, \tag{5.3.23}$$

with constant $b_0 = C_3$ remaining arbitrary.

Now, our task will be to determine the behaviour of function $f_{22}(\eta)$. This function satisfies the homogenous version of equation (5.3.20), namely,

$$f_{22}''' = \frac{1}{8}\lambda_0\eta^3 f_{22}'' - \lambda_0\left(\alpha - \frac{1}{4}\right)\eta^2 f_{22}' + \lambda_0\left(2\alpha - \frac{3}{4}\right)\eta f_{22}.$$
(5.3.24)

It should be solved with the initial conditions

$$f_{22}(0) = 0, \qquad f_{22}'(0) = 1, \qquad f_{22}''(0) = 0.$$
(5.3.25)

Let us express $f_{22}(\eta)$ in the form of the power series. In view of (5.3.25), the first term of the series should be written as

$$f_{22}(\eta) = \eta + \cdots .$$
(5.3.26)

Using (5.3.26) on the right-hand side of (5.3.24), we find

$$f_{22}''' = \lambda_0\left(\alpha - \frac{1}{2}\right)\eta^2 + \cdots .$$

The integration of this equation with conditions (5.3.25) yields the second term in (5.3.26):

$$f_{22}(\eta) = \eta + \frac{\lambda_0}{5!}(2\alpha - 1)\eta^5 + \cdots .$$

This procedure can be repeated, which leads to a conclusion that the power series for $f_{22}(\eta)$ has the form

$$f_{22}(\eta) = \sum_{n=0}^{\infty} c_n\eta^{4n+1},$$
(5.3.27)

where $c_0 = 1$.

A recurrence equation for the coefficients c_n of the series is obtained by substituting $c_n\eta^{4n+1}$ into the right-hand side of equation (5.3.24) and $c_{n+1}\eta^{4n+5}$ into its left-hand side. We find that

$$c_{n+1} = \frac{\lambda_0}{32}\frac{\left[n - (2\alpha - 1)\right]\left(n - \frac{1}{4}\right)}{\left(n + \frac{3}{4}\right)\left(n + \frac{5}{4}\right)(n + 1)}c_n.$$

Now, using mathematical induction, it can be shown that

$$c_n = -\left(\frac{\lambda_0}{32}\right)^n \frac{(1 - 2\alpha)_n}{(5/4)_n\,(4n - 1)\,n!}.$$

Here, $(a)_n$ denotes a quantity defined as

$$(a)_n = a(a + 1)(a + 2)\ldots(a + n - 1), \qquad (a)_0 = 1.$$
(5.3.28)

It may be expressed in terms of the Euler gamma function,[4]

$$(a)_n = \frac{\Gamma(a + n)}{\Gamma(a)}.$$

[4]Here we use a well-known property of the gamma function, $z\Gamma(z) = \Gamma(z + 1)$.

Consequently,

$$c_n = -\frac{\Gamma(5/4)}{\Gamma(1-2\alpha)} \left(\frac{\lambda_0}{32}\right)^n \frac{\Gamma(n+1-2\alpha)}{\Gamma(n+5/4)(4n-1)\,n!}. \tag{5.3.29}$$

It remains to substitute (5.3.29) back into (5.3.27), and we will have

$$f_{22}(\eta) = -\frac{\Gamma(5/4)}{\Gamma(1-2\alpha)} \sum_{n=0}^{\infty} \left(\frac{\lambda_0}{32}\right)^n \frac{\Gamma(n+1-2\alpha)}{\Gamma(n+5/4)(4n-1)\,n!}\, \eta^{4n+1}. \tag{5.3.30}$$

Now, our task will be to determine the asymptotic behaviour of $f_{22}(\eta)$ as $\eta \to \infty$. For this purpose, we express $f_{22}(\eta)$ in terms of Kummer's function $M(a,b,z)$, whose properties are well known.[5] Remember that Kummer's function is the solution of the confluent hypergeometric equation

$$z\frac{d^2w}{dz^2} + (b-z)\frac{dw}{dz} - aw = 0,$$

that is regular at point $z = 0$. In fact, $M(a,b,z)$ may be represented by the Taylor series

$$M(a,b,z) = 1 + \frac{a}{b}z + \frac{(a)_2}{(b)_2 2!}z^2 + \cdots + \frac{(a)_n}{(b)_n n!}z^n + \cdots$$

$$= \frac{\Gamma(b)}{\Gamma(a)} \sum_{n=0}^{\infty} \frac{\Gamma(a+n)}{\Gamma(b+n)\,n!} z^n, \tag{5.3.31}$$

which is convergent at any finite point z in the complex plane. We can varify that the series (5.3.31) for Kummer's function $M(a,b,z)$ may be converted into the series (5.3.30) for function $f_{22}(\eta)$ with the help of the following integral transformation:

$$f_{22}(\eta) = \eta - \eta^2 \int_0^{\eta} \xi^{-2} \left[M\left(1-2\alpha, \frac{5}{4}, \frac{\lambda_0}{32}\xi^4\right) - 1 \right] d\xi. \tag{5.3.32}$$

Generally, Kummer's function grows exponentially

$$M(a,b,z) = \frac{\Gamma(b)}{\Gamma(a)} e^z z^{a-b} + \cdots \tag{5.3.33}$$

as z tends to infinity along a ray that lies in the right half ($\Re\{z\} > 0$) of the complex plane z. However, function $f_{22}(\eta)$ should not be exponentially large at the outer edge of region 2a. An exponential growth of $f_{22}(\eta)$ would preclude the matching of solutions in region 2a with the solution in the main part of the boundary layer.[6] This contradiction is resolved by the fact that equation (5.3.33) cannot be used for

$$a = -m, \qquad m = 0,\,1,\,2\ldots, \tag{5.3.34}$$

when $\Gamma(a) = 0$. If a assumes one of the values in (5.3.34), then $(a)_{m+1}$ and all the subsequent members of the sequence (5.3.28) vanish, reducing the Taylor series (5.3.31)

---

[5] See, for example, Abramowitz and Stegun (1965).
[6] See Section 1.4.2 in Part 2 of this book series.

to a polynomial of degree m. Setting $a = 1 - 2\alpha$ in (5.3.34), leads to a conclusion that the sought eigenvalues are

$$\alpha = \frac{m+1}{2}, \qquad m = 0,\ 1,\ 2\dots. \tag{5.3.35}$$

5.3.2 Goldstein's singularity

Let us now return to the expansion (5.3.14). As any other asymptotic expansion, it should obey the rule that each subsequent term in (5.3.14) should be smaller than the previous one. This requirement is satisfied if $\alpha > 3/4$. Consequently, the first eigenvalue is

$$\alpha = 1.$$

The coefficient a_0 in the eigenfunction (5.3.19) depends on the distribution of the velocity $U_e(x)$ at the outer edge of the boundary layer on the entire interval $x \in [0, x_s]$. We will see that through special adjustment of $U_e(x)$ the coefficient a_0 may be made zero. However, in general, $a_0 \neq 0$, and the solution (5.3.14), (5.3.19), (5.3.23), (5.3.32) in region $2a$ assumes the form

$$\Psi(x, Y) = (-s)^{3/4} \frac{1}{6}\lambda_0 \eta^3 + (-s) f_1(\eta) + (-s)^{5/4} f_2(\eta) + \cdots, \tag{5.3.36}$$

where

$$f_1(\eta) = \frac{1}{2}a_0 \eta^2, \qquad f_2(\eta) = \frac{1}{2}b_0 \eta^2 - \frac{a_0^2}{240}\eta^5. \tag{5.3.37}$$

In addition to the terms shown, the expansion (5.3.36) contains the sum of an infinite number of successive eigenfunctions, and also includes additional terms produced by the higher-order terms in the expansion of the pressure gradient (5.3.7). All of these, however, are small compared to $(-s)^{5/4} f_2(\eta)$, and can be disregarded.

The solution (5.3.36), (5.3.37) does not satisfy the boundary condition (5.3.4d) at the outer edge of the boundary layer. Therefore, in addition to viscous region $2a$, it is necessary to consider the main part of the boundary layer (shown as region $2b$ in Figure 5.5), where the asymptotic analysis of the boundary-layer equations (5.3.4) is based on the limit

$$Y = O(1), \qquad s = x - x_s \to 0-.$$

The form of the asymptotic expansion of the stream function $\Psi(x, Y)$ in this region $2b$ may be determined with the help of the following standard procedure based on the

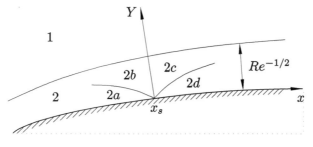

Fig. 5.5: Asymptotic regions' layout near the point of zero skin friction.

principle of matching of asymptotic expansions. If we substitute (5.3.37) into (5.3.36) and change the variables $\eta = Y/(-s)^{1/4}$, then we find that

$$\Psi(x, Y) = \left\{ \frac{1}{6}\lambda_0 Y^3 - \frac{a_0^2}{240}Y^5 + \cdots \right\} + (-s)^{1/2}\left\{ \frac{1}{2}a_0 Y^2 + \cdots \right\} + O[(-s)^{3/4}].$$

This suggests that the solution in region $2b$ should be sought in the form

$$\Psi(x, Y) = \Psi_{00}(Y) + (-s)^{1/2}\Psi_{01}(Y) + \cdots \quad \text{as} \quad s \to 0-, \qquad (5.3.38)$$

where functions $\Psi_{00}(Y)$ and $\Psi_{01}(Y)$ are such that

$$\left. \begin{aligned} \Psi_{00}(Y) &= \frac{1}{6}\lambda_0 Y^3 - \frac{a_0^2}{240}Y^5 + \cdots, \\ \Psi_{01}(Y) &= \frac{1}{2}a_0 Y^2 + \cdots \end{aligned} \right\} \quad \text{as} \quad Y \to 0. \qquad (5.3.39)$$

With (5.3.38), the four terms in the boundary-layer equation (5.3.4a) are calculated as

$$\frac{\partial \Psi}{\partial Y}\frac{\partial^2 \Psi}{\partial x \partial Y} = -\frac{1}{2}(-s)^{-1/2}\Psi'_{00}\Psi'_{01} + \cdots, \qquad \frac{\partial \Psi}{\partial x}\frac{\partial^2 \Psi}{\partial Y^2} = -\frac{1}{2}(-s)^{-1/2}\Psi_{01}\Psi''_{00} + \cdots,$$

$$\frac{dp_e}{dx} = \lambda_0 + \cdots, \qquad \frac{\partial^3 \Psi}{\partial Y^3} = \Psi'''_{00} + \cdots.$$

Clearly, the pressure gradient and the viscous term can be disregarded as $s \to 0-$, and equation (5.3.4a) reduces to

$$\Psi'_{00}\Psi'_{01} - \Psi_{01}\Psi''_{00} = 0.$$

The above equation may be written in the form

$$\left(\frac{\Psi_{01}}{\Psi'_{00}} \right)' = 0. \qquad (5.3.40)$$

Integrating (5.3.40) with initial conditions (5.3.39), we have

$$\Psi_{01} = \frac{a_0}{\lambda_0}\Psi'_{00}(Y). \qquad (5.3.41)$$

Here, $\Psi'_{00}(Y)$ is the velocity profile in the boundary layer at the point of zero skin friction, $x = x_s$. It is obtained by a numerical solution of the boundary-layer equation (5.3.4a) from the stagnation point $x = 0$ to the point $x = x_s$. The substitution of (5.3.41) back into (5.3.38) renders the solution in region $2b$ in the form

$$\Psi(x, Y) = \Psi_{00}(Y) + (-s)^{1/2}\frac{a_0}{\lambda_0}\Psi'_{00}(Y) + \cdots \quad \text{as} \quad s \to 0-. \qquad (5.3.42)$$

Now, we can calculate the velocity components (V_τ, V_n) in region $2b$. This is done by substituting (5.3.42) into (5.3.3) and then into (5.3.2). We have

$$V_\tau = \Psi'_{00}(Y) + \cdots, \qquad V_n = Re^{-1/2}(-s)^{-1/2}\frac{a_0}{2\lambda_0}\Psi'_{00}(Y) + \cdots.$$

We see that V_n develops a singularity as $x \to x_s - 0$. Singular behaviour is also displayed by other fluid-dynamic functions. Let us, for example, calculate the skin friction. For this purpose, we have to use the solution (5.3.36), (5.3.37) in region 2*a*. The substitution of (5.3.36), (5.3.37) together with (5.3.15) into (5.3.6) yields

$$\tau_w = (-s)^{1/2}a_0 + O\left[(-s)^{3/4}\right] \quad \text{as} \quad s \to 0-.$$

Since the skin friction τ_w is always positive upstream of the point $x = x_s$ (see Figure 5.4), we can claim that $a_0 > 0$.

The square root form of the singularity at the position of zero skin friction was first predicted by Landau and Lifshitz (1944). Four years later, Goldstein (1948) confirmed these predictions based on a more rigorous analysis of the boundary-layer equations, hence the name *Goldstein's singularity*. Goldstein also demonstrated that if the flow develops the singularity, then the solution cannot be extended downstream of the point of zero skin friction. Indeed, let us try to construct the solution behind $x = x_s$. When performing this task we need to keep in mind that the solution to the boundary-layer equation (5.3.4a) is sought in the class of continuous functions. This means that the leading-order term

$$\Psi(x, Y) = \Psi_{00}(Y) + \cdots \quad \text{as} \quad s \to 0+, \quad Y = O(1) \tag{5.3.43}$$

of the asymptotic expansion of the stream function $\Psi(x, Y)$ in region 2*c* (see Figure 5.5) should coincide with the leading-order term in the solution (5.3.38) for region 2*b*. It then follows from (5.3.39) that at the bottom of region 2*c*

$$\frac{\partial \Psi}{\partial Y} = \frac{1}{2}\lambda_0 Y^2 + \cdots. \tag{5.3.44}$$

By using (5.3.44) in (5.3.11), we can see that the thickness of the viscous region 2*d* (see Figure 5.5) may be estimated as

$$Y \sim s^{1/4}.$$

Thus, the asymptotic analysis of the boundary-layer equation (5.3.4a) in region 2*d* should be performed using the limit

$$s = x - x_s \to 0+, \qquad \xi = \frac{Y}{s^{1/4}} = O(1).$$

We shall seek the stream function in region 2*d* in the form of the asymptotic expansion

$$\Psi(x, Y) = s^{3/4}\frac{1}{6}\lambda_0\xi^3 + s^{\mu}\hat{f}_1(\xi) + s^{2\mu - 3/4}\hat{f}_2(\xi) + \cdots, \tag{5.3.45}$$

with the exponent μ unknown in advance. Functions $\hat{f}_1(\xi)$ and $\hat{f}_2(\xi)$ are determined in the same way as the functions $f_1(\eta)$ and $f_2(\eta)$ in the solution (5.3.14) for region 2*a*. We have

$$\hat{f}_1 = \frac{1}{2}\hat{a}_0\xi^2, \tag{5.3.46}$$

$$\hat{f}_2 = \frac{\hat{a}_0^2}{2\lambda_0}\xi^2 \int\limits_0^{\xi} \zeta^{-2}\left[M\left(1 - 2\mu, \frac{5}{4}, -\frac{\lambda_0}{32}\zeta^4\right) - 1\right]d\zeta + \frac{1}{2}\hat{b}_0\xi^2, \tag{5.3.47}$$

where $\hat{a}_0$ and $\hat{b}_0$ are arbitrary constants.

It remains to perform the matching of the solutions in regions 2c and 2d. For this purpose, the asymptotic behaviour of function $\hat{f}_2(\xi)$ at the outer edge of region 2d should be determined. It is known that the asymptotic behaviour of Kummer's function $M(a, b, z)$ depends on a choice of the ray along which z tends to infinity in the complex z-plane. If the ray lies in the left half-plane ($\Re\{z\} < 0$), then

$$M(a, b, z) = \frac{\Gamma(b)}{\Gamma(b - a)}(-z)^{-a}\left[1 + O\left(\frac{1}{z}\right)\right]. \tag{5.3.48}$$

By using (5.3.48) in (5.3.47), we find that

$$\hat{f}_2 = \frac{\hat{a}_0^2}{2\lambda_0}\frac{\Gamma(5/4)}{\Gamma(1/4 + 2\mu)}\left(\frac{\lambda_0}{32}\right)^{2\mu-1}\frac{\xi^{8\mu-3}}{8\mu - 5} + \cdots \quad \text{as} \quad \xi \to \infty. \tag{5.3.49}$$

Now, we can substitute (5.3.46) together with (5.3.49) into (5.3.45), and change the variable $\xi = Y/s^{1/4}$. As a result, we find that at the outer edge of region 2d

$$\Psi = \frac{1}{6}\lambda_0 Y^3 + \frac{\hat{a}_0^2}{2\lambda_0}\frac{\Gamma(5/4)}{\Gamma(1/4 + 2\mu)}\left(\frac{\lambda_0}{32}\right)^{2\mu-1}\frac{Y^{8\mu-3}}{8\mu - 5} + \cdots. \tag{5.3.50}$$

The solution in region 2c is given by (5.3.43). Substituting the first of equations (5.3.39) into (5.3.43), we see that at the 'bottom' of region 2c

$$\Psi = \frac{1}{6}\lambda_0 Y^3 - \frac{a_0^2}{240}Y^5 + \cdots. \tag{5.3.51}$$

According to the principle of matching of asymptotic expansions, the right-hand sides in (5.3.50) and (5.3.51) should coincide with one another. This is only possible if $\mu = 1$ and[7]

$$\hat{a}_0^2 = -a_0^2. \tag{5.3.52}$$

Here, constant a_0 is found as a part of the solution to the boundary-layer equation (5.3.4a) in the region that extends from the front stagnation point $x = 0$ to the point of zero skin friction $x = x_s$. To find constant $\hat{a}_0$, we should use equation (5.3.52). As this equation does not allow for a real solution, we have to conclude that the solution of the boundary-layer equation (5.3.4a) cannot be extended beyond the line $x = x_s$.

5.3.3   Weak singularity

Remember that the behaviour of the boundary layer at the leading edge of the aerofoil depends on parameter k. The latter is defined by the angle of attack and the profile shape; see equation (5.2.5). If k exceeds the critical value $k_0 = 1.1575$, then the solution terminates at the point of zero skin friction, where Goldstein's singularity is encountered. If, on the other hand, k is smaller than k_0, then the skin friction stays positive and the solution exists for all $x > 0$. In the latter case, the skin friction $\tau_w(x)$

---

[7]To simplify the coefficient in the second term in (5.3.50) we used a well-known property of the gamma function: $\Gamma(1 + z) = z\Gamma(z)$.

develops a minimum whose value decreases as parameter k approaches k_0; see Figure 5.4. Finally, when k reaches its critical value k_0, the minimal skin friction becomes zero, which happens at point $x_0 = 8.265$.

Let us denote the solution of the boundary-layer problem (5.3.4) for $k = k_0$ by $\Psi_0(x, Y)$. The behaviour of this solution near the point $x = x_0$ will now be studied. We start by noting that the procedure used in Section 5.3.2 for constructing the solution of the boundary-layer equation (5.3.4a) near the point of zero skin friction remains applicable for $\Psi_0(x, Y)$. However, we need to bear in mind that the function $\Psi_0(x, Y)$ may be thought of as the limiting solution of (5.3.4a) as $k \to k_0 - 0$. Since for all $k < k_0$ the solution continues through the line $x = x_0$, the same should be true for the limiting solution $\Psi_0(x, Y)$. In view of (5.3.52), the latter is only possible if the coefficient a_0 in the first eigenfunction (5.3.37) in expansion (5.3.36) is zero.

This conjecture is supported by the numerical solution of boundary-layer equation (5.3.4a). We have seen that for all $k > k_0$, Goldstein's singularity develops in the solution at the point of zero skin friction $x = x_s$. However, as the parameter k decreases, the point of zero friction x_s moves downstream, approaching x_0. Simultaneously, the singularity becomes weaker, that is, the coefficient a_0 decreases. At $k = k_0$ it turns zero, and the first eigenfunction in (5.3.36) disappears.

The second eigenvalue

$$\alpha = \frac{3}{2} \tag{5.3.53}$$

corresponds to $m = 2$ in (5.3.35). With (5.3.53), the solution in region $2a$ (see Figure 5.5) assumes the form

$$\Psi_0(x, Y) = (-s)^{3/4} \frac{1}{6} \lambda_0 \eta^3 + (-s)^{3/2} f_1(\eta)$$
$$+ (-s)^{7/4} F_1(\eta) + (-s)^{9/4} f_2(\eta) + \cdots \quad \text{as} \quad s \to 0-, \tag{5.3.54a}$$

where

$$\left. \begin{aligned} f_1 &= \frac{1}{2} a_0 \eta^2, \\ F_1 &= -\frac{1}{6} \lambda_1 \eta^3 + \frac{2\lambda_0 \lambda_1}{7!} \eta^7, \\ f_2 &= \frac{1}{2} b_0 \eta^2 - \frac{a_0^2}{5!} \eta^5 + \frac{\lambda_0 a_0^2}{8!} \eta^9. \end{aligned} \right\} \tag{5.3.54b}$$

To maintain uniformity in notations we still denote the coefficient in the eigenfunction by a_0 as before. However, it should be kept in mind that now this is the coefficient multiplying the second eigenfunction. The appearance of an additional term $(-s)^{7/4} F_1(\eta)$ in the expansion (5.3.54a) for $\Psi_0(x, Y)$ is related to the linear term, $\lambda_1 s$, in the Taylor expansion of the pressure gradient (5.3.7).

The solution in region $2b$ (see Figure 5.5) is expressed by the asymptotic expansion

$$\Psi_0(x, Y) = \Psi_{00}(Y) + (-s)\Psi_{01}(Y) + \cdots \quad \text{as} \quad s \to 0-. \tag{5.3.55}$$

By substituting (5.3.55) into the boundary-layer equation (5.3.4a), we find that

$$-\Psi_{00}' \Psi_{01}' + \Psi_{01} \Psi_{00}'' = -\lambda_0 + \Psi_{00}''',$$

or, equivalently,

$$\left(\frac{\Psi_{01}}{\Psi'_{00}}\right)' = \frac{\lambda_0 - \Psi'''_{00}}{(\Psi'_{00})^2}.$$

Integration of this equation yields

$$\Psi_{01}(Y) = \Psi'_{00}(Y)\left[C - \int\limits_0^Y \frac{\Psi'''_{00}(Y') - \lambda_0}{[\Psi'_{00}(Y')]^2}\, dY'\right]. \tag{5.3.56}$$

Matching the solution (5.3.55), (5.3.56) in region 2b with the solution (5.3.54) in region 2a shows, firstly, that constant C in (5.3.56) is

$$C = \frac{a_0}{\lambda_0}, \tag{5.3.57}$$

and, secondly, that

$$\Psi_{00}(Y) = \frac{1}{6}\lambda_0 Y^3 + \frac{2\lambda_0\lambda_1}{7!}Y^7 + \frac{\lambda_0 a_0^2}{8!}Y^9 + \cdots \quad \text{as} \quad Y \to 0. \tag{5.3.58}$$

This completes the construction of the solution before the point of zero skin friction, $x = x_0$. It is interesting to notice that substitution of (5.3.57) into (5.3.56) and then into (5.3.55) yields an asymptotic representation of $\Psi_0(x, Y)$,

$$\Psi_0(x, Y) = \Psi_{00}(Y) + s\Psi'_{00}(Y)\left[\int\limits_0^Y \frac{\Psi'''_{00}(Y') - \lambda_0}{[\Psi'_{00}(Y')]^2}\, dY' - \frac{a_0}{\lambda_0}\right] + \cdots, \tag{5.3.59}$$

which proves to be valid both in region 2b and region 2a; see Problem 1 in Exercises 13.

Now, we shall show that this solution can be continued downstream of point $x = x_0$. We start with region 2c, where $\Psi_0(x, Y)$ is sought in the form

$$\Psi_0(x, Y) = \Psi_{00}(Y) + s\widehat{\Psi}_{01}(Y) + \cdots \quad \text{as} \quad s \to 0+. \tag{5.3.60}$$

Function $\widehat{\Psi}_{01}(Y)$ is found in the same way as function $\Psi_{01}(Y)$ in (5.3.55). We have

$$\widehat{\Psi}_{01}(Y) = \Psi'_{00}(Y)\left[\widehat{C} + \int\limits_0^Y \frac{\Psi'''_{00}(Y') - \lambda_0}{[\Psi'_{00}(Y')]^2}\, dY'\right]. \tag{5.3.61}$$

In region 2d the solution is represented by the asymptotic expansion

$$\Psi_0(x, Y) = s^{3/4}\frac{1}{6}\lambda_0\xi^3 + s^{3/2}\hat{f}_1(\xi)$$
$$+ s^{7/4}\widehat{F}_1(\xi) + s^{9/4}\hat{f}_2(\xi) + \cdots \quad \text{as} \quad s \to 0+, \tag{5.3.62a}$$

with $\xi = Y/s^{1/4}$. Substituting (5.3.62a) into the boundary-layer equation (5.3.4a) and solving the resulting equations for $\hat{f}_1(\xi)$, $\widehat{F}_1(\xi)$, and $\hat{f}_2(\xi)$ with the no-slip conditions on the aerofoil surface (5.3.4c), we have

$$\hat{f}_1 = \frac{1}{2}\hat{a}_0\xi^2,$$

$$\widehat{F}_1 = \frac{1}{6}\lambda_1\xi^3 + \frac{2\lambda_0\lambda_1}{7!}\xi^7,$$

$$\hat{f}_2 = \frac{1}{2}\hat{b}_0\xi^2 + \frac{\hat{a}_0^2}{5!}\xi^5 + \frac{\lambda_0\hat{a}_0^2}{8!}\xi^9.$$

$$(5.3.62b)$$

It remains to match the asymptotic expansions (5.3.60) and (5.3.62) which represent Ψ_0 in regions $2c$ and $2d$, respectively. Remember that this is the procedure that was impossible to perform when dealing with the solution with Goldstein's singularity. The matching resulted in equation (5.3.52), which could not be solved for $\hat{a}_0$ in terms of real numbers. Now, instead of (5.3.52) we obtain

$$\hat{a}_0^2 = a_0^2. \qquad (5.3.63)$$

The matching also shows that constant $\widehat{C}$ in (5.3.61) is given by

$$\widehat{C} = \frac{\hat{a}_0}{\lambda_0}. \qquad (5.3.64)$$

The substitution of (5.3.64) back into (5.3.61) and then into (5.3.60) results in the equation

$$\Psi_0(x, Y) = \Psi_{00}(Y) + s\Psi_{00}'(Y)\left[\int_0^Y \frac{\Psi_{00}'''(Y') - \lambda_0}{\left[\Psi_{00}'(Y')\right]^2}\, dY' + \frac{\hat{a}_0}{\lambda_0}\right] + \cdots, \qquad (5.3.65)$$

which, similar to (5.3.59), may be used not only in the main part of the boundary layer (region $2c$), but also in the viscous sublayer (region $2d$).

Equation (5.3.63) shows that the solution of the boundary-layer problem (5.3.4), which is unique before the point of zero friction $x = x_0$, can be continued through this point in two ways. The first is given by

$$\hat{a}_0 = -a_0.$$

In this case, the solution (5.3.59) upstream of $x = x_0$ and the solution (5.3.65) downstream of this point may be expressed by a single equation

$$\Psi_0(x, Y) = \Psi_{00}(Y) + s\Psi_{00}'(Y)\left[\int_0^Y \frac{\Psi_{00}'''(Y') - \lambda_0}{\left[\Psi_{00}'(Y')\right]^2}\, dY' - \frac{a_0}{\lambda_0}\right] + O(s^2), \qquad (5.3.66)$$

which is valid in the entire vicinity of point $x = x_0$. It follows from (5.3.66) that the skin friction

$$\tau_w = \left.\frac{\partial^2\Psi_0}{\partial Y^2}\right|_{Y=0} = -a_0 s + O(s^2), \qquad (5.3.67)$$

changes sign at this point, which shows that a region of reverse flow forms downstream of $x = x_0$.

However, there exists a second branch of the solution with

$$\hat{a}_0 = a_0.$$

In this case, by combining (5.3.59) and (5.3.65) together, we have

$$\Psi_0 = \Psi_{00}(Y) + \Psi_{00}'(Y)\left[\frac{a_0}{\lambda_0}|s| + s\int_0^Y \frac{\Psi_{00}'''(Y') - \lambda_0}{\left[\Psi_{00}'(Y')\right]^2}\,dY'\right] + O(s^2). \qquad (5.3.68)$$

This solution has a singularity at $x = x_0$. In particular, let us calculate the angle $\vartheta = \arctan(V_n/V_\tau)$ made by the streamlines with the aerofoil surface. It follows from (5.3.2), (5.3.3), and (5.3.68) that

$$\vartheta = Re^{-1/2}\Theta, \qquad (5.3.69a)$$

where

$$\Theta = -\frac{\partial\Psi_0/\partial x}{\partial\Psi_0/\partial Y} = -\frac{a_0}{\lambda_0}\text{sign}(s) + \int_0^Y \frac{\lambda_0 - \Psi_{00}'''(Y')}{\left[\Psi_{00}'(Y')\right]^2}\,dY' + O(s). \qquad (5.3.69b)$$

We can see that Θ has a discontinuity at $x = x_0$. It further follows from (5.3.68) that the skin friction

$$\tau_w = a_0|s| + O(s^2) \quad \text{as} \quad s \to 0 \qquad (5.3.70)$$

has a minimum characteristic of the limiting solution of the boundary-layer equation (5.3.4a) as $k \to k_0 - 0$; see Figure 5.4.

As far as the smooth solution, given by (5.3.67), is concerned, we have encountered it in a number of examples when the boundary-layer separation proceeds through the viscous-inviscid interaction. In particular, Figure 2.18 shows a smooth behaviour of the flow near the separation point in the case of self-induced separation of the boundary layer in supersonic flow. The correspondent solution for subsonic flow is shown in Figure 2.29.

5.3.4   Formation of the singularity in the boundary layer

Let us now see what happens if the parameter k does not coincide with its critical value k_0, but the difference $\Delta k = k - k_0$ is small. In this case, the velocity at the outer edge of the boundary layer (5.2.6) may be represented by the Taylor expansion

$$U_e(x, k) = U_{e,0}(x) + \Delta k\, U_{e,1}(x) + O\left[(\Delta k)^2\right], \qquad (5.3.71)$$

where

$$U_{e,0}(x) = U_e(x, k_0), \qquad U_{e,1}(x) = \left.\frac{\partial U_e(x, k)}{\partial k}\right|_{k=k_0}.$$

Using the Bernoulli equation

$$\frac{1}{2}U_e^2 + p_e = \frac{1}{2},$$

we can see that the pressure in the boundary layer, $p_e(x)$, is also representable by the Taylor expansion

$$p_e(x) = p_0(x) + \Delta k\, p_1(x) + \cdots . \tag{5.3.72}$$

Here

$$p_0 = \frac{1}{2}\left(1 - U_{e,0}^2\right), \qquad p_1 = -U_{e,0}U_{e,1}.$$

Guided by (5.3.71) and (5.3.72), we seek the solution of the boundary-layer problem (5.3.4) in the form

$$\Psi = \Psi_0(x,Y) + \Delta k\,\Psi_1(x,Y) + \cdots \quad \text{as} \quad \Delta k \to 0. \tag{5.3.73}$$

The leading-order term $\Psi_0(x,Y)$ has been studied in detail in Section 5.3.3. We shall now consider function $\Psi_1(x,Y)$. The substitution of (5.3.73) together with (5.3.71) and (5.3.72) into (5.3.4) yields

$$\frac{\partial\Psi_0}{\partial Y}\frac{\partial^2\Psi_1}{\partial x\partial Y} + \frac{\partial^2\Psi_0}{\partial x\partial Y}\frac{\partial\Psi_1}{\partial Y} - \frac{\partial\Psi_0}{\partial x}\frac{\partial^2\Psi_1}{\partial Y^2} - \frac{\partial^2\Psi_0}{\partial Y^2}\frac{\partial\Psi_1}{\partial x} = -\frac{dp_1}{dx} + \frac{\partial^3\Psi_1}{\partial Y^3}, \tag{5.3.74a}$$

$$\Psi_1 = 0 \qquad\qquad \text{at} \quad x = 0, \tag{5.3.74b}$$

$$\Psi_1 = \frac{\partial\Psi_1}{\partial Y} = 0 \quad \text{at} \quad Y = 0, \tag{5.3.74c}$$

$$\frac{\partial\Psi_1}{\partial Y} = U_{e,1}(x) \qquad \text{at} \quad Y = \infty. \tag{5.3.74d}$$

The solution of the boundary-value problem (5.3.74) near point $x = x_0$ may be constructed in the same way as it was for the leading-order term $\Psi_0(x,Y)$. We start with region $2a$ (see Figure 5.5), where the asymptotic expansion of function $\Psi_1(x,Y)$ is sought in the form

$$\Psi_1(x,Y) = (-s)^\beta g_1(\eta) + (-s)^{\beta+3/4} g_2(\eta) + \cdots \quad \text{as} \quad s \to 0-, \tag{5.3.75}$$

with

$$\eta = \frac{Y}{(-s)^{1/4}}. \tag{5.3.76}$$

The leading-order term in (5.3.75) represents an eigenfunction of the local problem; parameter β is the eigenvalue to be determined.

The substitution of (5.3.75) together with (5.3.54) into (5.3.74a) results in the following equation for $g_1(\eta)$:

$$g_1''' - \frac{1}{8}\lambda_0\eta^3 g_1'' + \frac{1}{2}\lambda_0\left(\beta + \frac{1}{4}\right)\eta^2 g_1' - \lambda_0\beta\eta g_1 = 0.$$

For any β, its solution, satisfying the no-slip conditions

$$g_1(0) = g_1'(0) = 0,$$

is written as

$$g_1 = \frac{1}{2} a_1 \eta^2,$$ (5.3.77)

where a_1 is an arbitrary constant. In order to find the eigenvalue β, we need to consider the second term in (5.3.75). Function $g_2(\eta)$ satisfies the equation

$$g_2''' - \frac{1}{8} \lambda_0 \eta^3 g_2'' + \frac{1}{2} \lambda_0 (\beta + 1) \eta^2 g_2' - \lambda_0 \left(\beta + \frac{3}{4} \right) \eta g_2 = - \left(\frac{\beta}{2} + \frac{1}{4} \right) a_0 a_1 \eta^2,$$ (5.3.78)

which should be solved with the no-slip conditions on the aerofoil surface:

$$g_2(0) = g_2'(0) = 0.$$

The solution to (5.3.78) may be constructed in the same way as it was for equation (5.3.20). We find that

$$g_2(\eta) = \frac{a_0 a_1}{\lambda_0} (\eta - g_{22}) + \frac{1}{2} b_1 \eta^2,$$ (5.3.79)

where function $g_{22}(\eta)$ may be represented by the power series

$$g_{22}(\eta) = - \sum_{n=0}^{\infty} \left(\frac{\lambda_0}{32} \right)^n \frac{(-\beta - 1/2)_n}{(5/4)_n \, (4n - 1) \, n!} \, \eta^{4n+1}.$$ (5.3.80)

Alternatively, it may be expressed in terms of the Kummer function:

$$g_{22}(\eta) = \eta - \eta^2 \int_0^\eta \xi^{-2} \left[M \left(-\beta - \frac{1}{2}, \frac{5}{4}, \frac{\lambda_0}{32} \xi^4 \right) - 1 \right] d\xi.$$ (5.3.81)

It follows from (5.3.81) that $g_{22}(\eta)$ grows exponentially as $\eta \to \infty$ for all values of β except

$$\beta = m - \frac{1}{2}, \qquad m = 0,\, 1,\, 2 \ldots.$$

Hence, the first eigenvalue is

$$\beta = -\frac{1}{2}.$$ (5.3.82)

With (5.3.82) all the coefficients in (5.3.80), except the first one ($n = 0$), are zeros, and we have

$$g_{22} = \eta.$$ (5.3.83)

By using (5.3.83) in (5.3.79), and substituting (5.3.79) together with (5.3.82) and (5.3.77) into (5.3.75), we arrive at a conclusion that in region $2a$ the solution of equation (5.3.74a) is written as

$$\Psi_1(x, Y) = (-s)^{-1/2} \frac{1}{2} a_1 \eta^2 + (-s)^{1/4} \frac{1}{2} b_1 \eta^2 + \cdots.$$ (5.3.84)

Now, let us turn to the main part of the boundary layer (region $2b$ in Figure 5.5). To predict the form of the solution in this region, we use the substitution of variables (5.3.76) to cast the solution (5.3.84) in region $2a$ in the form

$$\Psi_1(x, Y) = (-s)^{-1}\frac{1}{2}a_1 Y^2 + (-s)^{-1/4}\frac{1}{2}b_1 Y^2 + \cdots .$$

The above equation suggests that in region $2b$, where $Y = O(1)$, the asymptotic expansion of $\Psi_1(x, Y)$ should be written as

$$\Psi_1(x, Y) = (-s)^{-1}\Psi_{11}(Y) + O\left[(-s)^{-1/4}\right] \quad \text{as} \quad s \to 0-, \tag{5.3.85}$$

where function $\Psi_{11}(Y)$ is such that

$$\Psi_{11}(Y) = \frac{1}{2}a_1 Y^2 + \cdots \quad \text{as} \quad Y \to 0. \tag{5.3.86}$$

The substitution of (5.3.85) and (5.3.55) into (5.3.74a) results in the following equation for Ψ_{11}:

$$\Psi'_{00}\Psi'_{11} - \Psi''_{00}\Psi_{11} = 0.$$

The solution of this equation, which satisfies the boundary condition (5.3.86), is easily found to be

$$\Psi_{11} = \frac{a_1}{\lambda_0}\Psi'_{00}(Y). \tag{5.3.87}$$

Here, we have used the fact that, according to (5.3.58),

$$\Psi_{00}(Y) = \frac{1}{6}\lambda_0 Y^3 + \cdots \quad \text{as} \quad Y \to 0. \tag{5.3.88}$$

It remains to substitute (5.3.87) back into (5.3.85), and we can conclude that in region $2b$

$$\Psi_1(x, Y) = (-s)^{-1}\frac{a_1}{\lambda_0}\Psi'_{00}(Y) + O\left[(-s)^{-3/4}\right] \quad \text{as} \quad s \to 0-. \tag{5.3.89}$$

Similar to (5.3.59), equation (5.3.89) proves to be valid not only in region $2b$, but also in region $2a$. This statement is easily verified by substituting (5.3.88) into (5.3.89) and comparing the resulting expression with (5.3.84).

Let us now return to expansion (5.3.73) for the stream function Ψ. As is any other asymptotic expansion, it is expected to exhibit proper ordering of the terms, namely, each subsequent term is expected be smaller than the previous one. The assumption that (5.3.73) satisfies this requirement has been used when deriving equation (5.3.74a), and hence lies in the foundation of the entire procedure employed for the analysis of $\Psi_1(x, Y)$. At each point (x, Y) situated upstream of point $x = x_0$, this assumption does hold, thanks to the smallness of Δk. However, as $x \to x_0 - 0$, function $\Psi_1(x, Y)$ develops a singularity, which leads to a violation of the supposed relationship between the terms in (5.3.73). Consequently, we have to reexamine the vicinity of point $x = x_0$ separately.

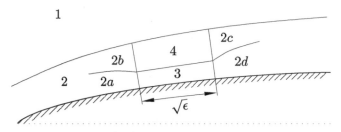

Fig. 5.6: The $O(\sqrt{\epsilon})$ vicinity of the point $x = x_0$.

To find the size of a new region that has to be introduced near $x = x_0$, we notice that the x-dependence of both $\Psi_0(x, Y)$ and $\Psi_1(x, Y)$ is determined by the eigenfunctions in the solutions (5.3.54), (5.3.84) for Ψ_0 and Ψ_1 in region 2a. By comparing the 'contributions' of the eigenfunctions in the asymptotic expansion (5.3.73):

$$(-s)^{3/2} \tfrac{1}{2} a_0 \eta^2 \sim \Delta k \, (-s)^{-1/2} a_1 \eta^2,$$

we find that the longitudinal extent of the new region is estimated as

$$|x - x_0| \sim \sqrt{\epsilon},$$

where $\epsilon = |\Delta k|$. The asymptotic structure of this region is shown in Figure 5.6. It is composed of two layers, the near-wall viscous sublayer (region 3) and the main part of the boundary layer (region 4).

Viscous sublayer (region 3)

Region 3 represents a continuation of region 2a into the $O(\sqrt{\epsilon})$ vicinity of the point $x = x_0$. Since $\eta = Y/(-s)^{1/4}$ is an order one quantity in region 2a, the thickness of region 3 may be estimated as

$$Y \sim |x - x_0|^{1/4} \sim \epsilon^{1/8}.$$

Consequently, the asymptotic analysis of the boundary-layer equation (5.3.4a) in region 3 should be based on the limit

$$x_* = \frac{x - x_0}{\epsilon^{1/2}} = O(1), \quad Y_* = \frac{Y}{\epsilon^{1/8}} = O(1), \quad \epsilon \to 0. \tag{5.3.90}$$

To predict the form of the asymptotic expansion of the stream function $\Psi(x, Y)$ in this region, we need to re-expand the solution in region 2a in terms of the variables (5.3.90). We start by substituting (5.3.54) and (5.3.84) into (5.3.73):

$$\Psi = (-s)^{3/4} \frac{1}{6} \lambda_0 \eta^3 + (-s)^{3/2} \frac{1}{2} a_0 \eta^2 + (-s)^{7/4} \left(-\frac{1}{6} \lambda_1 \eta^3 + \frac{2\lambda_0\lambda_1}{7!} \eta^7 \right)$$

$$+ (-s)^{9/4} \left(\frac{1}{2} b_0 \eta^2 - \frac{a_0^2}{5!} \eta^5 + \frac{\lambda_0 a_0^2}{8!} \eta^9 \right)$$

$$+ \Delta k \left[(-s)^{-1/2} \frac{1}{2} a_1 \eta^2 + (-s)^{1/4} \frac{1}{2} b_1 \eta^2 \right] + \cdots . \tag{5.3.91}$$

We then note that

$$(-s) = x_0 - x = \epsilon^{1/2}(-x_*), \qquad \eta = \frac{Y}{(-s)^{1/4}} = \frac{Y_*}{(-x_*)^{1/4}}, \tag{5.3.92}$$

and express Δk in the form

$$\Delta k = \epsilon \, \text{sign}(\Delta k). \tag{5.3.93}$$

The substitution of (5.3.92), (5.3.93) into (5.3.91) yields

$$\Psi = \epsilon^{3/8} \frac{1}{6} \lambda_0 Y_*^3$$
$$+ \epsilon^{6/8} \left[(-x_*) \frac{1}{2} a_0 Y_*^2 + (-x_*)^{-1} \text{sign}(\Delta k) \frac{1}{2} a_1 Y_*^2 \right]$$
$$+ \epsilon^{7/8} \left[\frac{1}{6} \lambda_1 x_* Y_*^3 + \frac{2 \lambda_0 \lambda_1}{7!} Y_*^7 \right]$$
$$+ \epsilon^{9/8} \left[\frac{a_0^2}{5!} x_* Y_*^5 + \frac{\lambda_0 a_0^2}{8!} Y_*^9 + (-x_*)^{7/4} \frac{1}{2} b_0 Y_*^2 \right.$$
$$\left. + (-x_*)^{-1/4} \text{sign}(\Delta k) \frac{1}{2} b_1 Y_*^2 \right] + \cdots . \tag{5.3.94}$$

This suggests that in region 3 the solution of the boundary layer equation (5.3.4a) should be sought in the form

$$\Psi = \epsilon^{3/8} \frac{1}{6} \lambda_0 Y_*^3 + \epsilon^{6/8} \Psi_1^*(x_*, Y_*) + \epsilon^{7/8} \left(\frac{1}{6} \lambda_1 x_* Y_*^3 + \frac{2 \lambda_0 \lambda_1}{7!} Y_*^7 \right)$$
$$+ \epsilon^{9/8} \Psi_2^*(x_*, Y_*) + \cdots . \tag{5.3.95}$$

It also follows from (5.3.94) that functions $\Psi_1^*(x_*, Y_*)$ and $\Psi_2^*(x_*, Y_*)$ should satisfy the following matching conditions with the solution in region 2a:

$$\Psi_1^* = \frac{1}{2} \left[a_0(-x_*) + \text{sign}(\Delta k) a_1 (-x_*)^{-1} \right] Y_*^2 + \cdots \qquad \text{as} \quad x_* \to -\infty, \tag{5.3.96}$$

and

$$\Psi_2^* = \frac{a_0^2}{5!} x_* Y_*^5 + \frac{\lambda_0 a_0^2}{8!} Y_*^9$$
$$+ \frac{1}{2} \left[b_0(-x_*)^{7/4} + \text{sign}(\Delta k) b_1 (-x_*)^{-1/4} \right] Y_*^2 + \cdots \qquad \text{as} \quad x_* \to -\infty. \tag{5.3.97}$$

Remember that we are dealing here with the classical formulation of the boundary-layer theory where the pressure distribution is known in advance. For k close to k_0 it is given by (5.3.72). Differentiating (5.3.72) with respect to x, we have

$$\frac{dp_e}{dx} = \frac{dp_0}{dx} + \Delta k \frac{dp_1}{dx} + \cdots . \tag{5.3.98}$$

In a small vicinity of point $x = x_0$, the first term dp_0/dx can be represented by the Taylor expansion (5.3.7). We further know that the second term dp_1/dx remains finite.

Hence, substituting (5.3.7) into (5.3.98) and expressing the results in terms of variables (5.3.90), we find that in region 3, the pressure gradient is given by

$$\frac{dp_e}{dx} = \lambda_0 + \epsilon^{1/2}\lambda_1 x_* + O(\epsilon). \tag{5.3.99}$$

Equations for functions Ψ_1^* and Ψ_2^* can now be deduced by substitution of (5.3.95) and (5.3.99) into the boundary-layer equation (5.3.4a). We find that the first of these is written as

$$\frac{1}{2}\lambda_0 Y_*^2 \frac{\partial^2 \Psi_1^*}{\partial x_* \partial Y_*} - \lambda_0 Y_* \frac{\partial \Psi_1^*}{\partial x_*} = \frac{\partial^3 \Psi_1^*}{\partial Y_*^3}. \tag{5.3.100a}$$

Equation (5.3.100a) should be solved with the no-slip conditions on the aerofoil surface:

$$\Psi_1^* = \frac{\partial \Psi_1^*}{\partial Y_*} = 0 \quad \text{at} \quad Y_* = 0. \tag{5.3.100b}$$

The solution to boundary-value problem (5.3.100) may be found using the method of Fourier transforms. With the Fourier transform $\breve{\Psi}_1$ of function $\Psi_1^*(x_*, Y_*)$ defined as

$$\breve{\Psi}_1(k, Y_*) = \int_{-\infty}^{\infty} \Psi_1^*(x_*, Y_*)e^{-ikx_*}\,dx_*, \tag{5.3.101}$$

equation (5.3.100a) and boundary conditions (5.3.100b) turn into

$$\frac{1}{2}ik\lambda_0 Y_*^2 \frac{d\breve{\Psi}_1}{dY_*} - ik\lambda_0 Y_* \breve{\Psi}_1 = \frac{d^3\breve{\Psi}_1}{dY_*^3}, \tag{5.3.102a}$$

$$\breve{\Psi}_1 = \frac{d\breve{\Psi}_1}{dY_*} = 0 \quad \text{at} \quad Y_* = 0. \tag{5.3.102b}$$

The general solution of equation (5.3.102a) is written as

$$\breve{\Psi}_1 = C_1\psi_{11} + C_2\psi_{12} + C_3\psi_{13}, \tag{5.3.103}$$

where ψ_{11}, ψ_{12}, and ψ_{13} are three complementary solutions of (5.3.102a). They may be chosen such that

$$\begin{aligned}
\psi_{11}(0) &= 1, & \psi_{11}'(0) &= 0, & \psi_{11}''(0) &= 0, \\
\psi_{12}(0) &= 0, & \psi_{12}'(0) &= 1, & \psi_{12}''(0) &= 0, \\
\psi_{13}(0) &= 0, & \psi_{13}'(0) &= 0, & \psi_{13}''(0) &= 1.
\end{aligned}$$

The first two solutions do not satisfy conditions (5.3.102b), and, therefore, we have to set $C_1 = C_2 = 0$ in (5.3.103). As far as the third solution is concerned, it appears to be

$$\psi_{13} = \frac{1}{2}Y_*^2,$$

which is verified by direct substitution into (5.3.102a).

Consequently, (5.3.103) reduces to

$$\breve{\Psi}_1 = \frac{1}{2} C_3(k) Y_*^2. \tag{5.3.104}$$

Applying the inverse Fourier transform to (5.3.104), we can conclude that[8]

$$\Psi_1^*(x_*, Y_*) = \frac{1}{2\pi} \int_{-\infty}^{\infty} \breve{\Psi}_1(k, Y_*) e^{ikx_*} dk = \frac{1}{2} A_*(x_*) Y_*^2, \tag{5.3.105}$$

where $A_*(x_*)$ is an arbitrary function of x_*, except it is known from (5.3.96) that

$$A_*(x_*) = a_0(-x_*) + \text{sign}(\Delta k) a_1 (-x_*)^{-1} + \cdots \quad \text{as} \quad x_* \to -\infty. \tag{5.3.106}$$

To find $A_*(x_*)$, we need to consider the next term in (5.3.95). Function Ψ_2^* satisfies the equation

$$\frac{1}{2} \lambda_0 Y_*^2 \frac{\partial^2 \Psi_2^*}{\partial x_* \partial Y_*} - \lambda_0 Y_* \frac{\partial \Psi_2^*}{\partial x_*} = \frac{\partial^3 \Psi_2^*}{\partial Y_*^3} - \frac{1}{2} A_* \frac{dA_*}{dx_*} Y_*^2. \tag{5.3.107a}$$

It should be solved subject to the no-slip conditions on the aerofoil surface:

$$\Psi_2^* = \frac{\partial \Psi_2^*}{\partial Y_*} = 0 \quad \text{at} \quad Y_* = 0. \tag{5.3.107b}$$

The solution to (5.3.107) may be also found with the help of the method of Fourier transforms. When using this method we have to keep in mind that the Fourier integral (5.3.101) converges if the function to which it is applied tends to zero as $|x_*| \to \infty$. It is easily seen from (5.3.97) that function Ψ_2^* does not satisfy this requirement. Keeping this in mind, we introduce a new unknown function Ψ_2 such that

$$\Psi_2^* = \Psi_2 + \frac{\lambda_0 a_0^2}{8!} Y_*^9 + \frac{a_0^2}{5!} x_* Y_*^5 + \frac{1}{2} B_*(x_*) Y_*^2 + G_*(x_*) Y_*. \tag{5.3.108}$$

Here, function $G_*(x_*)$ is defined as

$$G_*(x_*) = \frac{A_*^2 - a_0^2 x_*^2 - 2 a_0 a_1 \text{sign}(\Delta k)}{2\lambda_0}. \tag{5.3.109}$$

According to (5.3.106), it tends to zero as $x_* \to -\infty$.

Notice that the solution of the boundary-value problem (5.3.107) is not unique. Indeed, if Ψ_2^* is a solution of (5.3.107), then $\Psi_2^* + \frac{1}{2} B_*(x_*) Y_*^2$ is also a solution, with $B_*(x_*)$ being an arbitrary function. Still, to satisfy condition (5.3.97) and to ensure that Ψ_2 in (5.3.108) tends to zero as $x_* \to -\infty$, we have to impose the following restriction on $B_*(x_*)$:

$$B_*(x_*) = b_0(-x_*)^{7/4} + \text{sign}(\Delta k) b_1 (-x_*)^{-1/4} + \cdots \quad \text{as} \quad x_* \to -\infty.$$

---

[8]This result is verified by direct substitution of (5.3.105) into (5.3.100a) and (5.3.100b).

The transformation (5.3.108) renders the boundary-value problem (5.3.107) in the form

$$\frac{1}{2}\lambda_0 Y_*^2 \frac{\partial^2 \Psi_2}{\partial x_* \partial Y_*} - \lambda_0 Y_* \frac{\partial \Psi_2}{\partial x_*} = \frac{\partial^3 \Psi_2}{\partial Y_*^3},$$

$$\Psi_2 = 0, \quad \frac{\partial \Psi_2}{\partial Y_*} = -G_*(x_*) \quad \text{at} \quad Y_* = 0.$$

In terms of the Fourier transforms, it is written as

$$\frac{1}{2}ik\lambda_0 Y_*^2 \frac{d\breve{\Psi}_2}{dY_*} - ik\lambda_0 Y_* \breve{\Psi}_2 = \frac{d^3 \breve{\Psi}_2}{dY_*^3}, \tag{5.3.110a}$$

$$\breve{\Psi}_2 = 0, \quad \frac{d\breve{\Psi}_2}{dY_*} = -\breve{G}(k) \quad \text{at} \quad Y_* = 0. \tag{5.3.110b}$$

Here, $\breve{\Psi}_2(k, Y_*)$ and $\breve{G}(k)$ are the Fourier transforms of functions Ψ_2 and G_*, respectively.

The general solution of equation (5.3.110a) may be expressed in the form

$$\breve{\Psi}_2 = C_1 \psi_{21} + C_2 \psi_{22} + C_3 \psi_{23}, \tag{5.3.111}$$

where ψ_{21}, ψ_{22}, and ψ_{23} are three complementary solutions of the equation (5.3.110a). We shall choose these using the initial conditions

$$\left.\begin{aligned}
\psi_{21}(0) = 1, \quad \psi_{21}'(0) = 0, \quad \psi_{21}''(0) = 0, \\
\psi_{22}(0) = 0, \quad \psi_{22}'(0) = 1, \quad \psi_{22}''(0) = 0, \\
\psi_{23}(0) = 0, \quad \psi_{23}'(0) = 0, \quad \psi_{23}''(0) = 1.
\end{aligned}\right\} \tag{5.3.112}$$

It then follows from (5.3.110b) that

$$C_1 = 0, \qquad C_2 = -\breve{G}(k). \tag{5.3.113}$$

Factor C_3 remains an arbitrary function of k, and the third complementary solution ψ_{23} is easily seen to be

$$\psi_{23} = \frac{1}{2}Y_*^2. \tag{5.3.114}$$

Substituting (5.3.113) and (5.3.114) back into (5.3.111), we have

$$\breve{\Psi}_2 = -\breve{G}(k)\psi_{22} + \frac{1}{2}C_3(k)Y_*^2. \tag{5.3.115}$$

Now, let us take a closer look at function ψ_{22}. The power series for this function may be obtained in the same way as it was with function $f_{22}(\eta)$ representing the second complementary solution of equation (5.3.20). We start with the initial conditions for ψ_{22} in (5.3.112). They show that the first term in the series is

$$\psi_{22} = Y_* + \cdots. \tag{5.3.116}$$

Using (5.3.116) on the left-hand side of (5.3.110a) yields

$$\frac{d^3\psi_{22}}{dY_*^3} = -\frac{1}{2}ik\lambda_0 Y_*^2.$$

Integration of this equation gives the second term,

$$\psi_{22} = Y_* - \frac{ik\lambda_0}{120}Y_*^5 + \cdots .$$

This procedure can be repeated for subsequent terms, which leads to a conclusion that the power series of ψ_{22} should be sought in the form

$$\psi_{22}(Y_*) = \sum_{n=0}^{\infty} c_n Y_*^{4n+1}. \tag{5.3.117}$$

Alternatively, we can write

$$\psi_{22}(Y_*) = Y_* + \sum_{n=0}^{\infty} c_{n+1} Y_*^{4n+5}. \tag{5.3.118}$$

By using (5.3.117) on the left-hand side of equation (5.3.110a), and (5.3.118) on the right-hand side, we find that the coefficients of the series (5.3.117) satisfy the following recurrence equation:

$$c_{n+1} = \frac{ik\lambda_0}{32}\frac{n-1/4}{(n+3/4)(n+5/4)(n+1)}c_n. \tag{5.3.119a}$$

It has to be solved starting with

$$c_0 = 1. \tag{5.3.119b}$$

By direct substitution, we can verify that the solution of (5.3.119) is

$$c_n = -\left(\frac{ik\lambda_0}{32}\right)^n \frac{\Gamma(5/4)}{\Gamma(n+5/4)(4n-1)n!}. \tag{5.3.120}$$

Substituting (5.3.120) back into (5.3.117), we have

$$\psi_{22}(Y_*) = -\Gamma(5/4)\sum_{n=0}^{\infty}\left(\frac{ik\lambda_0}{32}\right)^n \frac{Y_*^{4n+1}}{\Gamma(n+5/4)(4n-1)n!}. \tag{5.3.121}$$

To determine the behaviour of $\psi_{22}(Y_*)$ at large values of Y_*, we express the right-hand side of (5.3.121) in terms of the Bessel function. It is known (see, for example Abramowitz and Stegun, 1965) that the Bessel function of the first kind and order ν may be represented by the power series

$$J_\nu(z) = \sum_{n=0}^{\infty} \frac{(-1)^n}{\Gamma(n+\nu+1)\,n!}\left(\frac{z}{2}\right)^{2n+\nu}, \tag{5.3.122}$$

that converges in the entire complex plane z except $z = \infty$. It is also known that in the limit, $|z| \to \infty$,

$$J_\nu(z) = \sqrt{\frac{2}{\pi z}} \left[\cos\left(z - \frac{1}{2}\nu\pi - \frac{1}{4}\pi\right) + e^{|\Im(z)|} O\left(\frac{1}{z}\right) \right]. \tag{5.3.123}$$

To reduce the power series in (5.3.122) to the power series in (5.3.121), we need to make a number of 'adjustments'. We choose $\nu = 1/4$, and write the first term in the series (5.3.122) separately:

$$J_{1/4}(z) = \frac{1}{\Gamma(5/4)} \left(\frac{z}{2}\right)^{1/4} + \sum_{n=1}^{\infty} \frac{(-1)^n}{\Gamma(n+5/4)\, n!} \left(\frac{z}{2}\right)^{2n+1/4}.$$

We can now see that

$$\int_0^z \left(\frac{2}{z}\right)^{3/2} \left[\frac{J_{1/4}(z)}{(z/2)^{1/4}} - \frac{1}{\Gamma(5/4)}\right] dz = 4 \sum_{n=1}^{\infty} \frac{(-1)^n}{\Gamma(n+5/4)(4n-1)n!} \left(\frac{z}{2}\right)^{2n-1/2},$$

and, therefore,

$$\sum_{n=0}^{\infty} \frac{(-1)^n}{\Gamma(n+5/4)(4n-1)n!} \left(\frac{z}{2}\right)^{2n+1/2} =$$

$$= \frac{z}{8} \int_0^z \left(\frac{2}{z}\right)^{3/2} \left[\frac{J_{1/4}(z)}{(z/2)^{1/4}} - \frac{1}{\Gamma(5/4)}\right] dz - \frac{1}{\Gamma(5/4)} \left(\frac{z}{2}\right)^{1/2}. \tag{5.3.124}$$

To find the asymptotic behaviour of the integral on the right-hand side of (5.3.124) at large values of z, we substitute (5.3.123) into (5.3.124), and restrict our attention to the leading-order exponentially growing term in the integrand. Using the integration by parts with[9]

$$u = \left(\frac{2}{z}\right)^{9/4}, \qquad dv = \cos\left(z - \frac{3}{8}\pi\right) dz,$$

$$du = -\frac{9}{8}\left(\frac{2}{z}\right)^{13/4} dz, \qquad v = \sin\left(z - \frac{3}{8}\pi\right),$$

we arrive at the conclusion that

$$\sum_{n=0}^{\infty} \frac{(-1)^n}{\Gamma(n+5/4)(4n-1)n!} \left(\frac{z}{2}\right)^{2n+1/2}$$

$$= \frac{1}{4\sqrt{\pi}} \left(\frac{2}{z}\right)^{5/4} \sin\left(z - \frac{3}{8}\pi\right) + \cdots \quad \text{as} \quad |z| \to \infty.$$

---

[9] See Section 1.1.2 in Part 2 of this book series.

It remains to set $z = \frac{1}{2} i\Omega^{1/2} Y_*^2$, with $\Omega = \frac{1}{2} ik\lambda_0$, and we can see that function ψ_{22}, given by (5.3.121), grows exponentially as $Y_* \to \infty$:

$$\psi_{22}(Y_*) = -\frac{\Gamma(5/4)}{2\sqrt{\pi} e^{i\pi/4} \Omega^{1/4}} \left(\frac{4}{i\Omega^{1/2} Y_*^2}\right)^{5/4} \sin\left(\frac{1}{2} i\Omega^{1/2} Y_*^2 - \frac{3}{8}\pi\right) + \cdots . \quad (5.3.125)$$

Thus, the exponential growth of the solution (5.3.115) for $\check{\Psi}_2$ can be suppressed if, and only if

$$\check{G}(k) = 0. \quad (5.3.126)$$

By applying the inverse Fourier transform to (5.3.126), we have $G_*(x_*) = 0$. Remember that function $G_*(x*)$ is defined by equation (5.3.109), whence

$$A_*^2 - a_0^2 x_*^2 - 2a_0 a_1 \text{sign}(\Delta k) = 0. \quad (5.3.127)$$

The solution of quadratic equation (5.3.127), which satisfies condition (5.3.106), is written as

$$A_* = a_0 \sqrt{x_*^2 + 2\frac{a_1}{a_0} \text{sign}(\Delta k)}. \quad (5.3.128)$$

This completes the flow analysis in region 3. To summarize the results, we substitute (5.3.105) into (5.3.95). Disregarding the higher-order terms, we have the solution for the stream function in region 3 in the form

$$\Psi = \epsilon^{3/8} \frac{1}{6} \lambda_0 Y_*^3 + \epsilon^{6/8} \frac{1}{2} A_*(x_*) Y_*^2 + \cdots , \quad (5.3.129)$$

where $A_*(x_*)$ is given by (5.3.128).

Main part of the boundary layer (region 4)

Let us now turn to region 4 (see Figure 5.6). The asymptotic analysis of the boundary-layer equation (5.3.4a) in this region is based on the limit

$$x_* = \frac{x - x_0}{\epsilon^{1/2}} = O(1), \qquad Y = O(1), \qquad \epsilon \to 0. \quad (5.3.130)$$

The form of the asymptotic expansion of the stream function Ψ in region 4 may be predicted by re-expanding the solution for region 2b (see Figure 5.6) in terms of variables (5.3.130). We start with the substitution of (5.3.59) and (5.3.89) into (5.3.73), which renders the solution in region 2b in the form

$$\Psi = \Psi_{00}(Y) + \Psi'_{00}(Y)\left[s\int_0^Y \frac{\Psi'''_{00} - \lambda_0}{(\Psi'_{00})^2} dY' - \frac{a_0 s - \Delta k a_1(-s)^{-1}}{\lambda_0}\right] + \cdots . \quad (5.3.131)$$

Now, we substitute (5.3.93) into (5.3.131) and take into account that $s = x - x_0 = \epsilon^{1/2} x_*$. We have

$$\Psi = \Psi_{00}(Y) + \epsilon^{1/2}\left\{\Psi'_{00}(Y)\left[x_* \int_0^Y \frac{\Psi'''_{00} - \lambda_0}{\left(\Psi'_{00}\right)^2}\, dY'\right.\right.$$
$$\left.\left. + \frac{a_0(-x_*) + a_1\mathrm{sign}(\Delta k)(-x_*)^{-1}}{\lambda_0}\right]\right\} + \cdots .$$

This suggests that in region 4 the stream function $\Psi(x, Y)$ has to be sought in the form

$$\Psi(x, Y) = \Psi_{00}(Y) + \epsilon^{1/2}\widetilde{\Psi}_1(x_*, Y) + \cdots , \qquad (5.3.132)$$

where $\widetilde{\Psi}_1$ should satisfy the following condition of matching with the solution in region 2b:

$$\widetilde{\Psi}_1 = \Psi'_{00}(Y)\left[x_* \int_0^Y \frac{\Psi'''_{00} - \lambda_0}{\left(\Psi'_{00}\right)^2}\, dY' + \frac{A_*(x_*)}{\lambda_0}\right] + \cdots \quad \text{as} \quad x_* \to -\infty. \qquad (5.3.133)$$

Here, we have used the fact that function $A_*(x_*)$ is represented at large negative x_* by (5.3.106).

We also need to formulate the condition of matching with the solution (5.3.129) in region 3. This is done by expressing (5.3.129) in terms of the variables of region 4. Since the scaled tangential coordinate x_* is common for regions 3 and 4, we only need to remember that the transformation of the normal coordinate is given by (5.3.90):

$$Y_* = \frac{Y}{\epsilon^{1/8}}. \qquad (5.3.134)$$

The substitution of (5.3.134) into (5.3.129) yields

$$\Psi = \frac{1}{6}\lambda_0 Y^3 + \epsilon^{1/2}\frac{1}{2}A_*(x_*)Y^2 + \cdots . \qquad (5.3.135)$$

Now, we turn to the asymptotic expansion (5.3.132) of the stream function in region 4. At the 'bottom' of region 4, the leading-order term in (5.3.132) may be simplified with the help of (5.3.58), which renders (5.3.132) in the form

$$\Psi(x, Y) = \frac{1}{6}\lambda_0 Y^3 + \epsilon^{1/2}\widetilde{\Psi}_1(x_*, Y) + \cdots . \qquad (5.3.136)$$

It remains to compare (5.3.135) with (5.3.136), and we can conclude that the sought matching condition is written as

$$\widetilde{\Psi}_1(x_*, Y) = \frac{1}{2}A_*(x_*)Y^2 + \cdots \quad \text{as} \quad Y \to 0. \qquad (5.3.137)$$

The equation for function $\widetilde{\Psi}_1(x_*, Y)$ is deduced by substituting (5.3.132) together with (5.3.99) into the boundary-layer equation (5.3.4a). We have

$$\Psi'_{00}\frac{\partial^2 \widetilde{\Psi}_1}{\partial x_* \partial Y} - \Psi''_{00}\frac{\partial \widetilde{\Psi}_1}{\partial x_*} = -\lambda_0 + \Psi'''_{00}. \qquad (5.3.138)$$

To solve equation (5.3.138), we cast it in the form

$$\frac{\partial^2}{\partial x_* \partial Y}\left(\frac{\widetilde{\Psi}_1}{\Psi'_{00}}\right) = \frac{\Psi'''_{00} - \lambda_0}{(\Psi'_{00})^2}. \tag{5.3.139}$$

It follows from (5.3.137) and (5.3.58) that

$$\left.\frac{\widetilde{\Psi}_1}{\Psi'_{00}}\right|_{Y=0} = \frac{A_*(x_*)}{\lambda_0}.$$

Therefore, integrating (5.3.139) with respect to Y, we have

$$\frac{\partial}{\partial x_*}\left(\frac{\widetilde{\Psi}_1}{\Psi'_{00}}\right) = \int_0^Y \frac{\Psi'''_{00}(Y') - \lambda_0}{[\Psi'_{00}(Y')]^2}\, dY' + \frac{A'_*(x_*)}{\lambda_0}. \tag{5.3.140}$$

We can also integrate (5.3.140) with respect to x_*. When performing this task, condition (5.3.133) has to be used. We find that

$$\widetilde{\Psi}_1(x_*, Y_*) = \Psi'_{00}(Y)\left[x_* \int_0^Y \frac{\Psi'''_{00}(Y') - \lambda_0}{[\Psi'_{00}(Y')]^2}\, dY' + \frac{A_*(x_*)}{\lambda_0}\right], \tag{5.3.141}$$

where function $A_*(x_*)$ is given by (5.3.128).

Substituting (5.3.141) back into (5.3.132), we have the solution in region 4 in the form

$$\Psi(x, Y) = \Psi_{00}(Y) + \epsilon^{1/2}\Psi'_{00}(Y)\left[x_* \int_0^Y \frac{\Psi'''_{00} - \lambda_0}{(\Psi'_{00})^2}\, dY' + \frac{A_*(x_*)}{\lambda_0}\right] + \cdots. \tag{5.3.142}$$

Through making use of (5.3.58), we can verify that in region 3, where $Y = \epsilon^{1/2}Y_*$, equation (5.3.142) reduces to (5.3.129), which signifies that (5.3.142) can be used not only in region 4, but also in region 3.

When discussing the properties of the solution, it is convenient to substitute (5.3.128) into (5.3.142) and to return to the original variable $x = x_0 + \epsilon^{1/2}x_*$. This yields

$$\Psi(x, Y) = \Psi_{00}(Y) + \Psi'_{00}(Y)\left[(x - x_0) \int_0^Y \frac{\Psi'''_{00} - \lambda_0}{(\Psi'_{00})^2}\, dY'\right.$$

$$\left. + \frac{a_0}{\lambda_0}\sqrt{(x - x_0)^2 + 2\frac{a_1}{a_0}\Delta k}\right] + \cdots. \tag{5.3.143}$$

The domain of existence of the solution is determined by the sign of the argument of the square root in (5.3.143). If $a_1\Delta k > 0$, then the solution exists for all x. The skin friction

$$\tau_w = \left.\frac{\partial^2 \Psi}{\partial Y^2}\right|_{Y=0} = a_0\sqrt{(x - x_0)^2 + 2\frac{a_1}{a_0}\Delta k} \tag{5.3.144}$$

remains positive everywhere, and reaches a minimum at point $x = x_0$. The value of the minimum decreases as $\Delta k \to 0$ according to the rule

$$\tau_w^{\min} = \sqrt{2a_0 a_1 \Delta k}. \tag{5.3.145}$$

However, if $a_1 \Delta k < 0$, then the solution exists only up to the point of zero friction,

$$x_s = x_0 - \sqrt{2\frac{|a_1 \Delta k|}{a_0}}.$$

Downstream of this point, the argument of the square root becomes negative. On approach to point $x = x_s$ the solution develops Goldstein's singularity. Indeed, it follows from (5.3.144) that

$$\tau = \left(8a_0^3 |a_1 \Delta k|\right)^{1/4} \sqrt{x_s - x} + \cdots \quad \text{as} \quad x \to x_s - 0.$$

Notice that the singularity becomes progressively weaker as $\Delta k \to 0$.

According to the numerical calculations (see Figure 5.4), the first of the situations described is realized for $\Delta k = k - k_0 < 0$ and the second for $\Delta k > 0$. This means that the constant a_1 is negative. Remember that in the case when $\Delta k = 0$, the boundary-layer equation (5.3.4a) admits two solutions. One of them, given by (5.3.66), passes smoothly through the point $x = x_0$, while the second solution (5.3.68) develops a singularity at this point. By comparing (5.3.66) and (5.3.68) with (5.3.143), we can see that this is the singular solution (5.3.68) that represents the limiting solution of the boundary-layer equation as $k \to k_0 - 0$.

To determine constants a_0 and a_1, we need to compare the analytical solution with the results of numerical calculations presented in Figure 5.4. Firstly, the skin friction distribution in Figure 5.4(b) calculated for $k = k_0$ should exhibit near $x = x_0$ the behaviour predicted by equation (5.3.70). Using this fact, we found that in the flow around parabolic leading edge of the aerofoil, $a_0 = 0.0085$. Then constant a_1 was found using equation (5.3.145). It appeared to be $a_1 = -1.24$.

5.4   Viscous-Inviscid Interaction

Up to this point, the flow analysis has been carried out in the framework of the classical boundary-layer theory. We started with the inviscid region, where the solution of the Euler equations was constructed using the impermeability condition on the aerofoil surface, that is, the existence of the boundary layer was completely ignored. As a result, we found the velocity distribution (5.2.6) on the aerofoil surface, while the pressure was calculated using the Bernoulli equation (5.3.5). Then, as a second step of Prandtl's hierarchical procedure, the boundary layer was analysed. For this purpose Prandtl's equation (5.3.4a) was used, with the pressure p_e and the velocity at the outer edge of the boundary layer U_e assumed unaffected by the presence of the boundary layer. This assumption is based on the observation that the boundary layer is only capable of causing an $O(Re^{-1/2})$ displacement of the streamlines from the aerofoil surface. Consequently, as long as the solution in the boundary layer remains regular, its influence on the pressure field is weak. It should be noted that the flow around a parabola at subcritical values of the parameter k represents a rare example of a flow for which the solution in the boundary layer remains regular along the entire surface, and the application of the hierarchical procedure does not lead to a contradiction.

The situation changes as the parameter k approaches the critical value k_0, and the solution in the boundary layer develops a singularity. Remember that when $k = k_0$, the streamlines in the boundary layer form a corner. The deflection angle θ at the corner may be calculated using equations (5.3.69). We have

$$\theta = Re^{-1/2}\left(\Theta\Big|_{x=x_0+0} - \Theta\Big|_{x=x_0-0}\right) = -Re^{-1/2}\frac{2a_0}{\lambda_0}. \tag{5.4.1}$$

It then follows from the thin aerofoil theory (see Problem 2 in Exercises 13) that the perturbations of the velocity components (V_τ, V_n) and of the pressure p, induced by the displacement effect of the boundary layer, are given by

$$V'_\tau = -Re^{-1/2}\frac{2a_0 U_0}{\pi\lambda_0}\ln\sqrt{(x-x_0)^2+y^2}+\cdots, \tag{5.4.2a}$$

$$V'_n = -Re^{-1/2}\frac{2a_0 U_0}{\pi\lambda_0}\left[\pi - \arctan\left(\frac{y}{x-x_0}\right)\right]+\cdots, \tag{5.4.2b}$$

$$p' = Re^{-1/2}\frac{2a_0 U_0^2}{\pi\lambda_0}\ln\sqrt{(x-x_0)^2+y^2}+\cdots. \tag{5.4.2c}$$

In particular, the pressure acting on the boundary layer is obtained by setting $y = 0$ in (5.4.2c):

$$p' = Re^{-1/2}\left[\frac{2a_0 U_0^2}{\pi\lambda_0}\ln|x-x_0|+O(1)\right] \quad \text{as} \quad x \to x_0. \tag{5.4.3}$$

Here, $U_0 = U_e(x_0)$, and λ_0 is the leading-order term in the pressure gradient (5.3.7). For an aerofoil with a parabolic nose, $U_0 = 1.286$ and $\lambda_0 = 0.024$.

It follows from (5.4.3) that the pressure gradient exhibits an unbounded growth:

$$\frac{dp'}{dx} = Re^{-1/2}\frac{2a_0 U_0^2}{\pi\lambda_0}\frac{1}{x-x_0}+\cdots. \tag{5.4.4}$$

Consequently, we can expect that there exists a vicinity of point $x = x_0$, where the pressure perturbations, induced by the displacement effect of the boundary layer, become large enough to start influencing the flow inside the boundary layer in the leading-order approximation, which then give rise to the process of viscous-inviscid interaction.

To determine the size of the interaction region, let us re-examine the procedure used to analyse the flow in the $O(\sqrt{\epsilon})$ vicinity of the singular point (see Figure 5.6). Remember that in regions 3 and 4 the dependence of the fluid-dynamic functions on the longitudinal coordinate x_* is expressed through function $A_*(x_*)$. The latter is determined by solving equation (5.3.107a). Thus, as soon as the induced pressure gradient becomes large enough 'to find its way' into this equation, the process of viscous-inviscid interaction begins.

The induced pressure gradient (5.4.4) can be compared with any term in equation (5.3.107a). Let us consider, for example, the first term on the right-hand side of

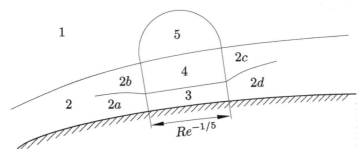

Fig. 5.7: The interaction region.

(5.3.107a). It is produced by the viscous term $\partial^3\Psi/\partial Y^3$ in equation (5.3.4a). In region 3 the stream function is represented by the asymptotic expansion (5.3.95). Being differentiated three times with respect to Y, it gives

$$\frac{\partial^3\Psi}{\partial Y^3} = \lambda_0 + \epsilon^{3/8}\frac{\partial^3\Psi_1^*}{\partial Y_*^3} + \epsilon^{1/2}\left(\lambda_1 x_* + \frac{\lambda_0\lambda_1}{12}Y_*^3\right) + \epsilon^{3/4}\frac{\partial^3\Psi_2^*}{\partial Y_*^3} + \cdots.$$

The fourth term on the right-hand side is a part of equation (5.3.107a). When comparing it with the induced pressure gradient (5.4.4),

$$\frac{Re^{-1/2}}{x - x_0} \sim \epsilon^{3/4}, \tag{5.4.5}$$

we have to take into account that in the region considered,

$$|x - x_0| \sim \sqrt{\epsilon}. \tag{5.4.6}$$

By solving (5.4.5) and (5.4.6) for ϵ and $|x - x_0|$, we find

$$\epsilon = |\Delta k| \sim Re^{-2/5}, \tag{5.4.7}$$

$$|x - x_0| \sim Re^{-1/5}. \tag{5.4.8}$$

Therefore, in what follows we assume that

$$k = k_0 + Re^{-2/5}k_1, \tag{5.4.9}$$

where constant k_1 remains an order one quantity as $Re \to \infty$. Our task is to study the flow in the interaction region which, apparently, has a three-tiered structure. In addition to regions 3 and 4, which lie inside the boundary layer (see Figure 5.6), we also need to introduce region 5 (see Figure 5.7). The latter is a part of the potential flow, and serves to convert the perturbations in the streamline shape into the perturbations of pressure.

Before proceeding further, we mention that the induced pressure gradient is only important in the interaction region. Outside this region the classical boundary-layer theory holds, which means that the solutions for regions 2a and 2b, constructed in Sections 5.3.3 and 5.3.4, are still valid.

5.4.1 Upper layer (region 5)

If we return to region 1, where X' and Y' are order one quantities (see Figure 5.3), and ignore the existence of the boundary layer, then the solution of the Euler equations for this region would be regular at point $x = x_0$. Keeping in mind that V_n is zero along the aerofoil contour, we can write the Taylor expansions for the velocity components and the pressure, centred at point $x = x_0$, as

$$V_\tau = U_0 + \left[a_1(x - x_0) + a_2 y \right]$$
$$+ \left[a_3(x - x_0)^2 + a_4(x - x_0)y + a_5 y^2 \right] + \cdots, \qquad (5.4.10a)$$

$$V_n = b_2 y + \left[b_4(x - x_0)y + b_5 y^2 \right] + \cdots, \qquad (5.4.10b)$$

$$p = P_{e0} + \left[c_1(x - x_0) + c_2 y \right]$$
$$+ \left[c_3(x - x_0)^2 + c_4(x - x_0)y + c_5 y^2 \right] + \cdots. \qquad (5.4.10c)$$

Remember that U_0 denotes the value of the tangential inviscid flow velocity at point $x = x_0$ on the aerofoil surface; P_{e0} is the corresponding value of the pressure. These are related to one another through the Bernoulli equation: $P_{e0} = \frac{1}{2}(1 - U_0^2)$.

Setting $y = 0$ in (5.4.10c) yields the pressure at the outer edge of the boundary layer:

$$p_e(x) = p\Big|_{y=0} = P_{e0} + c_1(x - x_0) + c_3(x - x_0)^2 + \cdots. \qquad (5.4.11)$$

By comparing (5.4.11) with (5.3.7), we can see that

$$c_1 = \lambda_0, \qquad c_3 = \frac{1}{2}\lambda_1. \qquad (5.4.12)$$

Equations (5.4.10) describe the flow immediately outside region 5. To account for the influence of the boundary layer on this flow, we need to superimpose (5.4.2) on (5.4.10). Keeping this in mind, we now turn to the flow analysis in region 5. The longitudinal extent of this region is given by (5.4.8). As any other potential flow of an incompressible fluid, the flow in region 5 is expected to be described by the Laplace equation. The principle of least degeneration, applied to this equation, requires the lateral size of region 5 to be comparable with its longitudinal size,

$$y \sim |x - x_0| \sim Re^{-1/5}.$$

This means that the asymptotic analysis of the Navier–Stokes equations (5.3.1) in region 5 has to be conducted based on the limit

$$x_* = \frac{x - x_0}{Re^{-1/5}} = O(1), \quad y_* = \frac{y}{Re^{-1/5}} = O(1), \quad Re \to \infty. \qquad (5.4.13)$$

To predict the form of the asymptotic expansions for the velocity components and pressure in region 5, we need to combine (5.4.10) and (5.4.2) and express the result in terms of the new variables (5.4.13). We have

$$V_\tau = U_0 + Re^{-1/5}(a_1 x_* + a_2 y_*) + Re^{-2/5}(a_3 x_*^2 + a_4 x_* y_* + a_5 y_*^2)$$
$$+ Re^{-1/2} \ln Re \frac{2a_0 U_0}{5\pi\lambda_0} - Re^{-1/2}\frac{2a_0 U_0}{\pi\lambda} \ln \sqrt{x_*^2 + y_*^2} + \cdots,$$

$$V_n = Re^{-1/5} b_2 y_* + Re^{-2/5}(b_4 x_* y_* + b_5 y_*^2)$$
$$- Re^{-1/2}\frac{2a_0 U_0}{\pi\lambda_0}\left[\pi - \arctan\left(\frac{y_*}{x_*}\right)\right] + \cdots,$$

$$p = P_{e0} + Re^{-1/5}(c_1 x_* + c_2 y_*) + Re^{-2/5}(c_3 x_*^2 + c_4 x_* y_* + c_5 y_*^2)$$
$$- Re^{-1/2} \ln Re \frac{2a_0 U_0^2}{5\pi\lambda_0} + Re^{-1/2}\frac{2a_0 U_0^2}{\pi\lambda} \ln \sqrt{x_*^2 + y_*^2} + \cdots.$$

This suggests that the solution in region 5 should be sought in the form

$$V_\tau = U_0 + Re^{-1/5}(a_1 x_* + a_2 y_*) + Re^{-2/5}(a_3 x_*^2 + a_4 x_* y_* + a_5 y_*^2)$$
$$+ Re^{-1/2} \ln Re \frac{2a_0 U_0}{5\pi\lambda_0} + Re^{-1/2}u_1^*(x_*, y_*) + \cdots, \qquad (5.4.14a)$$

$$V_n = Re^{-1/5} b_2 y_* + Re^{-2/5}(b_4 x_* y_* + b_5 y_*^2) + Re^{-1/2}v_1^*(x_*, y_*) + \cdots, \qquad (5.4.14b)$$

$$p = P_{e0} + Re^{-1/5}(c_1 x_* + c_2 y_*) + Re^{-2/5}(c_3 x_*^2 + c_4 x_* y_* + c_5 y_*^2)$$
$$- Re^{-1/2} \ln Re \frac{2a_0 U_0^2}{5\pi\lambda_0} + Re^{-1/2}p_1^*(x_*, y_*) + \cdots, \qquad (5.4.14c)$$

where functions u_1^*, v_1^*, and p_1^* satisfy the following matching conditions with the solution in region 1:

$$\left.\begin{aligned} u_1^* &= -\frac{2a_0 U_0}{\pi\lambda_0} \ln \sqrt{x_*^2 + y_*^2} + \cdots, \\ v_1^* &= -\frac{2a_0 U_0}{\pi\lambda_0}\left[\pi - \arctan\left(\frac{y_*}{x_*}\right)\right] + \cdots, \\ p_1^* &= \frac{2a_0 U_0^2}{\pi\lambda_0} \ln \sqrt{x_*^2 + y_*^2} + \cdots \end{aligned}\right\} \quad \text{as} \quad x_*^2 + y_*^2 \to \infty. \qquad (5.4.15)$$

The equations for these functions are deduced by substituting (5.4.14) into the Navier-Stokes equations (5.3.1), and working with the $O(Re^{-3/10})$ terms. We have

$$U_0 \frac{\partial u_1^*}{\partial x_*} = -\frac{\partial p_1^*}{\partial x_*}, \qquad U_0 \frac{\partial v_1^*}{\partial x_*} = -\frac{\partial p_1^*}{\partial y_*}, \qquad \frac{\partial u_1^*}{\partial x_*} + \frac{\partial v_1^*}{\partial y_*} = 0. \qquad (5.4.16)$$

The set of equations (5.4.16) is easily reduced to a single equation for function p_1^*. We start by eliminating $\partial u_1^*/\partial x_*$ from the first and third equations in (5.4.16). This results in

$$U_0 \frac{\partial v_1^*}{\partial y_*} = \frac{\partial p_1^*}{\partial x_*}, \qquad U_0 \frac{\partial v_1^*}{\partial x_*} = -\frac{\partial p_1^*}{\partial y_*}. \qquad (5.4.17)$$

Now, we eliminate v_1^* by cross-differentiating equations (5.4.17). We find that the pressure p_1^* satisfies the Laplace equation:

$$\frac{\partial^2 p_1^*}{\partial x_*^2} + \frac{\partial^2 p_1^*}{\partial y_*^2} = 0. \tag{5.4.18}$$

We will be using the method of Fourier transforms to solve equation (5.4.18). This method is applicable to functions that tend to zero as $|x_*| \to \infty$. It follows from (5.4.15) that p_1^* does not belong to this category. Therefore, we shall differentiate (5.4.18) with respect to x_*

$$\frac{\partial^2}{\partial x_*^2}\left(\frac{\partial p_1^*}{\partial x_*}\right) + \frac{\partial^2}{\partial y_*^2}\left(\frac{\partial p_1^*}{\partial x_*}\right) = 0, \tag{5.4.19}$$

and treat $\partial p_1^*/\partial x_*$ as the sought function. The 'far-field' boundary condition for this function is written as

$$\frac{\partial p_1^*}{\partial x_*} = \frac{2a_0 U_0^2}{\pi \lambda_0}\frac{x_*}{x_*^2 + y_*^2} \to 0 \quad \text{as} \quad x_*^2 + y_*^2 \to \infty. \tag{5.4.20}$$

Equation (5.4.19) also requires a boundary condition at $y_* = 0$. It will be formulated in the course of the flow analysis in region 4; see Figure 5.7.

5.4.2 Viscous sublayer (region 3)

When analysing the interactive flow regimes with the parameter k in the range (5.4.7), we can write the independent variables (5.3.90) in region 3 as

$$x_* = \frac{x - x_0}{Re^{-1/5}}, \qquad Y_* = \frac{y}{Re^{-11/20}}. \tag{5.4.21}$$

Remember that (x, y) are the 'body-fitted' coordinates introduced as shown in Figure 5.3.

To predict the form of the asymptotic expansion of the stream function in region 3, we again cast the solution in region $2a$ in terms of variables (5.4.21). When the flow was analysed in the frame of classical boundary-layer theory (see Section 5.3.4), this procedure led to equation (5.3.94). Now instead of (5.3.92) and (5.3.93), we have to use

$$(-s) = Re^{-1/5}(-x_*), \qquad \eta = \frac{Y_*}{(-x_*)^{1/4}}, \qquad \Delta k = k - k_0 = Re^{-2/5}k_1,$$

which, being substituted into (5.3.91), yield

$$\psi = Re^{-1/2}\Psi = Re^{-13/20}\frac{1}{6}\lambda_0 Y_*^3$$
$$+ Re^{-16/20}\left[(-x_*)\frac{1}{2}a_0 Y_*^2 + (-x_*)^{-1}\frac{1}{2}k_1 a_1 Y_*^2\right]$$
$$+ Re^{-17/20}\left(\frac{1}{6}\lambda_1 x_* Y_*^3 + \frac{2\lambda_0 \lambda_1}{7!}Y_*^7\right)$$
$$+ Re^{-19/20}\left[\frac{a_0^2}{5!}x_* Y_*^5 + \frac{\lambda_0 a_0^2}{8!}Y_*^9 + (-x_*)^{7/4}\frac{1}{2}b_0 Y_*^2\right.$$
$$\left. + (-x_*)^{-1/4}\frac{1}{2}k_1 b_1 Y_*^2\right] + \cdots .$$

This suggests that the asymptotic expansion of the stream function ψ in region 3 should be sought in the form

$$\psi = Re^{-13/20}\frac{1}{6}\lambda_0 Y_*^3 + Re^{-16/20}\Psi_1^*(x_*, Y_*)$$

$$+ Re^{-17/20}\left(\frac{1}{6}\lambda_1 x_* Y_*^3 + \frac{2\lambda_0\lambda_1}{7!}Y_*^7\right) + Re^{-19/20}\Psi_2^*(x_*, Y_*) + \cdots, \qquad (5.4.22)$$

where functions $\Psi_1^*(x_*, Y_*)$ and $\Psi_2^*(x_*, Y_*)$ are such that

$$\left.\begin{aligned} \Psi_1^* &= \frac{1}{2}\left[a_0(-x_*) + k_1 a_1(-x_*)^{-1}\right]Y_*^2 + \cdots, \\ \Psi_2^* &= \frac{a_0^2}{5!}x_* Y_*^5 + \frac{\lambda_0 a_0^2}{8!}Y_*^9 \\ &+ \frac{1}{2}\left[b_0(-x_*)^{7/4} + k_1 b_1(-x_*)^{-1/4}\right]Y_*^2 + \cdots \end{aligned}\right\} \quad \text{as} \quad x_* \to -\infty. \qquad (5.4.23)$$

We further represent the pressure in region 3 by the asymptotic expansion

$$p = P_{e0} + Re^{-1/5}\lambda_0 x_* + Re^{-2/5}\frac{1}{2}\lambda_1 x_*^2$$

$$- Re^{-1/2}\ln Re\frac{2a_0 U_0^2}{5\pi\lambda_0} + Re^{-1/2}P^*(x_*, Y_*) + \cdots, \qquad (5.4.24)$$

the form of which is obtained by setting $y_* = 0$ in (5.4.14c) and using (5.4.12).

We are now ready to formulate the equations of fluid motion in region 3. We start by substituting (5.4.22), (5.4.21) into (5.3.2). This leads to the following expressions of the velocity components in region 3,

$$V_\tau = Re^{-2/20}\frac{1}{2}\lambda_0 Y_*^2 + Re^{-5/20}\frac{\partial\Psi_1^*}{\partial Y_*}$$

$$+ Re^{-6/20}\left(\frac{1}{2}\lambda_1 x_* Y_*^2 + \frac{2\lambda_0\lambda_1}{6!}Y_*^6\right) + Re^{-8/20}\frac{\partial\Psi_2^*}{\partial Y_*} + \cdots, \qquad (5.4.25a)$$

$$V_n = -Re^{-12/20}\frac{\partial\Psi_1^*}{\partial x_*} - Re^{-13/20}\frac{1}{6}\lambda_1 Y_*^3 - Re^{-15/20}\frac{\partial\Psi_2^*}{\partial x_*} + \cdots. \qquad (5.4.25b)$$

We then substitute (5.4.25) together with (5.4.24) into the Navier–Stokes equations (5.3.1). We find from the x-momentum equation (5.3.1a) that the equation (5.3.100a) for Ψ_1^* retains its form:

$$\frac{1}{2}\lambda_0 Y_*^2\frac{\partial^2\Psi_1^*}{\partial x_*\partial Y_*} - \lambda_0 Y_*\frac{\partial\Psi_1^*}{\partial x_*} = \frac{\partial^3\Psi_1^*}{\partial Y_*^3}.$$

Its solution, satisfying the no-slip conditions

$$\Psi_1^* = \frac{\partial\Psi_1^*}{\partial Y_*} = 0 \quad \text{at} \quad Y_* = 0,$$

is written as

$$\Psi_1^* = \frac{1}{2}A_*(x_*)Y_*^2. \tag{5.4.26}$$

At this stage, the function $A_*(x_*)$ remains arbitrary; we can only claim that in view of (5.4.23),

$$A_*(x_*) = a_0(-x_*) + k_1 a_1(-x_*)^{-1} + \cdots \quad \text{as} \quad x_* \to -\infty. \tag{5.4.27}$$

Equation (5.3.107a) for function Ψ_2^* now acquires an additional term, the induced pressure gradient:

$$\frac{1}{2}\lambda_0 Y_*^2 \frac{\partial^2 \Psi_2^*}{\partial x_* \partial Y_*} - \lambda_0 Y_* \frac{\partial \Psi_2^*}{\partial x_*} = \frac{\partial^3 \Psi_2^*}{\partial Y_*^3} - \frac{1}{2}A_* \frac{dA_*}{dx_*}Y_*^2 - \frac{\partial P^*}{\partial x_*}. \tag{5.4.28}$$

The substitution of (5.4.24), (5.4.25) into the y-momentum equation (5.3.1b) shows that P^* does not change across region 3, that is

$$\frac{\partial P^*}{\partial Y_*} = 0. \tag{5.4.29}$$

5.4.3 Main part of the boundary layer (region 4)

When the parameter k belongs to the range (5.4.7), the asymptotic expansion (5.3.132) of the stream function in region 4 is written as

$$\psi = Re^{-1/2}\Psi_{00}(Y) + Re^{-7/10}\widetilde{\Psi}_1(x_*,Y) + \cdots . \tag{5.4.30}$$

Here,

$$x_* = \frac{x - x_0}{Re^{-1/5}}, \qquad Y = \frac{y}{Re^{-1/2}}.$$

By analogy with the pressure (5.4.24) in region 3, we seek the asymptotic expansion for the pressure in region 4 in the form

$$p = P_{e0} + Re^{-1/5}\lambda_0 x_* + Re^{-2/5}\frac{1}{2}\lambda_1 x_*^2$$
$$- Re^{-1/2}\ln Re\frac{2a_0 U_0^2}{5\pi\lambda_0} + Re^{-1/2}\widetilde{P}(x_*,Y) + \cdots . \tag{5.4.31}$$

The substitution of (5.4.30) into (5.3.2), yields

$$V_\tau = \Psi_{00}'(Y) + Re^{-1/5}\frac{\partial\widetilde{\Psi}_1}{\partial Y} + \cdots , \qquad V_n = -Re^{-1/2}\frac{\partial\widetilde{\Psi}_1}{\partial x_*} + \cdots . \tag{5.4.32}$$

We then substitute (5.4.32) and (5.4.31) into the x-momentum equation (5.3.1a). We find that the equation (5.3.138) retains its form.[10] Since the boundary conditions

---

[10]The induced pressure gradient $\partial P^*/\partial x_*$ is simply too weak to affect the flow in region 4.

(5.3.133) and (5.3.137) also remain unchanged, we can use for $\widetilde{\Psi}_1$ the solution given by (5.3.141):

$$\widetilde{\Psi}_1(x_*, Y_*) = \Psi'_{00}(Y)\left[x_* + \int_0^Y \frac{\Psi'''_{00} - \lambda_0}{\left(\Psi'_{00}\right)^2}\, dY' + \frac{A_*(x_*)}{\lambda_0}\right]. \tag{5.4.33}$$

It follows from (5.4.32) and (5.4.33) that the angle made by the streamlines with the aerofoil contour is calculated as

$$\vartheta = \frac{V_n}{V_\tau} = Re^{-1/2}\left[\int_0^Y \frac{\lambda_0 - \Psi'''_{00}}{\left(\Psi'_{00}\right)^2}\, dY' - \frac{1}{\lambda_0}\frac{dA_*}{dx_*}\right] + \cdots .$$

It changes with Y, but the curvature of the streamlines

$$\frac{\partial \vartheta}{\partial x} = -Re^{-3/10}\frac{1}{\lambda_0}\frac{d^2 A_*}{dx_*^2} + \cdots \tag{5.4.34}$$

stays constant across region 4.

To complete the flow analysis in region 4, we substitute (5.4.31) and (5.4.32) into the y-momentum equation (5.3.1b). By restricting our attention to the leading-order terms, we find that

$$\frac{\partial \widetilde{P}}{\partial Y} = \kappa(x_0)\left[\Psi'_{00}(Y)\right]^2.$$

Here, $\kappa(x_0)$ is the curvature of the aerofoil contour at point $x = x_0$. We see that while the pressure $\widetilde{P}$ changes with Y, the pressure gradient $\partial \widetilde{P}/\partial x_*$ remains constant across region 4:

$$\frac{\partial}{\partial Y}\left(\frac{\partial \widetilde{P}}{\partial x_*}\right) = 0.$$

5.4.4   Viscous-inviscid interaction problem

In order to describe the flow behaviour in the interaction region, we need to solve equation (5.4.28) in region 3,

$$\frac{1}{2}\lambda_0 Y_*^2 \frac{\partial^2 \Psi_2^*}{\partial x_* \partial Y_*} - \lambda_0 Y_* \frac{\partial \Psi_2^*}{\partial x_*} = \frac{\partial^3 \Psi_2^*}{\partial Y_*^3} - \frac{1}{2}A_*\frac{dA_*}{dx_*}Y_*^2 - \frac{\partial P^*}{\partial x_*}, \tag{5.4.35}$$

subject to the no-slip conditions on the aerofoil surface

$$\Psi_2^* = \frac{\partial \Psi_2^*}{\partial Y_*} = 0 \quad \text{at} \quad Y_* = 0, \tag{5.4.36}$$

and the matching condition (5.4.23) with the solution in region $2a$,

$$\Psi_2^* = \frac{a_0^2}{5!} x_* Y_*^5 + \frac{\lambda_0 a_0^2}{8!} Y_*^9$$
$$+ \frac{1}{2}\left\{ b_0(-x_*)^{7/4} + k_1 b_1 (-x_*)^{-1/4} \right\} Y_*^2 + \cdots \quad \text{as} \quad x_* \to -\infty. \qquad (5.4.37)$$

Since the pressure gradient $\partial P^*/\partial x_*$ in (5.4.35) is not known, we also need to solve equation (5.4.19) in region 5:

$$\frac{\partial^2}{\partial x_*^2}\left(\frac{\partial p_1^*}{\partial x_*} \right) + \frac{\partial^2}{\partial y_*^2}\left(\frac{\partial p_1^*}{\partial x_*} \right) = 0. \qquad (5.4.38)$$

The solution to (5.4.38) has to satisfy the far-field attenuation condition (5.4.20):

$$\frac{\partial p_1^*}{\partial x_*} \to 0 \quad \text{as} \quad x_*^2 + y_*^2 \to \infty, \qquad (5.4.39)$$

and a matching condition with the solution in region 4, which will now be deduced.
We know that the curvature of the streamlines in region 4 is given by (5.4.34):

$$\frac{\partial \vartheta}{\partial x} = -Re^{-3/10} \frac{1}{\lambda_0} \frac{d^2 A_*}{dx_*^2} + \cdots . \qquad (5.4.40)$$

Setting $y_* = 0$ in (5.4.14a) and (5.4.14b), we can easily find that, at the 'bottom' of region 5, the angle made by the streamlines with the aerofoil surface is calculated as

$$\vartheta = \frac{V_n}{V_\tau}\bigg|_{y_*=0} = Re^{-1/2} \frac{v_1^*(x_*,0)}{U_0} + \cdots . \qquad (5.4.41)$$

The differentiation of (5.4.41) with respect to x results in

$$\frac{\partial \vartheta}{\partial x} = Re^{-3/10} \frac{1}{U_0} \frac{\partial v_1^*}{\partial x_*}\bigg|_{y_*=0} + \cdots . \qquad (5.4.42)$$

By comparing (5.4.42) with (5.4.40), we conclude that the sought matching condition is written as

$$\frac{\partial v_1^*}{\partial x_*}\bigg|_{y_*=0} = -\frac{U_0}{\lambda_0} \frac{d^2 A_*}{dx_*^2}.$$

It remains to reformulate this condition for the pressure gradient. For this purpose the second equation in (5.4.17) is used. Differentiating this equation with respect to x_* and setting $y_* = 0$, we have

$$\frac{\partial}{\partial y_*}\left(\frac{\partial p_1^*}{\partial x_*} \right) = \frac{U_0^2}{\lambda_0} \frac{d^3 A_*}{dx_*^3} \quad \text{at} \quad y_* = 0. \qquad (5.4.43)$$

This completes the formulation of the viscous-inviscid interaction problem. To summarize, in order to describe the flow in the interaction region, we need to solve simultaneously equation (5.4.35) subject to the boundary conditions (5.4.36), (5.4.37) and equation (5.4.38) subject to the boundary conditions (5.4.39), (5.4.43). We start by expressing the interaction problem in a more convenient form. In the viscous sublayer (region 3), we introduce a new unknown function Ψ_2 defined by the equation

$$\Psi_2^* = \Psi_2 + \frac{\lambda_0 a_0^2}{8!}Y_*^9 + \frac{a_0^2}{5!}x_* Y_*^5 + \frac{1}{2}B_*(x_*)Y_*^2 + G_*(x_*)Y_*,$$

with

$$G_*(x_*) = \frac{A_*^2 - a_0^2 x_*^2 - 2k_1 a_0 a_1}{2\lambda_0},$$

and perform the affine transformations

$$\Psi_2 = \frac{a_0^{11/10} U_0^{9/5}}{\lambda_0^{17/10}}\bar{\Psi}, \qquad A_* = \frac{a_0^{3/5} U_0^{4/5}}{\lambda_0^{1/5}}A, \qquad \frac{\partial P^*}{\partial x_*} = \frac{a_0^{7/5} U_0^{6/5}}{\lambda_0^{4/5}}R(X),$$

$$G_* = \frac{a_0^{6/5} U_0^{8/5}}{\lambda_0^{7/5}}G(X), \qquad x_* = \frac{U_0^{4/5}}{a_0^{2/5}\lambda_0^{1/5}}X, \qquad Y_* = \frac{U_0^{1/5}}{a_0^{1/10}\lambda_0^{3/10}}\bar{Y}.$$

As a result, equation (5.4.35) and boundary conditions (5.4.36), (5.4.37) turn into

$$\frac{1}{2}\bar{Y}^2\frac{\partial^2\bar{\Psi}}{\partial X\partial\bar{Y}} - \bar{Y}\frac{\partial\bar{\Psi}}{\partial X} = \frac{\partial^3\bar{\Psi}}{\partial\bar{Y}^3} - R(X), \tag{5.4.44a}$$

$$\bar{\Psi} = 0, \quad \frac{\partial\bar{\Psi}}{\partial\bar{Y}} = -G(X) \quad \text{at} \quad \bar{Y} = 0, \tag{5.4.44b}$$

$$\bar{\Psi} \to 0 \quad \text{as} \quad X \to -\infty. \tag{5.4.44c}$$

Here,

$$G(X) = \frac{1}{2}\left(A^2 - X^2 + 2a\right), \tag{5.4.45}$$

with parameter a given by

$$a = k_1\frac{(-a_1)\lambda_0^{2/5}}{a_0^{1/5}U_0^{8/5}}. \tag{5.4.46}$$

In the new variables, condition (5.4.27) is written as

$$A(X) = (-X) - a(-X)^{-1} + \cdots \quad \text{as} \quad X \to -\infty, \tag{5.4.47}$$

which means that $G(X)$ tends to zero as $X \to -\infty$.

In the upper tier (region 5), we use the affine transformations

$$\frac{\partial p_1^*}{\partial x_*} = \frac{a_0^{7/5} U_0^{6/5}}{\lambda_0^{4/5}}r(X,\bar{y}), \qquad x_* = \frac{U_0^{4/5}}{a_0^{2/5}\lambda_0^{1/5}}X, \qquad y_* = \frac{U_0^{4/5}}{a_0^{2/5}\lambda_0^{1/5}}\bar{y}.$$

They render the equation (5.4.38) and boundary conditions (5.4.39), (5.4.43) in the form

$$\frac{\partial^2 r}{\partial X^2} + \frac{\partial^2 r}{\partial \bar{y}^2} = 0, \tag{5.4.48a}$$

$$r \to 0 \quad \text{as} \quad X^2 + \bar{y}^2 \to \infty, \tag{5.4.48b}$$

$$\left.\frac{\partial r}{\partial \bar{y}}\right|_{\bar{y}=0} = \frac{dQ}{dX}, \tag{5.4.48c}$$

where $Q(X)$ denotes the second derivative of function $A(X)$:

$$Q(X) = A''(X).$$

Finally, taking into account that the pressure gradient does not change across the middle tier (region 4), we can write

$$\left. r \right|_{\bar{y}=0} = R(X). \tag{5.4.49}$$

Notice that equations (5.4.44)–(5.4.49), which describe the flow in the interaction region, involve a single controlling parameter (5.4.46) that measures the deviation of the angle of attack from its critical value. Since $a_1 < 0$, the parameter a increases with k_1, and hence, with the angle of attack.

To solve the equations of viscous-inviscid interaction, we use the method of Fourier transforms. We start with the inviscid flow in region 5. It is easily seen that the Fourier transform renders equation (5.4.48a) and boundary conditions (5.4.48b), (5.4.48c) in the form

$$\frac{d^2 \breve{r}}{d\bar{y}^2} - k^2 \breve{r} = 0, \tag{5.4.50a}$$

$$\breve{r} = 0 \quad \text{at} \quad \bar{y} = \infty, \tag{5.4.50b}$$

$$\frac{d\breve{r}}{d\bar{y}} = ik\breve{Q} \quad \text{at} \quad \bar{y} = 0. \tag{5.4.50c}$$

Here, $\breve{r}(k,\bar{y})$ stands for the Fourier transform of the function $r(X,\bar{y})$:

$$\breve{r}(k,\bar{y}) = \int\limits_{-\infty}^{\infty} r(X,\bar{y})e^{-ikX}\,dX,$$

and $\breve{Q}(k)$ denotes the Fourier transform of the function $Q(X)$.

The general solution of the equation (5.4.50a) is written as

$$\breve{r} = C_1 e^{k\bar{y}} + C_2 e^{-k\bar{y}}. \tag{5.4.51}$$

To satisfy boundary condition (5.4.50b), we have to set $C_1 = 0$ for all $k > 0$, and $C_2 = 0$ for all $k < 0$. This reduces (5.4.51) to

$$\breve{r} = C e^{-|k|\bar{y}}. \tag{5.4.52}$$

The substitution of (5.4.52) into (5.4.50c) yields

$$C = -\frac{ik}{|k|}\breve{Q}. \tag{5.4.53}$$

It remains to substitute (5.4.53) back into (5.4.52), and we have the solution of the boundary-value problem (5.4.50) in the form

$$\check{r} = -\frac{ik}{|k|}\check{Q}e^{-|k|\bar{y}}. \tag{5.4.54}$$

Now, consider the flow in the viscous sublayer (region 3). The Fourier transform converts the boundary-value problem (5.4.44) into

$$\frac{1}{2}ik\bar{Y}^2\frac{d\check{\Psi}}{d\bar{Y}} - ik\bar{Y}\check{\Psi} = \frac{d^3\check{\Psi}}{d\bar{Y}^3} - \check{R}, \tag{5.4.55a}$$

$$\check{\Psi} = 0, \quad \frac{d\check{\Psi}}{d\bar{Y}} = -\check{G}(k) \quad \text{at} \quad \bar{Y} = 0. \tag{5.4.55b}$$

Here, $\check{\Psi}(k,\bar{Y})$, $\check{G}(k)$, and $\check{R}(k)$ are the Fourier transforms of functions $\bar{\Psi}(X,\bar{Y})$, $G(X)$, and $R(X)$, respectively.

Unlike (5.3.110a), equation (5.4.55a) is not homogeneous. Its general solution is written as

$$\check{\Psi} = C_1\phi_{21} + C_2\phi_{22} + C_3\phi_{23} + \phi_{2p}, \tag{5.4.56}$$

where ϕ_{2p} is a particular solution of the equation (5.4.55a) and ϕ_{21}, ϕ_{22}, and ϕ_{23} are the complementary solutions of the homogeneous part of (5.4.55a). These may be chosen such that

$$\left.\begin{array}{lll} \phi_{21}(0) = 1, & \phi'_{21}(0) = 0, & \phi''_{21}(0) = 0, \\ \phi_{22}(0) = 0, & \phi'_{22}(0) = 1, & \phi''_{22}(0) = 0, \\ \phi_{23}(0) = 0, & \phi'_{23}(0) = 0, & \phi''_{23}(0) = 1, \\ \phi_{2p}(0) = 0, & \phi'_{2p}(0) = 0, & \phi''_{2p}(0) = 0. \end{array}\right\} \tag{5.4.57}$$

It then follows from boundary conditions (5.4.55b) that $C_1 = 0$ and $C_2 = -\check{G}(k)$. Factor C_3 remains arbitrary, and the third complementary solution is easily seen to be $\psi_{23} = \frac{1}{2}\bar{Y}^2$. These turn (5.4.56) into

$$\check{\Psi} = -\check{G}(k)\phi_{22} + \frac{1}{2}C_3(k)\bar{Y}^2 + \phi_{2p}. \tag{5.4.58}$$

Let us consider function ϕ_{22} first. The behaviour of its counterpart, ψ_{22}, in the solution (5.3.111) of equation (5.3.110a) has been studied in detail in Section 5.3.4. In particular, we found that the asymptotic behaviour of ψ_{22} at the outer edge of region 3 is given by (5.3.125). It is easily seen that the homogenous part of (5.4.55a) is obtained by setting $\lambda_0 = 1$ in (5.3.110a). Consequently, to find the asymptotic behaviour of $\phi_{22}(\bar{Y})$ at large values of $\bar{Y}$, one simply needs to set $\lambda_0 = 1$ in (5.3.125), which requires $\Omega = \frac{1}{2}ik\lambda_0$ to be substituted with $\Omega = \frac{1}{2}ik$. We have

$$\phi_{22}(\bar{Y}) = -\frac{\Gamma(5/4)}{2\sqrt{\pi}e^{i\pi/4}\Omega^{1/4}}\left(\frac{4}{i\Omega^{1/2}\bar{Y}^2}\right)^{5/4}\sin\left(\frac{1}{2}i\Omega^{1/2}\bar{Y}^2 - \frac{3}{8}\pi\right) + \cdots. \tag{5.4.59}$$

Now, we turn to the particular solution $\phi_{2p}(\bar{Y})$. Function $\phi_{2p}(\bar{Y})$ satisfies the equation (5.4.55a). To obtain the power series for this function, we start by set $\bar{Y} = 0$ in (5.4.55a). We have

$$\frac{d^3\phi_{2p}}{d\bar{Y}^3} = \check{R}. \tag{5.4.60}$$

Integration of (5.4.60) with initial conditions given by (5.4.57) yields the first term of the power series:

$$\phi_{2p}(\bar{Y}) = \frac{1}{6}\check{R}\bar{Y}^3. \tag{5.4.61}$$

We then use (5.4.61) to calculate the left-hand side in (5.4.55a), and perform the integration again. This results in

$$\phi_{2p}(\bar{Y}) = \frac{1}{6}\check{R}\bar{Y}^3 + \frac{2}{7!}ik\check{R}\bar{Y}^7.$$

This procedure may be repeated for subsequent terms, leading to a conclusion that the power series of function ϕ_{2p} should be sought in the form

$$\phi_{2p}(\bar{Y}) = \frac{1}{6}\check{R}\bar{Y}^3 + \sum_{n=0}^{\infty} c_{n+1}\bar{Y}^{7+4n}. \tag{5.4.62}$$

Alternatively, we can write (5.4.62) as

$$\phi_{2p}(\bar{Y}) = \sum_{n=0}^{\infty} c_n\bar{Y}^{3+4n}. \tag{5.4.63}$$

Using (5.4.63) on the left-hand side of the equation (5.4.55a), and (5.4.62) on the right-hand side, yields the following recurrence equation for the coefficients c_n:

$$c_{n+1} = \frac{ik}{32}\frac{n+1/4}{(n+7/4)(n+5/4)(n+3/2)}c_n.$$

It can be easily verified by induction that the solution of this equation is written as

$$c_n = \frac{\check{R}}{6}\left(\frac{ik}{32}\right)^n \frac{\Gamma(3/2)\Gamma(7/4)}{\Gamma(n+3/2)\Gamma(n+7/4)(4n+1)}. \tag{5.4.64}$$

It remains to substitute (5.4.64) back into (5.4.63), and we will have[11]

$$\phi_{2p}(\bar{Y}) = \frac{\sqrt{\pi}}{16}\check{R}\Gamma(3/4)\sum_{n=0}^{\infty}\left(\frac{ik}{32}\right)^n \frac{\bar{Y}^{4n+3}}{\Gamma(n+3/2)\Gamma(n+7/4)(4n+1)}. \tag{5.4.65}$$

---

[11] Here it is taken into account that

$$\Gamma(3/2) = \tfrac{1}{2}\sqrt{\pi} \quad \text{and} \quad \Gamma(7/4) = \tfrac{3}{4}\Gamma(3/4).$$

We shall now show that the function $\psi_{2p}(\bar{Y})$ may be expressed in terms of the Struve function $H_\nu(z)$. It is known that the Struve function may be represented by the power series

$$H_\nu(z) = \sum_{n=0}^{\infty} \frac{(-1)^n}{\Gamma(n+3/2)\Gamma(n+\nu+3/2)} \left(\frac{z}{2}\right)^{2n+\nu+1}, \qquad (5.4.66)$$

that converges for all finite values of z. It is also known that

$$H_\nu(z) = \sqrt{\frac{2}{\pi z}} \sin\left(z - \frac{1}{2}\nu\pi - \frac{1}{4}\pi\right) + \cdots \quad \text{as} \quad |z| \to \infty. \qquad (5.4.67)$$

We choose the order of the Struve function to be $\nu = \frac{1}{4}$, and then it follows from (5.4.66) that

$$\sum_{n=0}^{\infty} \frac{(-1)^n}{\Gamma(n+3/2)\Gamma(n+7/4)(4n+1)} \left(\frac{z}{2}\right)^{2n+3/2} = \frac{z}{8} \int_0^z \left(\frac{2}{z}\right)^{7/4} H_{1/4}(z)\, dz. \quad (5.4.68)$$

By using (5.4.67) in the integral on the right-hand side of (5.4.68) and performing the integration by parts with

$$u = \left(\frac{2}{z}\right)^{9/4}, \qquad dv = \sin\left(z - \frac{3}{8}\pi\right) dz,$$

$$du = -\frac{9}{8}\left(\frac{2}{z}\right)^{13/4} dz, \qquad v = -\cos\left(z - \frac{3}{8}\pi\right),$$

we find that

$$\sum_{n=0}^{\infty} \frac{(-1)^n}{\Gamma(n+3/2)\Gamma(n+7/4)(4n+1)} \left(\frac{z}{2}\right)^{2n+3/2}$$

$$= -\frac{1}{4\sqrt{\pi}}\left(\frac{2}{z}\right)^{5/4} \cos\left(z - \frac{3}{8}\pi\right) + \cdots \quad \text{as} \quad |z| \to \infty. \qquad (5.4.69)$$

If we now set

$$\Omega = \frac{1}{2}ik, \qquad z = \frac{1}{2}i\Omega^{1/2}\bar{Y}^2, \qquad (5.4.70)$$

then equation (5.4.69) will turn into

$$\sum_{n=0}^{\infty} \left(\frac{ik}{32}\right)^n \frac{\bar{Y}^{4n+3}}{\Gamma(n+3/2)\Gamma(n+7/4)(4n+1)}$$

$$= -\frac{2}{\sqrt{\pi}e^{i3\pi/4}\Omega^{3/4}} \left(\frac{4}{i\Omega^{1/2}\bar{Y}^2}\right)^{5/4} \cos\left(\frac{1}{2}i\Omega^{1/2}\bar{Y}^2 - \frac{3}{8}\pi\right) + \cdots. \qquad (5.4.71)$$

It remains to substitute (5.4.71) into (5.4.65), and we can conclude that at large values of $\bar{Y}$,

$$\phi_{2p}(\bar{Y}) = -\frac{\check{R}\,\Gamma(3/4)}{8e^{i3\pi/4}\Omega^{3/4}}\left(\frac{4}{i\Omega^{1/2}\bar{Y}^2}\right)^{5/4}\cos\left(\frac{1}{2}i\Omega^{1/2}\bar{Y}^2 - \frac{3}{8}\pi\right) + \cdots. \qquad (5.4.72)$$

To progress further we need to define more precisely the way an analytic branch of the square root of $\frac{1}{2}ik$ is calculated in the second equation in (5.4.70). For this purpose we make a branch cut in the complex plane k along positive imaginary semi-axis (see Figure 5.8a). Expressing k in the form $k = re^{i\vartheta}$, and using (5.4.70), we see that for k lying on the real axis,

$$z = \begin{cases} \dfrac{1}{2}\sqrt{\dfrac{k}{2}}\,e^{i3\pi/4}\bar{Y}^2 & \text{if } k > 0, \\[3mm] \dfrac{1}{2}\sqrt{\dfrac{(-k)}{2}}\,e^{i\pi/4}\bar{Y}^2 & \text{if } k < 0. \end{cases}$$

This means that as $\bar{Y}$ tends to infinity, the corresponding point in the complex z-plane runs along one of the rays shown in Figure 5.8(b); which one depends on the sign of k. On both rays $\Re\{iz\} < 0$, whence

$$\left.\begin{aligned} \sin\left(z - \frac{3}{8}\pi\right) &= \frac{e^{i(z-3\pi/8)} - e^{-i(z-3\pi/8)}}{2i} = -\frac{1}{2i}e^{-i(z-3\pi/8)} + \cdots, \\ \cos\left(z - \frac{3}{8}\pi\right) &= \frac{e^{i(z-3\pi/8)} + e^{-i(z-3\pi/8)}}{2} = \frac{1}{2}e^{-i(z-3\pi/8)} + \cdots. \end{aligned}\right\} \qquad (5.4.73)$$

By substituting (5.4.73) into (5.4.72) and (5.4.59), and then into (5.4.58), we find that at the outer edge of region 3 (see Figure 5.7),

$$\check{\Psi} = -\frac{1}{4e^{i3\pi/4}\Omega^{1/4}}\left[\frac{\Gamma(5/4)}{\sqrt{\pi}}\check{G}(k) + \frac{\Gamma(3/4)}{4\Omega^{1/2}}\check{R}(k)\right]\left(\frac{4}{i\Omega^{1/2}\bar{Y}^2}\right)^{5/4}e^{-i(z-3\pi/8)} + \cdots.$$

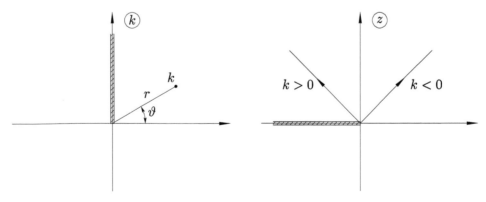

(a) Complex plane of the Fourier variable k.   (b) Complex plane z.

Fig. 5.8: Graphical illustration of equations (5.4.70).

To ensure that the solution in region 3 can be matched with the solution in region 4, we have to suppress the exponential growth of the function $\breve{\Psi}$, which is done by setting

$$\frac{\Gamma(5/4)}{\sqrt{\pi}}\breve{G}(k) + \frac{\Gamma(3/4)}{4\Omega^{1/2}}\breve{R}(k) = 0. \qquad (5.4.74)$$

Equation (5.4.74) relates the Fourier transform of the pressure gradient $R(X)$ in region 3 with the Fourier transform of function $G(X)$. The latter is related to the function $A(X)$ by means of equation (5.4.45). The second relationship between $A(X)$ and $R(X)$ is given by the solution (5.4.54) in the upper tier (region 5). The solutions in regions 3 and 5 are linked to one another through equation (5.4.49). By writing this equation in terms of the Fourier transforms, and using (5.4.54), we find that

$$\breve{R}(k) = \breve{r}\Big|_{\bar{y}=0} = -\frac{ik}{|k|}\breve{Q}(k). \qquad (5.4.75)$$

The Fourier transform of the pressure gradient, $\breve{R}(k)$, may be easily eliminated from (5.4.75) and (5.4.74), which leads to[12]

$$\breve{G}(k) = \frac{\sqrt{\pi}}{2}\Lambda\frac{(ik)^{1/2}}{|k|}\breve{Q}(k). \qquad (5.4.76)$$

Here, Λ is a constant given by

$$\Lambda = \frac{\Gamma(3/4)}{\sqrt{2}\,\Gamma(5/4)}.$$

Applying the inverse Fourier transformation to equation (5.4.76), renders it in the form

$$\frac{1}{2}\left(A^2 - X^2 + 2a\right) = \frac{\sqrt{\pi}}{2}\Lambda\frac{1}{2\pi}\int_{-\infty}^{\infty}\frac{(ik)^{1/2}}{|k|}\breve{Q}(k)e^{ikX}\,dk. \qquad (5.4.77)$$

Here, $\breve{Q}(k)$ is the Fourier transform of $A''(X)$:

$$\breve{Q}(k) = \int_{-\infty}^{\infty}A''(\xi)e^{-ik\xi}\,d\xi. \qquad (5.4.78)$$

If we substitute (5.4.78) into the integral on the right-hand side of (5.4.77) and change the order of integration, then we find that

$$A^2 - X^2 + 2a = \frac{\Lambda}{2\sqrt{\pi}}\int_{-\infty}^{\infty}A''(\xi)I(X,\xi)\,d\xi, \qquad (5.4.79)$$

where

$$I(X,\xi) = \int_{-\infty}^{\infty}\frac{(ik)^{1/2}}{|k|}e^{ik(X-\xi)}\,dk. \qquad (5.4.80)$$

---

[12]Remember that $\breve{G}(k)$ is the Fourier transform of the function $G(X) = \frac{1}{2}\left(A^2 - X^2 + 2a\right)$.

It may be shown (see Problem 3 in Exercises 13) that

$$I(X,\xi) = \begin{cases} 0 & \text{if } \xi < X, \\ \dfrac{2\sqrt{\pi}}{\sqrt{\xi - X}} & \text{if } \xi > X. \end{cases} \tag{5.4.81}$$

The substitution of (5.4.81) into (5.4.79) results in the following integro-differential equation for function $A(X)$:

$$A^2 - X^2 + 2a = \Lambda \int_X^\infty \frac{A''(\xi)}{\sqrt{\xi - X}} \, d\xi. \tag{5.4.82}$$

This equation does not allow for further simplification, and has to be solved numerically. When performing the calculations we need to use appropriate boundary conditions for $A(X)$. The first of these is given by (5.4.47):

$$A(X) = (-X) - a(-X)^{-1} + \cdots \quad \text{as} \quad X \to -\infty. \tag{5.4.83}$$

It represents the condition of matching with the solution in the boundary layer upstream of the interaction region. Expressing (5.3.70) in terms of function $A(X)$ we can write the condition of matching with the solution downstream of the interaction region as

$$A(X) = X + \cdots \quad \text{as} \quad X \to \infty. \tag{5.4.84}$$

For more detailed analysis of the asymptotic behaviour of $A(X)$ at large values of $|X|$ see Problem 4 in Exercises 13.

5.4.5 Numerical results

The results of the calculations are presented in Figures 5.9–5.12.[13] Figure 5.9 shows how function $A(X)$ changes as the parameter a is gradually increased. Note that the function $A(X)$ plays a dual role in the marginal separation theory. Firstly, through equation (5.4.34), it defines the shape of the streamlines in region 4, and, therefore, similar to the corresponding function in the classical triple-deck theory, it may be termed the displacement function. However, in the marginal separation theory, it also appears to be proportional to the shear stress on the aerofoil surface. Indeed, according to (5.4.22), (5.4.26), the two-term asymptotic expansion of the stream function in region 3 is written as

$$\psi = Re^{-13/20} \frac{1}{6} \lambda_0 Y_*^3 + Re^{-16/20} \frac{1}{2} A_*(x_*) Y_*^2 + \cdots, \qquad y = Re^{-11/20} Y_*.$$

Consequently, using (5.4.21), the dimensionless skin friction may be calculated as

$$\tau_w = \frac{1}{\sqrt{Re}} \frac{\partial^2 \psi}{\partial y^2}\bigg|_{y=0} = Re^{-1/5} A_*(x_*) = Re^{-1/5} \frac{a_0^{3/5} U_0^{4/5}}{\lambda_0^{1/5}} A(X).$$

---

[13]The author is grateful to Prof. S. Braun of Vienna TU for providing the data for these figures.

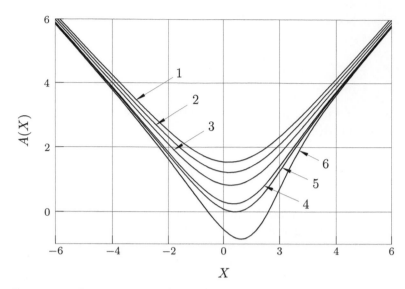

Fig. 5.9: Solutions of the equation (5.4.82) on the upper branch of the fundamental curve: 1) $a = -0.5$; 2) $a = 0.0$; 3) $a = 0.5$; 4) $a = 1.0$; 5) $a = a_s = 1.139$; 6) $a = a_c = 1.330$.

Graph 2 in Figure 5.9 is plotted for $a = 0$, which corresponds to the critical value of the angle of attack, as estimated based on the classical boundary-layer theory. When the viscous-inviscid interaction is ignored, the Prandtl equations yield a singular solution for $a = 0$, with the skin friction given by (5.3.70). We see that the interaction acts to smooth out the singularity. The minimal skin friction is lifted, and the τ_w appears to be positive for all values of $X \in (-\infty, \infty)$. For graph 5, the parameter a has been adjusted in such a way that the minimal skin friction returns back to zero to capture the incipience of the separation. We found that this happens at point $X = 0.406$ when the parameter a reaches the value $a_s = 1.139$. Graph 6 is plotted for the critical value of the parameter $a_c = 1.330$. It shows a region of negative A between $X = -0.566$ and $X = 1.605$, which is occupied by the separation bubble. Interestingly, the solution does not exist beyond $a = a_c$.

This important result is illustrated by Figure 5.10, where the so-called *fundamental curve* is displayed. This curve shows the entire set of admissible solutions of the marginal separation theory. It is constructed in the following way. Given a, the solution of the boundary-value problem (5.4.82)–(5.4.84) yields a distribution of the shear stress $A(X)$ along the aerofoil surface. Each such solution is represented by a point on the fundamental curve, which is obtained by taking the value of $A(X)$ at $X = 0$, and plotting it against the parameter a. The numbered circles on the fundamental curve represent the corresponding graphs in Figure 5.9. Notice that for $a = a_s$ (graph 5), the minimal skin friction is zero, but $A(0)$ is still positive.

When the parameter a reaches its critical value a_c, the fundamental curve turns back to form the second branch of the solutions. Figure 5.11 compares the solutions on the upper and lower branches for $a = a_s$. The first of these represents the flow with the

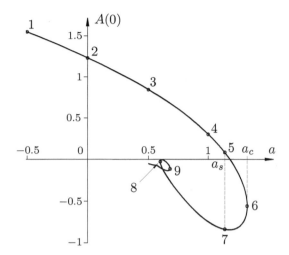

Fig. 5.10: The fundamental curve.

separation region just about to appear; in the second, the separation region is already well developed. In fact, the length of the separation region grows monotonically as an 'observer' follows the fundamental curve from the point 5 towards the critical point 6, and then all the way along the lower branch. This trend is demonstrated by Figure 5.12, where, in addition to the solution at point 7, the solutions at points 8 and 9, which lie on a small loop on the lower branch, are shown. Despite the fact that parameter a is still rather large on the loop, the solution already shows an asymptotic behaviour characteristic of small a. When $a = 0$, the equation (5.4.82) admits two solutions:

$$A = -X \quad \text{and} \quad A = X. \tag{5.4.85}$$

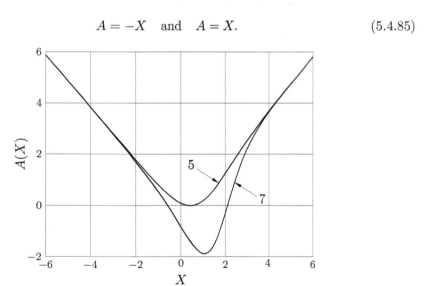

Fig. 5.11: Comparison of the solutions on the upper and lower branches for $a = a_s = 1.139$.

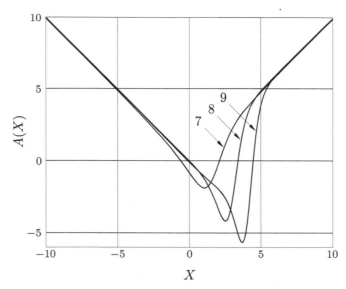

Fig. 5.12: Solutions on the lower branch of the fundamental curve: 7) $a = a_s = 1.139$; 8) $a = 0.600$; 9) $a = 0.680$.

Neither of these satisfies both boundary conditions (5.4.83), (5.4.84), but they are clearly visible in Figure 5.12 as two fragments of the solution, connected to one another through a sharp 'jump' in a region that becomes progressively shorter and moves to the right as $a \to 0-$.

In conclusion, according to the marginal separation theory, the flow near the leading edge of a thin aerofoil exhibits a *hysteresis* behaviour. It should be noted that hysteresis is routinely observed in experiments with aerofoils. These are normally conducted in such a way that the angle of attack is at first increased slowly enough to keep the flow quasi-steady, and then, after achieving the *aerofoil stall*, it is gradually decreased. Normally, the angle of attack at which the separation region forms on the upper surface of the aerofoil does not coincide with the angle of attack at which the flow returns back to attached form. As a result, the graph of the lift force versus the angle of attack assumes the shape of a hysteresis curve. Within the hysteresis loop, two flow states become possible for each value of the angle of attack. The choice between them depends on the history of the development of the flow.

Experimental observations of the short separation bubbles show that these are formed in a smooth manner without abrupt change of the flow field, which is only possible if the solution remains on the upper branch of the fundamental curve when the parameter a passes through a_s. Of course, the flow cannot continue to change smoothly after parameter a reaches the critical value a_c. The non-existence of the solution to (5.4.82)–(5.4.84) for $a > a_c$ suggests that the flow has to undergo a sudden change, known from experiments as the *bubble bursting*.

In order to calculate the values of the angle of attack at the incipience of the short separation bubble and at the bubble bursting, we need to return to equation (5.2.5).

By combining it with (5.4.9) and (5.4.46), we find that the scaled angle of attack $\alpha_* = \alpha/\varepsilon$ is given by

$$\alpha_* = \frac{1}{\sqrt{2}}\left[k_0 + Re^{-2/5}\frac{a_0^{1/5}U_0^{8/5}}{(-a_1)\lambda_0^{2/5}}\,a\right] + \frac{1}{2\pi}\int\limits_0^1 \frac{Y_+'(\zeta) + Y_-'(\zeta)}{\sqrt{\zeta(1-\zeta)}}\,d\zeta. \qquad (5.4.86)$$

Remember that constants k_0, U_0, λ_0, a_0, and a_1 are found by solving the Prandtl's equations of the classical boundary-layer theory. For an aerofoil with parabolic nose

$$k_0 = 1.1575, \quad U_0 = 1.286, \quad \lambda_0 = 0.024, \quad a_0 = 0.0085, \quad a_1 = -1.24. \qquad (5.4.87)$$

Using (5.4.87) in equation (5.4.86), renders it in the form

$$\alpha_* = 0.8185 + 1.4610 \cdot Re^{-2/5}a + \frac{1}{2\pi}\int\limits_0^1 \frac{Y_+'(\zeta) + Y_-'(\zeta)}{\sqrt{\zeta(1-\zeta)}}\,d\zeta.$$

If we assume, for example, that the aerofoil is symmetric ($Y_- = -Y_+$), then we will have

$$\alpha_* = 0.8185 + 1.4610 \cdot Re^{-2/5}a + \cdots \quad \text{as} \quad Re \to \infty. \qquad (5.4.88)$$

Setting $a = a_s = 1.139$ in (5.4.88) gives the angle of attack at moment of the formation of the short separation bubble:

$$\alpha_s = \varepsilon\left(0.8185 + 1.6641 \cdot Re^{-2/5} + \cdots\right). \qquad (5.4.89)$$

If we choose $a = a_c = 1.33$, then we will find that the bubble bursting has to be expected at the angle of attack

$$\alpha_c = \varepsilon\left(0.8185 + 1.9431 \cdot Re^{-2/5} + \cdots\right). \qquad (5.4.90)$$

When applying these results to real flows, we need to keep in mind that the marginal separation theory relies on the assumption of laminar flow. This assumption is well justified for attached flows near the leading edge of a thin aerofoil because the characteristic length scale here is not the aerofoil cord L, but a much smaller quantity, the radius of the aerofoil nose $r = L\varepsilon^2$. Under conditions typical of aerodynamic applications, the Reynolds number $Re = V_\infty r/\nu$ is not large enough for the attached boundary layer to become turbulent. In fact, the flow is observed to remain laminar even after the short separation bubble is formed. However, separated flows are known to be less stable and undergoes a rather rapid transition to turbulence before the reattachment point. The transition has a significant effect on the behaviour of the separation bubble. For example, when the turbulence is enhanced by introducing an additional acoustic noise in the wind tunnel test section, it always delays the bubble bursting. This explains why formula (5.4.89) proves to be fairly accurate (see Hsiao and Pauley, 1994), while (5.4.90) underestimates the critical angle of attack α_c.

Exercises 13

1. Consider the boundary-layer flow that has the 'weak singularity' at the point of zero skin friction $x = x_0$. It is known that upstream of this point, the boundary layer splits into two regions: the viscous sublayer (region $2a$), and the main part of the boundary layer (region $2b$). You may use without proof the fact that in region $2b$ the stream function is given by equation (5.3.59):

$$\Psi_0(x,Y) = \Psi_{00}(Y) + s\Psi_{00}'(Y)\left[\int_0^Y \frac{\Psi_{00}'''(Y') - \lambda_0}{\left[\Psi_{00}'(Y')\right]^2}\,dY' - \frac{a_0}{\lambda_0}\right] + \cdots. \qquad (5.4.91)$$

Consider the 'bottom' of region $2b$, where

$$\Psi_{00}(Y) = \frac{1}{6}\lambda_0 Y^3 + \frac{2\lambda_0\lambda_1}{7!}Y^7 + \frac{\lambda_0 a_0^2}{8!}Y^9 + \cdots. \qquad (5.4.92)$$

Substitute (5.4.92) into (5.4.91) and express the resulting equation in terms of the variable $\eta = Y/(-s)^{1/4}$ of region $2a$. Observe that this reduces (5.4.91) to

$$\Psi_0(x,Y) = (-s)^{3/4}\frac{1}{6}\lambda_0\eta^3 + (-s)^{3/2}\frac{1}{2}a_0\eta^2$$
$$+ (-s)^{7/4}\left(-\frac{1}{6}\lambda_1\eta^3 + \frac{2\lambda_0\lambda_1}{7!}\eta^7\right)$$
$$+ (-s)^{9/4}\left(-\frac{a_0^2}{5!}\eta^5 + \frac{\lambda_0 a_0^2}{8!}\eta^9\right) + \cdots,$$

which coincides with the solution (5.3.54) in region $2a$.

2. Consider a two-dimensional steady incompressible inviscid fluid flow near a corner point O on a rigid body surface as shown in Figure 5.13. Assume that the wall deflection angle θ is small, and represent the solution of the Euler equations in the form of asymptotic expansions

$$u = U_0 + \theta u_1(x,y) + \cdots, \quad v = \theta v_1(x,y) + \cdots, \quad p = \theta p_1(x,y) + \cdots. \qquad (5.4.93)$$

Here u, v, and p are the non-dimensional velocity components and pressure; U_0 denotes the value of the tangential velocity u that would be observed in the flow if the angle θ was zero.

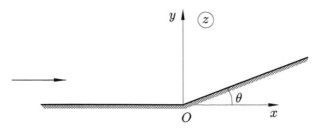

Fig. 5.13: Flow past a corner point of a rigid body surface.

By substituting (5.4.93) into the Euler equations

$$u\frac{\partial u}{\partial x} + v\frac{\partial u}{\partial y} = -\frac{\partial p}{\partial x},$$

$$u\frac{\partial v}{\partial x} + v\frac{\partial v}{\partial y} = -\frac{\partial p}{\partial y},$$

$$\frac{\partial u}{\partial x} + \frac{\partial v}{\partial y} = 0,$$

prove that $f(z) = p_1 + iU_0v_1$ is an analytic function of the complex variable $z = x + iy$, and try to find the solution for $f(z)$ near the corner in the form

$$p_1 + iU_0v_1 = (C_r + iC_i)\ln z + (D_r + iD_i) + \cdots \quad \text{as} \quad z \to 0. \tag{5.4.94}$$

Set $z = re^{i\vartheta}$ in (5.4.94), and using the impermeability condition on the body surface upstream and downstream of the corner point O,

$$\frac{v}{u} = \begin{cases} \tan\theta \approx \theta & \text{at} \quad y = 0,\ x > 0, \\ 0 & \text{at} \quad y = 0,\ x < 0, \end{cases}$$

show that

$$C_r = -\frac{U_0^2}{\pi}, \qquad C_i = 0, \qquad D_i = U_0^2.$$

Hence, deduce that in a small vicinity of the corner point

$$\left.\begin{aligned} p &= -\theta\frac{U_0^2}{\pi}\ln r + \cdots, \\ u &= \theta\frac{U_0}{\pi}\ln r + \cdots, \\ v &= \theta U_0\left(1 - \frac{\vartheta}{\pi}\right) + \cdots \end{aligned}\right\} \quad \text{as} \quad r \to 0. \tag{5.4.95}$$

3. Consider integral (5.4.80):

$$I(X,\xi) = \int\limits_{-\infty}^{\infty} \frac{(ik)^{1/2}}{|k|}e^{ik(X-\xi)}dk, \tag{5.4.96}$$

where an analytic branch of $(ik)^{1/2}$ is defined by introducing a branch cut along the positive imaginary semi-axis (see Figure 5.14), and expressing k in the exponential form $k = se^{i\vartheta}$, where $\vartheta \in \left(-\frac{3}{2}\pi, \frac{1}{2}\pi\right)$.
   Split integral (5.4.96) into two,

$$I(X,\xi) = \int\limits_{-\infty}^{0} \frac{(ik)^{1/2}}{-k}e^{ik(X-\xi)}dk + \int\limits_{0}^{\infty} \frac{(ik)^{1/2}}{k}e^{ik(X-\xi)}dk, \tag{5.4.97}$$

and assume, first, that $\xi < X$. In this case, when calculating the first integral I_1 in (5.4.97), change the path of integration from the negative real semi-axis (contour

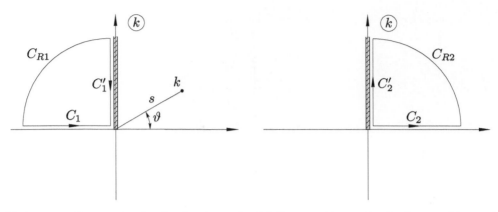

(a) Contour of integration for the first integral in (5.4.97).

(b) Contour of integration for the second integral in (5.4.97).

Fig. 5.14: Calculation of the integrals in (5.4.97) for $\xi < X$.

C_1 in Figure 5.14a) to the positive imaginary semi-axis C_1'. Prove that the integral along the quarter-circle C_{R1} tends to zero as the radius R of the quarter-circle tends to infinity.

When performing the integration along C_1', introduce a new integration variable s, such that $k = is$, and show that the adopted rule of calculation of $(ik)^{1/2}$ gives

$$(ik)^{1/2} = -is^{1/2}.$$

Hence, show that

$$I_1 = -i \int_0^\infty \frac{e^{-(X-\xi)s}}{\sqrt{s}} \, ds.$$

When calculating the second integral, I_2, in (5.4.97), change the contour of integration as shown in Figure 5.14(b). Show that

$$I_2 = i \int_0^\infty \frac{e^{-(X-\xi)s}}{\sqrt{s}} \, ds.$$

Hence, conclude that

$$I(X,\xi) = I_1 + I_2 = 0 \quad \text{for} \quad \xi < X.$$

For the case when $\xi > X$, change the contour of integration for the two integrals in (5.4.97) as indicated in Figure 5.15, and show that

$$I_1 = I_2 = \frac{\sqrt{\pi}}{\sqrt{\xi - X}}.$$

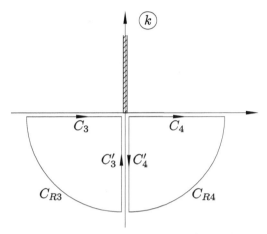

Fig. 5.15: Calculation of the integrals in (5.4.97) for $\xi > X$.

4. Consider the equation (5.4.82) of the marginal separation theory:

$$A^2 - X^2 + 2a = \Lambda \int_X^\infty \frac{A''(\xi)}{\sqrt{\xi - X}}\, d\xi. \tag{5.4.98}$$

Remember that its solution has to satisfy boundary conditions (5.4.83), (5.4.84):

$$A(X) = (-X) - a(-X)^{-1} + \cdots \quad \text{as} \quad X \to -\infty, \tag{5.4.99}$$
$$A(X) = X + \cdots \quad\quad\quad\quad\ \text{as} \quad X \to \infty. \tag{5.4.100}$$

Your task is to find the next order terms in (5.4.99) and (5.4.100). You may perform this task in the following steps:

(a) First, assume that $X \to \infty$. Notice that the right-hand side of equation (5.4.98) tends to zero as $X \to \infty$, and argue that the first two terms of the asymptotic expansion of $A(X)$ may be written as

$$A(X) = X - \frac{a}{X} + \cdots \quad \text{as} \quad X \to \infty. \tag{5.4.101}$$

Differentiate (5.4.101) twice, and substitute the result into the integral on the right-hand side of (5.4.98). Perform the integration using the following substitutions: $\xi = Xs$, then $s = t^2 + 1$ and, finally, $t = \tan\theta$. Conclude that

$$A^2 - X^2 + 2a = -\frac{3}{4}a\Lambda\pi X^{-5/2} + \cdots \quad \text{as} \quad X \to \infty. \tag{5.4.102}$$

Using (5.4.102), find two additional terms in (5.4.101).

(b) Now, assume that $X \to -\infty$. Write the integral on the right-hand side of (5.4.98) as

$$\int_X^\infty \frac{A''(\xi)}{\sqrt{\xi - X}}\, d\xi = \int_X^{-\Delta} \frac{A''(\xi)}{\sqrt{\xi - X}}\, d\xi + \int_{-\Delta}^{\Delta} \frac{A''(\xi)}{\sqrt{\xi - X}}\, d\xi + \int_{\Delta}^\infty \frac{A''(\xi)}{\sqrt{\xi - X}}\, d\xi, \tag{5.4.103}$$

where parameter Δ is such that

$$\Delta \ll (-X), \qquad \Delta \gg (-X)^{1/3}. \tag{5.4.104}$$

Start with the middle integral. Notice that, thanks to the first inequality in (5.4.104), it may be approximated as

$$I_2 \approx \int_{-\Delta}^{\Delta} \frac{A''(\xi)}{\sqrt{-X}} \, d\xi = \frac{1}{\sqrt{-X}} \int_{-\Delta}^{\Delta} A''(\xi) \, d\xi = \frac{1}{\sqrt{-X}} \Big[A'(\Delta) - A'(-\Delta) \Big].$$

Keeping in mind that Δ is large, calculate $A'(\Delta)$ and $A'(-\Delta)$ using (5.4.100) and (5.4.99), respectively, and conclude that

$$I_2 = \frac{2}{\sqrt{-X}} + \cdots . \tag{5.4.105}$$

When evaluating the first integral in (5.4.103), calculate $A''(\xi)$ with the help of (5.4.99). Then use the fact that $(-\xi) \le \Delta$ everywhere in the integration interval, and show that

$$|I_1| \le \frac{2a}{\Delta^3} \int_{X}^{-\Delta} \frac{d\xi}{\sqrt{\xi - X}} = \frac{4a}{\Delta^3} \sqrt{(-X) - \Delta}.$$

Finally, consider the third integral in (5.4.103):

$$I_3 = \int_{\Delta}^{\infty} \frac{A''(\xi)}{\sqrt{\xi - X}} \, d\xi. \tag{5.4.106}$$

Use asymptotic expansion (5.4.101) to calculate $A''(\xi)$ in (5.4.106), and introduce a new integration variable t through the substitution $\xi = (-X)(t^2 - 1)$. This leads to

$$I_3 = -\frac{4a}{(-X)^{5/2}} \int_{\sqrt{1+\Delta/(-X)}}^{\infty} \frac{dt}{(t-1)^3 (t+1)^3}.$$

Notice that $t + 1 > 2$ everywhere in the integration interval. Hence, deduce that

$$|I_3| \le \frac{4a}{(-X)^{5/2}} \int_{\sqrt{1+\Delta/(-X)}}^{\infty} \frac{dt}{(t-1)^3 8} \approx \frac{a}{4(-X)^{1/2}} \frac{1}{\Delta^2}.$$

Argue that under conditions (5.4.104), integrals I_1 and I_3 are much smaller than I_2. Substitute (5.4.105) into (5.4.103) and then into (5.4.98). Conclude that

$$A^2 - X^2 + 2a = \frac{2\Lambda}{\sqrt{-X}} + \cdots \quad \text{as} \quad X \to -\infty.$$

5. Show that the equation of marginal separation theory

$$A^2 - X^2 + 2a = \Lambda \int_X^\infty \frac{A''(\xi)}{\sqrt{\xi - X}} \, d\xi$$

may be inverted to be written as

$$A'(X) = 1 - \frac{1}{\pi\Lambda} \int_X^\infty \frac{A^2(\xi) - \xi^2 + 2a}{\sqrt{\xi - X}} \, d\xi.$$

When performing this task you may use without proof equation (5.4.76):

$$\breve{G}(k) = \frac{\sqrt{\pi}}{2} \Lambda \frac{(ik)^{1/2}}{|k|} \breve{Q}(k), \tag{5.4.107}$$

where $\breve{G}(k)$ and $\breve{Q}(k)$ are Fourier transforms of functions $G(X) = \frac{1}{2}(A^2 - X^2 + 2a)$ and $Q(X) = A''$, respectively.

Start by introducing a new function $S(X)$ such that $S = G'(X)$, and deduce from equation (5.4.107) that

$$\breve{Q} = \frac{2}{\sqrt{\pi}\Lambda} \frac{|k|}{(ik)^{3/2}} \breve{S}. \tag{5.4.108}$$

Apply the inverse Fourier transform to (5.4.108), and by using the technique described in Problem 3, show that

$$A''(X) = -\frac{1}{2\pi\Lambda} \int_X^\infty \frac{S(\xi)}{\sqrt{\xi - X}} \, d\xi. \tag{5.4.109}$$

Now, consider the integral

$$F(X) = \int_X^\infty \frac{G(\xi)}{\sqrt{\xi - X}} \, d\xi. \tag{5.4.110}$$

Perform the integration by parts in (5.4.110), using (5.4.102), and show that

$$F(X) = -2 \int_X^\infty \sqrt{\xi - X} S(\xi) \, d\xi. \tag{5.4.111}$$

Then differentiate (5.4.111) with respect to X, and deduce that

$$F'(X) = \int_X^\infty \frac{S(\xi)}{\sqrt{\xi - X}} \, d\xi. \tag{5.4.112}$$

Finally, substitute (5.4.112) into (5.4.109) and integrate the resulting equation using the fact that, according to (5.4.100), $A' = 1$ at $X = \infty$.

References

Abramowitz, M. and Stegun, I. A. (1965). *Handbook of Mathematical Functions* (third edn). Dover Publications.

Ackerberg, R. C. (1970). Boundary-layer separation at a free streamline. Part 1. Two-dimensional flow. *J. Fluid Mech.*, **44**(2), 211–225.

Batchelor, G. K. (1956). On steady laminar flow with closed streamlines at large Reynolds number. *J. Fluid Mech.*, **1**(2), 177–190.

Blasius, H. (1908). Grenzschichten in flüssigkeiten mit kleiner reibung. *Z. Mathematik Physik*, **56**, 1–37.

Bodonyi, R. J. and Kluwick, A. (1998). Transonic trailing-edge flow. *Q. J. Mech. Appl. Math.*, **51**, 297–310.

Braun, S. and Kluwick, A. (2004). Unsteady three-dimensional marginal separation caused by surface mounted obstacles and/or local suction. *J. Fluid Mech.*, **514**, 121–152.

Braun, S. and Scheichl, S. (2014). On recent developments in marginal separation theory. *Phil. Trans. Roy. Soc.*, **A 372**, doi: 10.1098/rsta.2013.0343.

Brillouin, M. (1911). Les surfaces de glissement d'Helmholtz et la résistance des fluides. *Ann. Chim. Phys.*, 8e sér. **23**, 145–230.

Brodetsky, S. (1923). Discontinuous fluid motion past circular and elliptic cylinders. *Proc. Roy. Soc. London*, **A 102**, 542–553.

Brown, S. N. (1985). Marginal separation of a three-dimensional boundary layer on a line of symmetry. *J. Fluid Mech.*, **158**, 95–111.

Brown, S. N. and Stewartson, K. (1970). Trailing-edge stall. *J. Fluid Mech.*, **42**(3), 561–584.

Chapman, D. R. (1950). Laminar mixing in a compressible fluid. *NACA Rep.* 958.

Chapman, D. R., Kuehn, D. M., and Larson, H. K. (1958). Investigation of separated flows in supersonic and subsonic streams with emphasis on the effect of transition. *NACA Rep.* 1356.

Chow, R. and Melnik, R. E. (1976). Numerical solution of the triple-deck equations for laminar trailing-edge stall. *Lect. Notes Phys.*, **59**, 135–144.

Crank, J. and Nicolson, P. (1947). A practical method for numerical evaluation of solutions of the partial differential equations of the heat-conduction type. *Proc. Camb. Phil. Soc.*, **43**(1), 50–67.

Dennis, S. C. R. and Chang, G. Z. (1969). Numerical investigation of the Navier–Stokes equations for steady two-dimensional flow. *Phys. Fluids*, **12**(12, Suppl. II), 88–93.

Dennis, S. C. R. and Dunwoody, J. (1966). The steady flow of a viscous fluid past a flat plate. *J. Fluid Mech.*, **24**(3), 577–595.

Dettman, J. W. (1965). *Applied Complex Variables*. Macmillan (reprinted 1984 by Dover Publications).

Falkner, V. M. and Skan, S. W. (1930). Some approximate solutions of the boundary-layer equations. *Aero. Res. Coun., Rep. Memor.* 1314.

Flachsbart, O. (1935). Der widerstand quer angeströmter rechteckplatten bei reynoldsschen zahlen 1000 bis 6000. *Z. Angew. Math. Mech.*, **15**, 32–37.

Fomina, I. G. (1983). On asymptotic flow theory near corner points on a solid surface. *Uch. zap. TsAGI*, **14**(5), 31–38.

Glauert, M. B. (1957). The flow past a rapidly rotating cylinder. *Proc. Roy. Soc. London*, **A 242**, 108–115.

Goldstein, S. (1930). Concerning some solutions of the boundary layer equations in hydrodynamics. *Proc. Camb. Phil. Soc.*, **26**(1), 1–30.

Goldstein, S. (1948). On laminar boundary-layer flow near a position of separation. *Q. J. Mech. Appl. Math.*, **1**(1), 43–69.

Hartree, D. R. (1939). A solution of the laminar boundary-layer equations for retarded flow. *Aero. Res. Coun. Rep. Memo.* 2426 (issued 1949).

Hiemenz, K. (1911). Die grenzschict neinem in den gleichformigen flussigkeitsstrom eingetauchten geraden Kreiszylinder. *Dingler's Polytechnic J.*, **326**, 321–410.

Howarth, L. (1938). On the solution of the laminar boundary layer equations. *Proc. Roy. Soc. London*, **A 164**, 547–579.

Hsiao, C. T. and Pauley, L. L. (1994). Comparison of the triple-deck theory, interactive boundary layer method, and Navier–Stokes computations for marginal separation. *Trans. ASME J. Fluids Eng.*, **116**, 22–28.

Jobe, C. E. and Burggraf, O. R. (1974). The numerical solution of the asymptotic equations of the trailing edge flow. *Proc. Roy. Soc. London*, **A 340**, 91–111.

Jones, B. M. (1934). Stalling. *J. Roy. Aero. Soc.*, **38**, 753–770.

Kirchhoff, G. (1869). Zur theorie freier flüssigkeitsstrahlen. *J. für die Reine und Angew. Math.*, **70**(4), 289–298.

Korolev, G. L. (1991). Interaction theory and non-uniqueness of separated flows around solid bodies. In V. V. Kozlov and A. V. Dovgal, Eds., *Separated Flows and Jets*, pp. 139–142. Springer, Berlin.

Korolev, G. L., Gajjar, J. S. B., and Ruban, A. I. (2002). Once again on the supersonic flow separation near a corner. *J. Fluid Mech.*, **463**, 173–199.

Kozlova, I. G. and Mikhailov, V. V. (1970). Strong viscous interaction on delta and swept wing. *Izv. Akad. Nauk SSSR, Mech. Zhidk. Gaza* (6), 94–99.

Kravtsova, M. A., Zametaev, V. B., and Ruban, A. I. (2005). An effective numerical method for solving viscous-inviscid interaction problems. *Phil. Trans. R. Soc. Lond. A*, **363**(1830), 1157–1167.

Landau, L. D. and Lifshitz, E. M. (1944). *Mechanics of Continuous Media*. Gostekhizdat, Moscow.

Levi-Civita, T. (1907). Scie e leggi di resistenza. *Rendconte de Circolo Matematico di Palermo*, **XXIII**, 1–37.

Liepmann, H. W., Roshko, A., and Dhawan, S. (1952). On reflection of shock waves from boundary layers. *NACA Rep.* 1100.

Lighthill, M. J. (1953). On boundary layers and upstream influence. II. Supersonic flows without separation. *Proc. Roy. Soc. London*, **A 217**, 478–507.

Messiter, A. F. (1970). Boundary-layer flow near the trailing edge of a flat plate. *SIAM J. Appl. Math.*, **18**(1), 241–257.

Negoda, V. V. and Sychev, Vik. V. (1986). The boundary layer on a rapidly rotating cylinder. *Izv. Akad. Nauk SSSR, Mech. Zhidk. Gaza* (5), 36–45.

Neiland, V. Ya. (1969*a*). Theory of laminar boundary layer separation in supersonic flow. *Izv. Akad. Nauk SSSR, Mech. Zhidk. Gaza* (4), 53–57.

Neiland, V. Ya. (1969*b*). On asymptotic theory that predicts heat transfer near corner points. *Izv. Akad. Nauk SSSR, Mech. Zhidk. Gaza* (5), 53–60.

Neiland, V. Ya. (1970*a*). On asymptotic theory of steady two-dimensional supersonic flows with separation region. *Izv. Akad. Nauk SSSR, Mech. Zhidk. Gaza* (3), 22–33.

Neiland, V. Ya. (1970*b*). Upstream propagation of perturbations in a boundary layer interacting with hypersonic flow. *Izv. Akad. Nauk SSSR, Mech. Zhidk. Gaza* (4), 40–49.

Neiland, V. Ya. (1971). Flow behind the boundary-layer separation point in a supersonic stream. *Izv. Akad. Nauk SSSR, Mech. Zhidk. Gaza* (3), 19–25.

Neiland, V. Ya., Bogolepov, V. V., Dudin, G. N., and Lipatov, I. I. (2008). *Asymptotic Theory of Supersonic Viscous Gas Flows*. Elsevier.

Neiland, V. Ya. and Sychev, V. V. (1966). Asymptotic solutions of the Navier–Stokes equations in regions of large local perturbations. *Izv. Akad. Nauk SSSR, Mech. Zhidk. Gaza* (4), 43–49.

Nishioka, M. and Miyagi, T. (1978). Measurements of velocity distributions in the laminar wake of a flat plate. *J. Fluid Mech.*, **84**(4), 705–715.

Oswatitsch, K. and Wieghardt, K. (1948). Theoretical analysis of stationary potential flows and boundary layers at high speed. *NACA Tech. Memo.* 1189.

Prandtl, L. (1904). Über flüssigkeitsbewegung bei sehr kleiner Reibung. In *Verh. III. Intern. Math. Kongr., Heidelberg*, pp. 484–491. Teubner, Leipzig, 1905.

Prandtl, L. and Tietjens, O. G. (1934). *Applied Hydro- and Aeromechanics*. McGraw-Hill (reprinted 1957 by Dover Publications).

Ruban, A. I. (1976). On the theory of laminar flow separation of a fluid from a corner point on a solid surface. *Uch. zap. TsAGI*, **7**(4), 18–28.

Ruban, A. I. (1981). A singular solution of the boundary-layer equations which can be extended continuously through the point of zero skin friction. *Izv. Akad. Nauk SSSR, Mech. Zhid. Gaza* (6), 42–52.

Ruban, A. I. (1982). Asymptotic theory of short separation bubbles at the leading edge of a thin airfoil. *Izv. Akad. Nauk SSSR, Mech. Zhid. Gaza* (1), 42–52.

Ruban, A. I. and Sychev, V. V. (1973). Hypersonic viscous gas flow past a small aspect ratio wing. *Uch. zap. TsAGI*, **4**(5), 18–25.

Ruban, A. I. and Turkyilmaz, I. (2000). On laminar separation at a corner point in transonic flow. *J. Fluid Mech.*, **423**, 345–380.

Ruban, A. I., Wu, X., and Pereira, R. M. S. (2006). Viscous-inviscid interaction in transonic Prandtl–Meyer flow. *J. Fluid Mech.*, **568**, 387–424.

Schlichting, H. (1933). Laminare strahlausbreitung. *Z. Angew. Math. Mech.*, **13**, 260–263.

Shidlovskii, V. P. (1977). The structure of the viscous fluid flow near the edge of a rotating disk. *Prikl. Mat. Mech.*, **41**(3), 464–472.

Smith, F. T. (1977). The laminar separation of an incompressible fluid streaming past a smooth surface. *Proc. Roy. Soc. London*, **A 356**, 443–463.

Smith, F. T. and Duck, P. W. (1977). Separation of jets or thermal boundary layers from a wall. *Q. J. Mech. Appl. Math.*, **30**, 143–156.

Smith, F. T. and Merkin, J. H. (1982). Triple-deck solutions for subsonic flow past humps, steps, concave or convex corners and wedged trailing edges. *Int. J. Comput. Fluids*, **10**(1), 7–25.

Stewartson, K. (1955). On the motion of a flat plate at high speed in a viscous compressible fluid. Steady motion. *J. Aeronaut. Sci..*, **22**, 303–309.

Stewartson, K. (1969). On the flow near the trailing edge of a flat plate. *Mathematika*, **16**(1), 106–121.

Stewartson, K. (1970). On laminar boundary layers near corners. *Q. J. Mech. Appl. Math.*, **23**(2), 137–152.

Stewartson, K. (1971). On laminar boundary layers near corners. Corrections and an addition. *Q. J. Mech. Appl. Math.*, **24**(3), 387–389.

Stewartson, K., Smith, F. T., and Kaups, K. (1982). Marginal separation. *Stud. Appl. Math.*, **67**(1), 45–61.

Stewartson, K. and Williams, P. G. (1969). Self-induced separation. *Proc. Roy. Soc. London*, **A 312**, 181–206.

Sychev, V. V. (1972). Laminar separation. *Izv. Akad. Nauk SSSR, Mech. Zhidk. Gaza* (3), 47–59.

Sychev, Vik. V. (1986). Singular solution of boundary-layer equations on a moving surface. *Izv. Akad. Nauk SSSR, Mech. Zhidk. Gaza* (2), 43–52.

Sychev, V. V., Ruban, A. I., Sychev, Vic. V., and Korolev, G. L. (1998). *Asymptotic Theory of Separated Flows*. Cambridge University Press.

Tani, I. (1964). Low-speed flows involving bubble separations. *Progr. Aeronaut. Sci.*, **5**, 70–103.

Timoshin, S. N. (1996). Concerning marginal singularities in the boundary-layer flow on a downstream-moving surface. *J. Fluid Mech.*, **308**, 171–194.

Tollmien, W. (1931). Grenzschichttheorie. *Handbuch der Experimentalphysik*, **4**(1), 241–287.

Villat, H. (1914). Sur la validité des solutions de certains problémes d'hydrodynamique. *J. Math. Pures Appl.*, 6e sér. **10**, 231–290.

Werlé, H. and Gallon, M. (1972). Contrôle d'ecoulements par jet transversal. *Aéronaut. Astronaut.*, **34**, 21–33.

Zametaev, V. B. (1986). Existence and non-uniqueness of local separation region in viscous jets. *Izv. Akad. Nauk SSSR, Mech. Zhidk. Gaza* (1), 38–45.

Index